COELIAC DISEASE AND GLUTEN-RELATED DISORDERS

COELIAC DISEASE AND GLUTEN-RELATED DISORDERS

Edited by

ANNALISA SCHIEPATTI

Researcher in Gastroenterology, Department of Internal Medicine and Medical Therapy, University of Pavia, Pavia, Italy

DAVID S. SANDERS

Professor of Gastroenterology, The University of Sheffield; NHS Consultant, The Academic Unit of Gastroenterology, Royal Hallamshire Hospital, UK

ACADEMIC PRESS

An imprint of Elsevier

elsevier.com/books-and-journals

Academic Press is an imprint of Elsevier
125 London Wall, London EC2Y 5AS, United Kingdom
525 B Street, Suite 1650, San Diego, CA 92101, United States
50 Hampshire Street, 5th Floor, Cambridge, MA 02139, United States
The Boulevard, Langford Lane, Kidlington, Oxford OX5 1GB, United Kingdom

Notices
Knowledge and best practice in this field are constantly changing. As new research and experience broaden
our understanding, changes in research methods, professional practices, or medical treatment may become
necessary.

Practitioners and researchers must always rely on their own experience and knowledge in evaluating and
using any information, methods, compounds, or experiments described herein. In using such information
or methods they should be mindful of their own safety and the safety of others, including parties for whom
they have a professional responsibility.

To the fullest extent of the law, neither the Publisher nor the authors, contributors, or editors, assume any
liability for any injury and/or damage to persons or property as a matter of products liability, negligence or
otherwise, or from any use or operation of any methods, products, instructions, or ideas contained in the
material herein.

British Library Cataloguing-in-Publication Data
A catalogue record for this book is available from the British Library

Library of Congress Cataloging-in-Publication Data
A catalog record for this book is available from the Library of Congress

ISBN: 978-0-12-821571-5

For Information on all Academic Press publications visit our
website at https://www.elsevier.com/books-and-journals

Publisher: Stacy Masucci
Acquisitions Editor: Stacy Masucci
Editorial Project Manager: Megan Ashdown
Production Project Manager: Selvaraj Raviraj
Cover Designer: Mark Rogers

Typeset by Aptara, New Delhi, India

Working together
to grow libraries in
developing countries

www.elsevier.com • www.bookaid.org

CONTENTS

Contributors

Daniel Agardh
Department of Clinical Sciences, Lund University, Malmö, Sweden

Julie Antvorskov
Rigshospitalet, The Bartholin Institute, Copenhagen, Denmark

Federico Biagi
Gastroenterology Unit, IRCCS Pavia, ICS Maugeri, University of Pavia, Italy

Margit Brottveit
Healthy Life Centre, Municipality of Nes, N-2150 Årnes, Norway

Antonio Carroccio
Unit of Internal Medicine, "V. Cervello" Hospital, Ospedali Riuniti "Villa Sofia-Cervello," Department of Health Promotion Sciences, Maternal and Infant Care, Internal Medicine and Medical Specialties (PROMISE), University of Palermo, Palermo, Italy

Ashish Chauhan
Department of Gastroenterology and Human Nutrition, All India Institute of Medical Sciences, New Delhi, India

Carolina Ciacci
Celiac Center at Department of Medicine, Surgery, Dentistry, Scuola Medica Salernitana, University of Salerno, Italy

Jette Frederiksen
University of Copenhagen, Department of Clinical Medicine, Faculty of Health Sciences, Copenhagen, Denmark; Rigshospitalet-Glostrup, Multiple Sclerosis Clinic, Department of Neurology, Glostrup, Denmark; Rigshospitalet, The Bartholin Institute, Copenhagen, Denmark

Marios Hadjivassiliou
Academic Department of Neurosciences, Sheffield Teaching hospitals NHS Trust and University of Sheffield, Royal Hallamshire Hospital, Sheffield, United Kingdom

Kaisa Hervonen
Celiac Disease Research Center, Faculty of Medicine and Health Technology, Tampere University, Tampere, Finland; Department of Dermatology, Tampere University Hospital, Tampere, Finland

Knud Josefsen
University of Copenhagen, Department of Clinical Medicine, Faculty of Health Sciences, Copenhagen, Denmark; Rigshospitalet-Glostrup, Multiple Sclerosis Clinic, Department of Neurology, Glostrup, Denmark; Rigshospitalet, The Bartholin Institute, Copenhagen, Denmark

Shiva Dahal-Koirala
K.G. Jebsen Coeliac Disease Research Centre, Institute of Clinical Medicine, University of Oslo, Oslo, Norway; Department of Immunology, University of Oslo, Oslo, Norway

Kalle Kurppa
Tampere University and Tampere University Hospital, Tampere, Finland

Daniel A. Leffler
Division of Gastroenterology, Department of Medicine, Beth Israel Deaconess Medical Center, Boston, MA, United States

Knut E.A. Lundin
KG Jebsen Coeliac Disease Research Centre, Institute of Clinical Medicine, Faculty of Medicine, University of Oslo, Oslo, Norway and Department of gastroenterology, Division of Surgery, inflammationa and transplantation, Oslo University Hospital Rikshospitalet, N-0372 Oslo, Norway

Govind K. Makharia
Department of Gastroenterology and Human Nutrition, All India Institute of Medical Sciences, New Delhi, India

Pasquale Mansueto
Department of Health Promotion Sciences, Maternal and Infant Care, Internal Medicine and Medical Specialties (PROMISE), University of Palermo, Palermo, Italy

Moschoula Passali
University of Copenhagen, Department of Clinical Medicine, Faculty of Health Sciences, Copenhagen, Denmark

Hugo A. Penny
Academic Unit of Gastroenterology, University of Sheffield, Sheffield, United Kingdom; Lydia Becker Institute of Inflammation and Immunology, University of Manchester, Manchester, United Kingdom

Mahendra Singh Rajput
Department of Gastroenterology and Human Nutrition, All India Institute of Medical Sciences, New Delhi, India

Anupam Rej
Academic Unit of Gastroenterology, Royal Hallamshire Hospital, Sheffield Teaching Hospital NHS Foundation Trust, Sheffield S10 2JF, United Kingdom

Timo Reunala
Celiac Disease Research Center, Faculty of Medicine and Health Technology, Tampere University, Tampere, Finland

Louise Fremgaard Risnes
K.G. Jebsen Coeliac Disease Research Centre, Institute of Clinical Medicine, University of Oslo, Oslo, Norway; Department of Immunology, Oslo University Hospital, Oslo, Norway

Teea Salmi
Celiac Disease Research Center, Faculty of Medicine and Health Technology, Tampere University, Tampere, Finland; Department of Dermatology, Tampere University Hospital, Tampere, Finland

David S. Sanders
Academic Unit of Gastroenterology, University of Sheffield, Sheffield, United Kingdom; Academic Department of Gastroenterology, Royal Hallamshire Hospital & University of Sheffield, United Kingdom

Annalisa Schiepatti
Gastroenterology Unit, IRCCS Pavia, ICS Maugeri, University of Pavia, Italy

Aurelio Seidita
Department for the Treatment and Study of Abdominal Diseases and Abdominal Transplantation, IRCCS–ISMETT (Istituto di Ricovero e Cura a Carattere Scientifico – Istituto Mediterraneo per i Trapianti e Terapie ad alta specializzazione), Palermo, Italy

Gry Skodje
Department of gastroenterology, Division of Medicine, Oslo University Hospital Ullevål, N-0450 Oslo, Norway

Ludvig M. Sollid
K.G. Jebsen Coeliac Disease Research Centre, Institute of Clinical Medicine, University of Oslo, Oslo, Norway; Department of Immunology, University of Oslo, Oslo, Norway; Department of Immunology, Oslo University Hospital, Oslo, Norway

Amelie Therrien
Division of Gastroenterology, Department of Medicine, Beth Israel Deaconess Medical Center, Boston, MA, United States

Shakira Yoosuf
Division of Gastroenterology, Department of Medicine, Beth Israel Deaconess Medical Center, Boston, MA, United States

Fabiana Zingone
Gastroenterology Unit, Department of Surgery, Oncology and Gastroenterology, University of Padua, Padua, Italy

Panagiotis Zis
Academic Department of Neurosciences, Sheffield Teaching hospitals NHS Trust and University of Sheffield, Royal Hallamshire Hospital, Sheffield, United Kingdom

CHAPTER 1

Epidemiology and clinical features of celiac disease in adults

Mahendra Singh Rajput, Ashish Chauhan, Govind K Makharia
Department of Gastroenterology and Human Nutrition, All India Institute of Medical Sciences, New Delhi, India

1.1 Introduction

Celiac disease (CeD) is an autoimmune disease due to hyperreactivity to gluten in genetically susceptible individuals [1,2]. Until a few decades ago, CeD was considered to be an uncommon disease and it was thought to be limited to individuals of European ancestry [3]. Advancement in the knowledge about this disease over the past three decades has led this to the limelight from a fairly uncommon disease before the 1980s. The key advancements that lead to widespread recognition of CeD are summarized below (Fig. 1.1). Firstly, making a diagnosis of CeD was very tedious and it required three sequential intestinal mucosal biopsies including one at baseline showing villous atrophy, a second one demonstrating histological improvement on a gluten-free diet (GFD), and a third biopsy showing worsening of villous abnormalities upon reintroduction of gluten into the diet [4]. In most current guidelines, the diagnostic criteria have been simplified and the diagnosis of CeD can be made on the basis of a combination of a positive celiac-specific serologic test and small intestinal biopsy specimens showing villous abnormalities [1,2]. Furthermore, a diagnosis of CeD can also be made based on the presence of high titer of anti-tissue transglutaminase antibody (x10 folds) even without demonstration of villous abnormalities [5].

Secondly, CeD was thought to be a disease in children and seen mainly by pediatricians, but now, CeD is diagnosed at any age group, including the elderly [5]. Thirdly, CeD was thought to affect only people of European origin and subsequently, CeD was recognized in Caucasians living in North America, Oceania, and even South America. Now, CeD is well-recognized in the non-Caucasian population including Africans and Asians. The evolution of CeD has been very interesting. Even in those countries where CeD was thought to be uncommon, CeD was found to be common once systematic population-based studies were conducted in those countries [6,7].

Fourthly, it was initially thought that gluten hypersensitivity in CeD is limited to the small intestinal mucosa and all other features are secondary to malabsorption caused by the disease, but it is now established that many of the features of gluten hypersensitivity such as dermatitis herpetiformis (DH) cannot be explained on the basis of malabsorption

Coeliac Disease and Gluten-Related Disorders.
DOI: https://doi.org/10.1016/B978-0-12-821571-5.00012-X

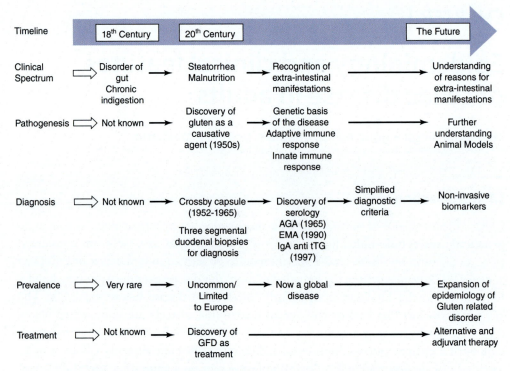

Figure 1.1 *Changing epidemiology and clinics of celiac disease through the different ages.* AGA- anti-gliadin antibodies, EMA- Endomysial antibody, anti-tTG - Antitransglutamiase, GFD - Gluten free diet.

alone. It is now been widely accepted that many other organs such as skin, brain, and bones are affected in CeD with or without of intestinal involvement [8–11].

Fifthly, the discovery of reliable serologic tests such as antitissue transglutaminase antibody (IgA tTG Ab), antiendomysial antibody (EMA) or antideamidated gliadin peptide antibody (anti-DGP Ab) have not only allowed screening of high-risk groups for CeD, but also made it possible to estimate the true prevalence of CeD in the general population. And lastly, an increase in the awareness about this disease amongst healthcare professionals and amongst general population also played a big role in unravelling this hidden disease (Fig. 1.1).

1.2 Initial epidemiological studies based on clinical symptoms

One of the oldest epidemiologic studies on CeD was conducted in 1950 when the diagnosis of CeD was based entirely on the presence of typical gastrointestinal symptoms. These studies established the cumulative incidence of the disease in England and Wales as 1 in 8000, and that in Scotland in 1 in 4000 [12]. With the introduction of more specific tests for malabsorption and advent of intestinal biopsy, the awareness about CeD greatly

Table 1.1 Seroprevalence and prevalence of biopsy-confirmed CeD in various continents; CeD (Celiac disease). From Global Prevalence of Celiac Disease: Systematic Review and Meta-analysis. (c. 2018). *Clinical Gastroenterology and Hepatology, 16*(6), 823-836. https://doi.org/10.1016/j.cgh.2017.06.037.

Continent	Seroprevalence (95% CI)	Prevalence of Bx confirmed CeD (95% CI)
Europe	1.3 (1.1-1.5)	0.8 (0.6-1.1)
North America	1.4 (0.7-2.2)	0.5
South America	1.3 (0.5-2.5)	0.4 (0.1-0.6)
Africa	1.1 (0.4-2.2)	0.5 (0.2-0.9)
Asia	1.8 (1.0-2.9)	0.6 (0.4-0.8)
Oceania	1.4 (1.4-1.8)	0.8 (0.2-1.7)

increased, which led to an increase in the incidence of the disease in the 1970s to 1 in 450 from Ireland, Scotland, and Switzerland [13,14] (Fig. 1.1).

1.3 Modern epidemiological study based on the serological test

In 1996, a multicenter study from Italy using three-layered strategy including clinical screening, serological tests, and intestinal biopsies in the school children gave birth to the modern epidemiology of CeD. Amongst 17201 healthy Italian students, the overall prevalence of CeD was found to be 1 in 184. More interestingly, only 1 of 7 was previously diagnosed as CeD, highlighting a big iceberg phenomenon, where clinically detectable patients were just a few and a large number of subjects remained undetected. Ever since epidemiological studies started appearing from various countries of the world which are summarized in the following sections [15].

1.4 Global prevalence of CeD

The assessment of the prevalence of CeD is accomplished as seroprevalence of CeD (a proportion of people having positive anti-tTG Ab and /or anti-endomysial Ab) or prevalence of biopsy-confirmed CeD, where intestinal mucosal biopsies show villous abnormalities of modified Marsh grade 2 or more along with a positive serological test.

1.5 Global seroprevalence of CeD

A recent systematic review and meta-analysis of population-based studies, including 275818 subjects have shown that a pooled global seroprevalence of CeD in the general population is 1.4% (95% CI 1·1%, 1·7%) [7] (Fig. 1.2). The seroprevalence of CeD varies from continent to continent, highest being in Europe and Asia (Table 1.1) Furthermore, the seroprevalence also varies from country to country, highest being in Algeria, Czech Republic, India, Israel, Mexico, Malaysia, Saudi-Arabia, Sweden, Portugal, and Turkey and lowest in Estonia, Germany, Iceland, Libya, Poland, Republic of San Marino, Spain, and Switzerland [7].

Figure 1.2 *Country-wise seroprevalence of CeD; Countries are stratified into 4 groups of percentiles representing the 0 to 25th percentile (light grey) to the 76th to 100th percentile (dark black).* Countries with highest percentiles of seroprevalence (76th to 100th) include Algeria, Czech Republic, India, Israel, Mexico, Malaysia, Saudi-Arabia, Sweden, Portugal, and Turkey with lowest percentiles of seroprevalence (0th -25th) include Estonia, Germany, Iceland, Libya, Poland, Republic of San Marino, Spain and Switzerland. From Global Prevalence of Celiac Disease: Systematic Review and Meta-analysis. (c. 2018). *Clinical Gastroenterology and Hepatology, 16*(6), 823-836. https://doi.org/10.1016/j.cgh.2017.06.037.

1.6 Global prevalence of biopsy-confirmed CeD

The global pooled prevalence of biopsy-confirmed CeD has been shown to be 0•7% (95% CI 0•5%, 0•9%) [7] (Fig. 1.3). On stratification of countries into quintiles based on the prevalence of biopsy-confirmed CD, countries with the highest prevalence (76th to 100th quintile) include Argentina, Egypt, Hungary, Finland, India, New-Zealand and Sweden and the countries with the lowest prevalence (0 to 25th quintile) include Brazil, Germany, Republic of San Marino, Russia, and Tunisia (Fig. 1.3) [7].

Most population-based epidemiological studies on CeD prevalence are based on serological data, and the diagnosis of CeD in all seropositive patients has not been confirmed by invasive small intestinal mucosal biopsies. Therefore, the global pooled prevalence of biopsy-confirmed CeD, which is 0.7% (95% CI: 0.5–0.9%), is lower than the seroprevalence.

The epidemiology of CeD is still evolving and population-based data is still not available from many countries.

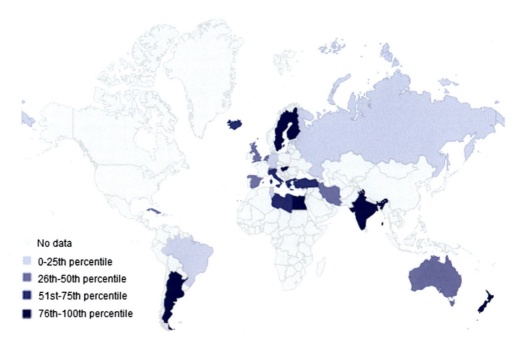

No data

0-25th percentile

26th-50th percentile

51st-75th percentile

76th-100th percentile

Figure 1.3 *Country-wise prevalence of biopsy-confirmed CeD; The prevalence was stratified into 4 groups of percentiles representing the 0 to 25th percentile (light grey) to the 76th to 100th percentile (dark black).* The countries with highest percentiles (76th to 100th) include Argentina, Egypt, Hungary, Finland, India, New-Zealand, and Sweden and the countries with lowest percentiles (0th -25th) include Brazil, Germany, Republic of San Marino, Russia and Tunisia. From Global Prevalence of Celiac Disease: Systematic Review and Meta-analysis. (c. 2018). *Clinical Gastroenterology and Hepatology, 16*(6), 823-836. https://doi.org/10.1016/j.cgh.2017.06.037.

1.7 Prevalence of CeD by gender and age

CeD is more common in females (0•6 %; 95% CI 0•5%, 0•8%) than in males (0•4%, 95% CI 0•3%, 0•5%) [7]. Furthermore, CeD is more common in pediatric age group compared to that in adults 0•9% (95% CI 0•6, 1.3) vs 0•5% (95% CI 0•3, 0•8). (P<0.001) [7].

1.8 Continent wise prevalence of CeD

1.8.1 CeD in Europe

Initial studies of CeD prevalence in the general population have been published from many European countries. In a multi-national European study including Finland, Germany, Italy and UK, 29,212 subjects were tested for CeD by anti–tTG antibody, and all those who had either a positive or a borderline titer of anti–tTG Ab were further tested for EMA in serum. The overall prevalence CeD in this multinational study was found to be 1.0% (95% CI 0.9 - 1.1). Interestingly, the prevalence of CeD was not uniform in the four participating European countries, despite sharing a similar distribution of causal factors (level of gluten intake and frequency of HLA-DQ2 and -DQ8 genotype). The prevalence of CeD was 2.0% (95% CI 2.0 – 2.8) in Finland, 1.5 (95% CI 1.1-1.9) in UK, 0.7% (95% CI 0.4–1.0) in Italy, 0.3% in Germany (95% CI 0.1–0.5) [16]. Interestingly another study from Germany including 12741 participants aged 1 to 17 years has shown a prevalence of CeD of 0.9% [17]. Furthermore, continent wise systemic review and meta-analysis of 33 studies have shown the seroprevalence and prevalence of biopsy-confirmed of CeD in Europe to be 1.3% (95% 1.1-1.5) and 0.8% (95% 0.6-1.1), respectively [7].

1.8.2 Celiac disease in America (North and South America)

CeD had been considered to be an uncommon disease in America till 2003, the results of a prevalence study reporting 1 in 133 Americans having CeD, led to more widespread recognition of this disease in the United States (USA) [18]. The systematic review and meta-analysis of population-based studies including 7 studies and 17,778 subjects have revealed that the seroprevalence of CeD in North America is 1.4% (95% CI 0.7-2.2). There is a paucity of data on the prevalence of CeD based on biopsy confirmation.

CeD is well known in those South American countries that are populated by individuals of European origin, such as Brazil. In a study including 4405 subjects from Brazil, the overall seroprevalence and prevalence of biopsy confirmed CeD was 3.6 and 3.41 per 1000, respectively. Prevalence in adult and children was 2.11 and 5.44 per 1000, respectively [19]. As per the systematic review of the studies from South America, the pooled seroprevalence and prevalence of biopsy-confirmed CeD was 1.3% (95% 0.5-2.5) (11 studies and 20245 subjects screened) and 0.4% (0.1-0.6) (5 studies and 16550 subjects), respectively [7].

1.8.3 Prevalence of CeD in Oceania

As in the European countries, a population-based study from Australia including 3011 subjects showed the seroprevalence and biopsy-confirmed prevalence of CeD to be 1/251 and 1/430, respectively [20]. A similar population-based study done in New Zealand including 1064 subjects has shown the prevalence of CeD to be 1.1% [21].

1.8.4 Prevalence of CeD in Africa

The pooled seroprevalence (7 studies and 15,775 subjects) and prevalence of biopsy-confirmed CeD (4 studies and 7902 subjects) in African continent was shown to be 1.1% (95% CI 0.4-2.2) and 0.5% (95% CI 0.2-0.9) [7]. The Saharawi population of Arab-Berber origin, originally living in Western Sahara, has been reported to have the highest prevalence of CeD in the world. A study of 990 Saharawi children, showed the prevalence of CeD in this population to be 5.6% [22]. The reasons for the high prevalence of CeD in the Saharawi population are likely to be a high rate of consanguinity and high gluten consumption in this population. Some other prevalence studies done in Africa have shown a prevalence of CeD to be 0.5% in Egypt, 0.8% in Libya, and 0.6% in Tunisia [23,24].

1.8.5 Prevalence of CeD in Asia

Until recent times, CeD was considered extremely rare in Asia, and patients presenting with celiac-like symptoms (chronic diarrhea, malnutrition, abdominal distention, etc) were diagnosed usually as having tropical sprue or kwashiorkor [25]. After the widespread availability of serologic tests, multiple screening studies performed in different Asian countries such as Turkey, Iran, Israel, Jordan and India have shown that CeD is not an uncommon disorder and it is often underdiagnosed in Asia [26]. Due to the heterogeneity of populations, genetics, economic condition and dietary habits, the epidemiology of CeD varies according to the geographical areas of Asia taken into consideration [26].

In India, CeD has been recognized mainly in the Northern part of India, where wheat is the predominant cereal consumed and a population-based study including 2879 subjects showed a prevalence of CeD to be 1.04% (1 in 96) [27]. Later, a pan-India study including 23,331 healthy adults from 3 different regions of India, showed a regional variation in the prevalence of CeD. While the age-adjusted seroprevalence of CeD in Northern, North Eastern regions were 1.23% and 0.87%, respectively, it was only 0.10% in the Southern region, showing a Northern and Southern region gradient in the prevalence [28]. Since the prevalence of HLA-DQ2 and DQ8 haplotype was similar in all the three regions, this regional difference in the prevalence was attributed to the differences in the wheat (gluten) eating pattern, which was highest in Northern part of India and lowest in Southern part of India [28].

In a systematic review and meta-analysis, the prevalence studies from Asia were segregated into studies from Middle East (Iran, Turkey, Saudi Arabia, Israel, Jordan) and South East Asia (data from India, Malaysia and Egypt). The pooled seroprevalence and prevalence of biopsy-confirmed CeD in the Middle East region and South East Region of Asia has been reported to be 1.6% (95% CI 1.2-2.1) and 0.6% (95% CI 0.4-0.8), and 2.6% (95% CI 0.3-7.2) and 0.8% (0.4-1.4), respectively, which were quite similar to that is reported from many European countries [7].

While the epidemiology of CeD in China was largely unknown until recent years, except for a small case series. In a cross-sectional study including 19,778 Chinese adolescents and young adults (age 16-25 years) from 27 geographic regions in China were screened for CeD using a combination of IgG anti-DGP and IgA anti-tTG Ab. More than 2% (2.19%) of them were detected to have a positive one of the antibody including 1.8%

for IgG anti DGP and 0.36% for IgA anti tTG. The prevalence of people with a positive antibody varied remarkably amongst different regions of China and it was 12 times higher in the Northern provinces, such as Shandong, Shaanxi, and Henan, where wheat was the staple diet [29]. In another recent study including 2277 in-patients with gastrointestinal symptoms in four major ethnic groups of Xinjiang Uyghur Autonomous Region of China (including 1391 Han, 608 Uyghur, 146 Kazakh and 132 Hui), the seroprevalence and prevalence of biopsy-confirmed CeD was observed to be 1.27% (95% CI, 0.81-1.73%), and 0.35% (95% Cl, 0.11-0.59%), respectively [30]. CeD was found to be three times more common in rural part with significantly higher wheat consumption compared to urban living subjects (3.16% vs 0.97%, P < 0.01). Interestingly, of 246 patients with diarrhea-predominant IBS in China, 2.85% were reported to have CeD [31]. These preliminary studies have established the foundation for exploration of the exact prevalence of CeD and regional geographical differences in the prevalence of CeD in China.

In a pilot study including 562 young healthy volunteers from Malaysia, the sero-prevalence of CeD was found to be 1.25% (95% CI 0.78-1.72%). In this multi-ethnic country, all three ethnic groups such as Malay (0.8%), Chinese (1.7%) and Indian (1.3%) were affected [32]. In a study from Japan including 2008 subjects, anti-tTG Ab was found to be in a high proportion (8%), none of them, however, was EMA positive and only one showed celiac-type alterations at the small intestinal biopsy [33]. Similarly, in a study including 1961 Vietnamese children the seroprevalence based on anti-tTG Ab was observed to be 1%, but none of them was EMA positive [34].

While there are also reports of CeD from other Asian countries such as Pakistan, Bangladesh, Singapore, there are no formal reports on CeD from Indonesia, Korea, Taiwan and many other Asian nations [35,26].

1.8.6 Increase in the global prevalence of CeD over time

A study from the United States has shown that the prevalence of CeD was only 0.2% in the year 1975, but over past 25-30 years, it has increased by five-folds [36]. The pooled prevalence of CeD has increased from 0.6% (95% CI, 0.5%–0.7%) during the period of 1991-2000 to 0.8% (95% CI, 0.5%–1%) during 2011-2016 [7]. Furthermore, the increase in the prevalence over time is not limited to the United States and Europe, but it has been reported also in other parts of the world [37].

Although the prevalence of CeD in the general population has increased, the disorder still remains heavily underdiagnosed. It has been estimated that approximately five to ten patients remain undiagnosed for one diagnosed patient with CeD. This happens because a large proportion of patients with CeD have either atypical symptoms or minimal symptoms or they can even be asymptomatic. The diagnostic rate of CeD depends mostly on the level of awareness of the physicians. Despite an increase in the prevalence of CeD, and presence of a large pool of patients with CeD globally, the majority of patients (83%–95%) in developed countries, and possibly even a higher number in developing countries, still remain undiagnosed [38,40–42].

1.8.7 Increase in the incidence of CeD

Not only the prevalence but the incidence of CeD has also increased throughout the Western world. The pooled average annual incidence of CeD has been estimated to be rising by 7.5% (95% CI: 5.8, 9.3) per year over the past several decades. A recent systematic review reported the pooled incidence of CeD in women and men to be 17.4 (95% CI: 13.7, 21.1) and 7.8 (95% CI: 6.3, 9.2) per 100,000 person-years, respectively [37]. The increase in the incidence in CeD is both due to improved diagnostics and awareness about the disease amongst the physicians and changes in our environment and eating practices [38].

1.8.8 Spectrum of clinical manifestations of CeD

CeD is now considered a systemic disorder and their clinical presentation may be with gastrointestinal symptoms, called "classical CeD" accounting for 50-60% of all cases and non-gastrointestinal symptoms called "non-classical CeD" accounting for 40-50% of cases [39]. While some patients with CeD have fully expressed disease and present early in life, while in others the disease is expressed in a milder form and hence does not come to clinical attention till adulthood. Additionally, there are situations where the clinical presentation of CeD is abrupt and explosive in a relatively asymptomatic individual, which partly may be due to either occurrence of a second hit or the development of a complication.

1.8.9 Gastrointestinal manifestations (Classical CeD)

Many adult patients with CeD present with classical GI symptoms including chronic or intermittent diarrhea, steatorrhea, abdominal bloating, flatulence, and weight loss [40]. Some of the patients with CeD may have mild GI manifestations such as altered bowel activity, abdominal pain/discomfort, and bloating, and with this symptom complex, they are likely to be labeled as having functional gastrointestinal diseases including irritable bowel syndrome (IBS). In a meta-analysis of 22 eligible studies including 6991 patients with IBS, Ford et al. reported that 3.3% (95 % CI 2.3%, 4.5%) of them had CeD [41]. As one can expect, the prevalence of CeD is highest amongst patients with IBS-diarrhea predominant (pooled prevalence of 5.4%, 95% CI 3.3-7.8%) and IBS-Mixed suggesting that all patients with IBS-D and IBS-M should be screened for CeD [41]. Furthermore, the seroprevalence of CeD in patients with functional dyspepsia is higher (7.9% vs 3.9%) as compared with controls [42].

Table 1.2 Prevalence of CeD amongst extra-intestinal manifestations and high-risk groups.

Condition	Prevalence CeD % (95% CI)	
Iron deficiency anaemia [50]	3.2 (95% CI 2.6-3.9)	
Short stature [58]	All cause	7.4 (95% CI 4.7-10.6)
	Idiopathic	11.6 (95% CI 4.1-22.2)
Infertility [98]	All cause	2.3 (95% CI 1.4-3.5)
	Idiopathic	3.2 (95% CI 2-4.9)
Type 1 diabetes [61]	6 (95% CI 5-6.9)	
Osteoporosis [69]	1.6 (95% CI 1.1-2.0)	
Autoimmune thyroid disorder [63]	1.4 (95% CI 1-1.8)	
Cryptogenic cirrhosis [76]	2.50%	
Autoimmune hepatitis [80]	3.50%	
Unexplained transaminases [70]	4 (95% CI 1-7)	

1.9 Extra-intestinal manifestations of CeD (Table 1.2 and Table 1.3)

1.9.1 Hematological manifestations

Anaemia is a common manifestation of CeD affecting approximately 12-69% in the western countries and 85-90% of Indian patients with CeD. Iron deficiency could be the sole manifestation of CeD even in the absence of diarrhea [48–56]. Iron deficiency is the commonest form of anaemia in CeD [43,44]. Although symptomatic B12 deficiency is uncommon, 8% to 41% of patients with CeD, however, have been detected to have vitamin B_{12} deficiency at the time of diagnosis [45–47]. The exact reason for deficiency of vitamin B12 is not known. However, it may be due to under-secretion of gastric acid, bacterial overgrowth, and occurrence of co-existent autoimmune gastritis. Folate deficiency has also been reported in patients with CeD [45,48]. Anemia in patients with CeD has been shown to respond very well to GFD and iron/vitamin supplementation [44,45]. Recent pieces of evidence suggest that CeD patients with anaemia have a significantly longer duration of symptoms, lower albumin levels, higher anti–tTG antibody level, and more often advanced enteropathy in comparison to CeD patients having normal hemoglobin [49].

On the other hand, 3.2% (95% CI 2.6% to 3.9%) of patients with iron deficiency anaemia (1 in 31) have CeD, as shown in a meta–analysis of 18 studies including 2998 patients with iron deficiency anemia [50].

The haematological manifestations of CeD are not limited to anaemia but other cell lines and coagulation cascade are also affected [45]. Up to 60% of CeD patients can have thrombocytosis but only a minority of patients have thrombocytopenia [52]. Thrombocytosis is believed to occur due to presence of an active inflammatory state, iron deficiency state and hyposplenism [45]. Venous thromboembolism has been well

Table 1.3 Extra-intestinal manifestations of CeD.

Systemic manifestations	Manifestation
Hematological	Anemia, both iron deficiency and B12 deficiency, Thrombocytosis
Endocrinological	Growth failure, short stature, Hypothyroidism Secondary hyperparathyroidism
Musculo-skeletal	Osteopenia and osteoporosis, osteomalacia, pathologic fractures, muscle atrophy, tetany
Hepatic	Elevated serum transminases, Autoimmune hepatitis Cryptogenic cirrhosis, Primary sclerosing cholangitis, Primary biliary cirrhosis
Metabolic	Metabolic syndrome, Obesity
Dermatological	Dermatitis herpetiformis Vitiligo, alopecia areata, cutaneous vasculitis, Urticaria, Atopic dermatitis, Psoriasis, Linear IgA dermatosis, Vitiligo, and Lupus rythematous
Reproductive	Delayed menarche and amenorrhea, Early menopause Infertility, recurrent abortion, Intrauterine growth retardation Low birth weight
Neuropsychiatric	Cerebellar ataxia, Peripheral neuropathy, seizures Autism spectrum disorder, Attention deficit hyperactive disorder, Anxiety disorder, Depression and Mood disorder Schizophrenia
Oro-dental	Dental enamel dysplasia, Recurrent aphthous ulcer Dry mouth and xerostomia

reported in CeD and there are reports describing co-occurrence of CeD and hepatic venous outflow tract obstruction [45,51,52].

1.9.2 Endocrinological manifestations

As CeD affects the early part of life, growth failure/retardation is an important manifestation of CeD. It is interesting to mention that while the height of men with CeD is slightly shorter, the women with CeD are taller than the respective control population [53]. The short stature is more common in those diagnosed during childhood and during adolescence than those diagnosed in adulthood. Such an event may occur because of multiple reasons. Those having the severe disease are more likely to be symptomatic and hence they come to clinical attention early, whereas those with mild disease may remain unnoticed in childhood and later diagnosed in adulthood. More importantly, the institution of a GFD in these patients is associated with early catch-up growth for the initial 2-3 years [54,55]. Early diagnosis and compliance with a GFD result in rapid recovery and patients may achieve normal adult height [56,57].

In a systematic review and meta-analysis involving 17 studies including 3759 patients with short stature, 7.4% (95% CI 4.7-10.6%) of all-cause short stature and 11.6% (95% CI 4.1-22.2%) of idiopathic short stature, respectively had biopsy-confirmed CeD [58]. The pathogenesis of short stature in patients with CeD is multi-factorial including malabsorption, lower levels of insulin like growth factor (IGF)-1, IGF-2, insulin like growth factor binding protein (IGFBP)-1 and IGFBP-3, partial insensitivity to the growth and concomitant hypothyroidism, hypogonadism and Turner's syndrome [59,60].

Association between CeD and type 1 diabetes (T1DM) and autoimmune thyroiditis is also well established. In a systematic review and meta-analysis, Elfstrom et al pooled the data from 27 studies involving 26,605 patients with T1DM and reported that 6% (95% CI 5-6.9%) of them have CeD. Thus, more than 1 in eighteen patients with T1DM have biopsy-confirmed CeD. The prevalence of CeD is lower in adults (2.7%, 95% CI 2.1-2.3%), than in children (6.2%, 95% CI 6.1-6.3%) with type 1 diabetes (P<0.001) [61].

As many as 10-15% of patients with CeD have co-existent clinical hypothyroidism or hyperthyroidism and conversely up to 2-4% of patients with autoimmune thyroid disorders have CeD [62]. In a recent systematic review and meta-analysis of 15 studies involving 6024 patients with autoimmune thyroid disorders, Roy et al reported that 1.4% (95% CI 1-1.8%) of them had biopsy-confirmed CeD [63]. The prevalence of CeD was lower in those with hypothyroidism (1.4%, 95% CI 1-1.9%) compared to those with hyperthyroidism (2.6%, 95% CI 0.7-4.4%) [63].

The evidences of association between other endocrinopathies such as Addison's disease, primary or secondary hyperparathyroidism, and autoimmune hypophysitis and CeD is not well established. A complete remission of autoimmune hypophysitis has been reported with a GFD [64].

1.9.3 Skeletal manifestations (osteopenia and osteoporosis)

CeD is associated with various degrees of osteopenia and osteoporosis. Several studies have demonstrated that low bone mineral density (BMD) is invariably present in adults and children with untreated CeD, regardless of clinical presentation [65-66]. In a systematic review and meta-analysis including 563 premenopausal women and men with CeD, the pooled prevalence of osteoporosis and osteopenia was found to be 14.4% [95%CI: 9–20.5%] and 39.6% [31.1–48.8%], respectively [67]. Not only there is an increase of the prevalence of osteoporosis in them, but they are also at a higher risk of developing bone fractures. A systematic review and meta-analysis including 20,995 CeD patients and 97,777 controls from eight studies published between 2000 and 2007, showed that patients with CeD have a 43% higher risk for developing non-traumatic fracture compared with people without CeD [68].

In a systematic review and meta-analysis of 3188 individuals with osteoporosis, Laszkowska, et al reported that 1.6% (1 in 62) individuals with osteoporosis have biopsy-confirmed CeD [69]. Low bone density in patients with CeD is multifactorial including malabsorption of calcium and vitamin D resulting in secondary hyperparathy-roidism, chronic inflammation, high cytokine levels that interfere with bone growth, and autoimmune factors. The skeletal fragility, as measured by microarchitecture by high-resolution computed tomography, is also diminished in individuals with newly diagnosed CeD [70]. The bone density tends to improve on a GFD, especially if detected early, strict adherence to GFD, and supplemented with calcium and vitamin D supplementation, thus signifying the importance of early detection and recognition of the bone disease in patients with CeD [70,71].

1.9.4 Hepatic manifestations

The spectrum of liver involvement in patients with CeD ranges from an asymptomatic elevation of serum transaminases to end-stage liver disease. A systematic review and meta-analysis showed that 27% (95% CI 13% to 44%) of newly diagnosed patients with CeD have elevated serum transaminases and they normalize in 63% to 90% of patients within 1 year of GFD [72].

Conversely, 6% (95% CI 3% to 10%) and 4% (95% CI 1% to 7%) of patients with cryptogenic hypertransaminasemia have a positive celiac serology and biopsy-confirmed CeD, respectively [72]. The persistence of liver cell injury, which is asymptomatic most of the time, can ultimately lead to cirrhosis of the liver if remain unattended and untreated. There have been multiple reports including clinical observations, case series and population-based cohort in past one-decade describing detection of CeD in patients having chronic liver disease (CLD) or cirrhosis [52,73,74]. In one of the first reports, Kaukinen et al had reported reversal of hepatic dysfunction after initiation of a GFD in four patients awaiting liver transplantation and eventually three of them were remitted from the liver transplantation list [75]. In a more recent prospective study, Wakim-Fleming, et al have reported CeD in 2.5% of 204 consecutive biopsy-proven patients with liver cirrhosis. They also reported improvement in liver function tests with the initiation of a GFD in these patients [76]. Furthermore, patients with CeD are also more likely to die from liver cirrhosis than the general population [74].

Patients with CeD has also been reported to co-exist with other autoimmune liver diseases such as autoimmune hepatitis (AIH), primary biliary cirrhosis (PBC) and primary sclerosing cholangitis (PSC). CeD has been reported in 4% to 6% of patients with autoimmune hepatitis [77–79]. Van-Gerven, et al reported that 3.5% of 410 patients with AIH were seropositive for CeD which was higher than 0.35% prevalence of CeD in the general Dutch population [80]. Similarly, 4-11% of patients with PBC were found to have co-existent CeD [77]. Conversely, patients with CeD have more than three times higher odds of having PBC compared to the occurrence of PBC in the general population (0.17% vs 0.05%) [81]. In another large population-based cohort of 13,818 CeD patients, Ludvigsson, et al have reported 4.5 times higher risk of having PSC amongst patients with CeD in comparison to that in the general population [73]. In patients having a combination of both autoimmune liver disorders and CeD, symptoms of one might precede the other, or even may co-exist at the time of diagnosis and therefore need careful evaluation [82]. Thinking of a diagnosis of CeD in patients with cirrhosis is challenging because of two reasons. Firstly, the symptoms of CeD such as diarrhea, fatigability, anaemia, failure to gain weight and height can also be caused by cirrhosis and secondly reliability of anti-tTG antibody positivity is questioned because of occurrence of false-positive anti-tTG antibody in those having in cirrhosis of the liver. Therefore, one should keep a high index of suspicion of CeD in patients with cirrhosis, especially cryptogenic, if they have chronic diarrhea, short stature and disproportionately severe anaemia or iron refractory anaemia [52,82].

The abovementioned evidence while suggesting an association between CeD and a spectrum of liver diseases, Jabobson, et al. have described histological changes in the liver of 37 patients with CeD having an increase in serum transaminases, with normal liver in 5, non-specific necro-inflammatory changes in 25, chronic hepatitis in 6, and PSC in one [83].

1.9.5 Metabolic manifestations

While thinking of CeD, a picture in mind often comes as a person with emaciation, short stature and low weight, as described in many texts books. That indeed was a description of the most advanced and long-standing patient with CeD. It is now appreciated that 8% to 44% of the patients with CeD are overweight or even obese at the time of diagnosis. Patients with CeD are also at higher risk of developing metabolic syndrome [84]. In a study, Tortora, et al have observed an increase in the prevalence of metabolic syndrome from 2% at the time of diagnosis to 30% after 12 months of GFD [85]. In another study including 44 treatment naïve patients with CeD, 5 (11.4%) had metabolic syndrome at the baseline, which further rose to 18.2% after one year of GFD. Furthermore, it has been reported that 26.3% of patients on a GFD for more than one year have metabolic syndrome [86]. Use of processed gluten-free foods, having a higher glycemic index and high saturated fat content, a state of hyperabsorption on a GFD and differential secretion pattern of brain-gut axis hormones may underlie weight gain and development of the metabolic syndrome in such patients [87,88]. Long-term use of GFD can lead to perpetual and progressive worsening of metabolic parameters and thus potentially predisposing to a higher risk for cardiovascular diseases. Therefore, it is imperative that we recognize metabolic abnormalities as a reality in patients with CeD and hence initiate appropriate preventive strategies, including counseling for physical activities and a well-balanced GFD.

1.9.6 Dermatological manifestations

Dermatitis herpetiformis (DH) is amongst the first extra-intestinal manifestations to be recognized in patients with CeD [89]. DH is characterized by clusters of papules and vesicles associated with intense pruritus, which are followed by erosions and excoriations.

The typical sites for DH lesions include extensor surfaces of upper and lower extremities, elbows, knees, scalp and buttocks. Histology with immunohistochemistry (IHC) or immunofluorescence (IF) of skin biopsies is diagnostic for DH which shows granular IgA deposits and neutrophil infiltrates in the papillary dermis [89]. It is interesting to note only approximately two-thirds of patients with DH have villous abnormalities and one-third of them have no enteropathy [90]. A questionnaire-based study including 1,138 biopsy-confirmed patients with CeD showed that the prevalence of DH in patients with CeD is 9.8% [91]. (Please refer to chapter 8 for a detailed description of dermatitis herpetiformis).

1.9.7 Reproductive manifestations

There is evidence to suggest that women with CeD have delayed menarche, early menopause, recurrent abortions, infertility, intrauterine growth retardation, and low birth weight (preterm and small for gestational age babies) [92,93]. Amenorrhea occurs in approximately one-third of women of childbearing age, and menarche is often delayed, typically by 1 year, in untreated subjects. The connection between female infertility and CeD was first described in 1970 by Morris et al, who described reversal of infertility with GFD in 3 females with infertility having CeD [94]. Based on the observations of population-based study, Zugna et al have suggested that women with CeD while having a comparable lifetime risk of infertility with the general population, their fertility is however reduced in the 2 years preceding the diagnosis of CeD and it is not uncommon for infertile women with CeD to become pregnant shortly after commencing GFD [95].

Conversely, a systematic review of cohort studies including 884 women with all-cause infertility and 623 women with unexplained infertility have reported that 2.3% (95% CI, 1.4-3.5) and 3.2% (95% CI 2%, 4.9%) of them have CeD, respectively [96]. A systematic review of case-control studies also has also shown that women with all-cause infertility have 3.5 times higher odds of having CeD in comparison with the controls (OR=3.5; 95% CI, 1.3-9; P <0.01). Similarly, women with "unexplained infertility" had 6 times higher odds of having CeD than controls (OR = 6; 95% CI, 2.4-14.6) [96].

1.9.8 Neuropsychiatric manifestations

While neurological conditions such as ataxia and peripheral neuropathy are well recognized, other neuropsychiatric manifestations such as migraine, epilepsy, dementia, cognitive impairment and depression are also recognized in the spectrum of clinical manifestations of CeD and gluten-related disorders. The term gluten ataxia was coined when patients with idiopathic sporadic ataxia were detected to have a positive anti-gliadin antibodies (either IgG or IgA or both) with or without the presence of enteropathy [39]. While cerebellar ataxia is being reported in 0-6% of patients with CeD [97], it is diagnosed more often while investigating a patient presenting predominantly with manifestations of cerebellar ataxia.

In addition to anti–gliadin Ab, antibodies to transglutaminase 6 (anti–TG-6) have been identified as a potential new serological marker for gluten ataxia [98]. Patients with gluten ataxia have not only clinical manifestations of ataxia, but approximately 60% of them also have demonstrable cerebellar atrophy on imaging and metabolic abnormalities on magnetic resonance spectroscopy [99]. (Please refer to chapter 11 for detail description).

1.9.9 Oral cavity and dental manifestations

The oral manifestations of CeD include cheilosis, angular cheilitis, recurrent aphthous ulcers (RAU), dry mouth and dental enamel defects (DED). It has been suggested that oral manifestation of CeD can help in detection of CeD, even in the absence of typical gastrointestinal manifestations [100]. While Aine, et al in 1990 reported DED in up to 69% of patients with CeD [101], recent systematic review and meta-analysis has suggested that 50% (95% CI 44-57%) of patients with CeD have DED. Furthermore, patients with CeD are at more than the two-fold higher risk of DED as compared to healthy people (RR: 2.31, 95% CI 1.71-3.12) [102]. The factors which lead to DED are multiple and include immune-mediated damage, nutritional disturbances and genetic factors [54]. Since dental development completes generally by 7 years of age, DED often sets in childhood-onset CeD and the teeth may remain normal if the disease starts later in life. Furthermore, DED is irreversible and do not regress with GFD [54]. DED is not really a benign disease and it can cause early colonization of cariogenic bacteria due to roughened tooth surfaces and hence enhance carcinogenesis. DED can also cause structural breakdown due to reduced enamel quantity and quality [103]. It increases dental hypersensitivity due to dentine exposure, tooth wear, and if anteriorly located, aesthetics can be a big concern for the patients [104].

RAU is self-limiting lesions and affects 5–50% of the general population [105]. RAU is present in 3% to 61% of patients with CeD and other gluten-related disorders such as non-celiac gluten sensitivity [106]. A recent meta-analysis including 21 studies has shown that patients with CeD are at more than three times higher risk of developing RAU (odds ratio 3.7, 95% CI 2.6, 5.3; P<0.00001) [107]. These lesions improve with GFD [100,108].

While DED and oral manifestations are so common in patients with CeD, the teeth are generally not examined by physicians and hence the dental manifestations are missed. It is therefore essential that physicians should train in the identification of basic dental lesions so that patients can be referred to dentists for the specific management.

1.10 Mortality in celiac disease

Both diagnosed and undiagnosed patients with CeD have been described to have high morbidity and mortality [109,110]. Large population-based studies have shown 1.3- to 2-fold higher mortality in them but these data are based on the studies conducted largely in the past when diagnosis of CeD was made infrequently and GFD was not available easily [111]. While there is still a debate on the increase in mortality in patients with CeD at present times when the awareness about the disease and diagnostic and therapeutic facility are easily available, even a recent study in Swedish population has shown a small (1.2 per 1000-person year) but significant increase in mortality in patients with CeD. Mortality risk was greatest after first year of diagnosis but persisted even after 10 year of the diagnosis [112]. This increased risk of mortality in them is has been attributed to the co-morbidities associated with CeD including malignancy [113] osteoporosis [114] type 1 diabetes [115], and other conditions [116]. (Please refer to chapter 5 for the complications of celiac disease).

In conclusion, CeD is now a global disease and it is a multisystemic disorder. The clinical spectrum of CeD is very heterogeneous. It is essential that we recognize the evolving spectrum of manifestations of CeD.

References

[1] P.H.R. Green, C. Cellier, Celiac disease, N. Engl. J. Med. 357 (2007) 1731–1743.

[2] A. Fasano, C. Catassi, Celiac disease, N. Engl. J. Med. 367 (2012) 2419–2426.

[3] E. Lionetti, C. Catassi, New clues in celiac disease epidemiology, pathogenesis, clinical manifestations, and treatment, Int. Rev. Immunol. 30 (2011) 219–231.

[4] Revised criteria for diagnosis of coeliac disease, Report of Working Group of European Society of Paediatric Gastroenterology and Nutrition, Arch. Dis. Child. 65 (1990) 909–911.

[5] P. Singh, S. Shergill, G.K. Makharia, Celiac disease in older adults, J. Gastrointestin. Liver Dis. 22 (2013) 357–362.

[6] P. Singh, S. Arora, A. Singh, T.A. Strand, G.K. Makharia, Prevalence of celiac disease in Asia: A systematic review and meta-analysis, J. Gastroenterol. Hepatol. 31 (2016) 1095–1101.

[7] P. Singh, A. Arora, T.A. Strand, D.A. Leffler, C. Catassi, P.H. Green, et al., Global Prevalence of Celiac Disease: Systematic Review and Meta-analysis, Clin. Gastroenterol. Hepatol. 16 (2018) 823–836 e2.

[8] M. Hadjivassiliou, D.S. Sanders, R.A. Grünewald, N. Woodroofe, S. Boscolo, D. Aeschlimann, Gluten sensitivity: from gut to brain, Lancet Neurol. 9 (2010) 318–330.

[9] E. Antiga, M. Caproni, I. Pierini, D. Bonciani, P. Fabbri, Gluten-free diet in patients with dermatitis herpetiformis: not only a matter of skin, Arch. Dermatol. 147 (2011) 988–989 author reply 989.

[10] M.N. Marsh, Gluten, major histocompatibility complex, and the small intestine. A molecular and immunobiologic approach to the spectrum of gluten sensitivity ('celiac sprue'), Gastroenterology 102 (1992) 330–354.

[11] G. Makharia, Where are Indian adult celiacs? Trop. Gastroenterol. 27 (2006) 1–3.

[12] L.S.P. Davidson, J.R. Fountain, Incidence of the sprue syndrome; with some observations on the natural history, Br. Med. J. 1 (1950) 1157–1161.

[13] J. Van Stirum, K. Baerlocher, A. Fanconi, E. Gugler, O. Tönz, D.H. Shmerling, The incidence of coeliac disease in children in Switzerland, Helv. Paediatr. Acta. 37 (1982) 421–430.

[14] M. Mylotte, B. Egan-Mitchell, C.F. McCarthy, B. McNicholl, Incidence of coeliac disease in the West of Ireland, Br. Med. J. 1 (1973) 703–705.

[15] C. Catassi, E. Fabiani, I.M. Rätsch, G.V. Coppa, P.L. Giorgi, R. Pierdomenico, et al., The coeliac iceberg in Italy. A multicentre antigliadin antibodies screening for coeliac disease in school-age subjects, Acta. Paediatr. Oslo. Nor. 1992 Suppl. 412 (1996) 29–35.

[16] K. Mustalahti, C. Catassi, A. Reunanen, E. Fabiani, M. Heier, S. McMillan, et al., The prevalence of celiac disease in Europe: results of a centralized, international mass screening project, Ann. Med. 42 (2010) 587–595.

[17] M.W. Laass, R. Schmitz, H.H. Uhlig, K.-P. Zimmer, M. Thamm, S. Koletzko, The Prevalence of Celiac Disease in Children and Adolescents in Germany, Dtsch Aerzteblatt Online (2015), doi:10.3238/arztebl.2015.0553.

[18] A. Fasano, I. Berti, T. Gerarduzzi, T. Not, R.B. Colletti, S. Drago, et al., Prevalence of celiac disease in at-risk and not-at-risk groups in the United States: a large multicenter study, Arch. Intern. Med. 163 (2003) 286–292.

[19] R. Pratesi, L. Gandolfi, S.G. Garcia, I.C. Modelli, P. Lopes de Almeida, A.L. Bocca, et al., Prevalence of coeliac disease: unexplained age-related variation in the same population, Scand. J. Gastroenterol. 38 (2003) 747–750.

[20] C.J. Hovell, J.A. Collett, G. Vautier, A.J. Cheng, E. Sutanto, D.F. Mallon, et al., High prevalence of coeliac disease in a population-based study from Western Australia: a case for screening? Med. J. Aust. 175 (2001) 247–250.

[21] H.B. Cook, M.J. Burt, J.A. Collett, M.R. Whitehead, C.M. Frampton, B.A. Chapman, Adult coeliac disease: prevalence and clinical significance, J. Gastroenterol. Hepatol. 15 (2000) 1032–1036.

[22] C. Catassi, I.M. Rätsch, L. Gandolfi, R. Pratesi, E. Fabiani, R. El Asmar, et al., Why is coeliac disease endemic in the people of the Sahara? Lancet London Engl. 354 (1999) 647–648.

[23] M. Abu-Zekry, D. Kryszak, M. Diab, C. Catassi, A. Fasano, Prevalence of celiac disease in Egyptian children disputes the east-west agriculture-dependent spread of the disease, J. Pediatr. Gastroenterol. Nutr. 47 (2008) 136–140.

[24] M. Ben Hariz, M. Kallel-Sellami, L. Kallel, A. Lahmer, S. Halioui, S. Bouraoui, et al., Prevalence of celiac disease in Tunisia: mass-screening study in schoolchildren, Eur. J. Gastroenterol. Hepatol. 19 (2007) 687–694.

[25] S.J. Baker, V.I. Mathan, Tropical enteropathy and tropical sprue, Am. J. Clin. Nutr. 25 (1972) 1047–1055.

[26] G.K. Makharia, C. Catassi, Celiac disease in Asia, Gastroenterol. Clin. North Am. 48 (2019) 101–113.

[27] G.K. Makharia, A.K. Verma, R. Amarchand, S. Bhatnagar, P. Das, A. Goswami, et al., Prevalence of celiac disease in the northern part of India: A community based study: Celiac disease in India, J. Gastroenterol. Hepatol. 26 (2011) 894–900.

[28] B.S. Ramakrishna, G.K. Makharia, K. Chetri, S. Dutta, P. Mathur, V. Ahuja, et al., Prevalence of Adult Celiac Disease in India: Regional Variations and Associations, Am. J. Gastroenterol. 111 (2016) 115–123.

[29] J. Yuan, C. Zhou, J. Gao, J. Li, F. Yu, J. Lu, et al., Prevalence of Celiac Disease Autoimmunity Among Adolescents and Young Adults in China, Clin. Gastroenterol. Hepatol. Off Clin. Pract. J. Am. Gastroenterol. Assoc. 15 (2017) 1572–1579 e1.

[30] C. Zhou, F. Gao, J. Gao, J. Yuan, J. Lu, Z. Sun, et al., Prevalence of coeliac disease in Northwest China: heterogeneity across Northern Silk road ethnic populations, Aliment. Pharmacol. Ther. 51 (2020) 1116–1129.

[31] G.J. Kou, J. Guo, X.L. Zuo, C.Q. Li, C. Liu, R. Ji, et al., Prevalence of celiac disease in adult Chinese patients with diarrhea-predominant irritable bowel syndrome: A prospective, controlled, cohort study, J. Dig. Dis. 19 (2018) 136–143.

[32] T.W.-C. Yap, W.-K. Chan, A.H.-R. Leow, A.N. Azmi, M.-F. Loke, J. Vadivelu, et al., Prevalence of serum celiac antibodies in a multiracial Asian population—a first study in the young Asian adult population of Malaysia, PloS One 10 (2015) e0121908.

[33] M. Fukunaga, N. Ishimura, C. Fukuyama, D. Izumi, N. Ishikawa, A. Araki, et al., Celiac disease in non-clinical populations of Japan, J. Gastroenterol. 53 (2018) 208–214.

[34] S. Zanella, L. De Leo, L. Nguyen-Ngoc-Quynh, B. Nguyen-Duy, T. Not, M. Tran-Thi-Chi, et al., Cross-sectional study of coeliac autoimmunity in a population of Vietnamese children, BMJ Open 6 (2016) e011173.

[35] G.K. Makharia, Celiac disease screening in southern and East Asia, Dig. Dis. Basel. Switz. 33 (2015) 167–174.

[36] C. Catassi, D. Kryszak, B. Bhatti, C. Sturgeon, K. Helzlsouer, S.L. Clipp, et al., Natural history of celiac disease autoimmunity in a USA cohort followed since 1974, Ann. Med. 42 (2010) 530–538.

[37] J.A. King, J. Jeong, F.E. Underwood, J. Quan, N. Panaccione, J.W. Windsor, et al., Incidence of Celiac Disease Is Increasing Over Time: A Systematic Review and Meta-analysis, Am. J. Gastroenterol. 115 (2020) 507–525.

[38] H. Okada, C. Kuhn, H. Feillet, J.-F. Bach, The 'hygiene hypothesis' for autoimmune and allergic diseases: an update, Clin. Exp. Immunol. 160 (2010) 1–9.

[39] J.F. Ludvigsson, D.A. Leffler, J.C. Bai, F. Biagi, A. Fasano, P.H.R. Green, et al., The Oslo definitions for coeliac disease and related terms, Gut. 62 (2013) 43–52.

[40] L.J. Virta, K. Kaukinen, P. Collin, Incidence and prevalence of diagnosed coeliac disease in Finland: results of effective case finding in adults, Scand. J. Gastroenterol. 44 (2009) 933–938.

[41] A.C. Ford, W.D. Chey, N.J. Talley, A. Malhotra, B.M.R. Spiegel, P. Moayyedi, Yield of diagnostic tests for celiac disease in individuals with symptoms suggestive of irritable bowel syndrome: systematic review and meta-analysis, Arch. Intern. Med. 169 (2009) 651–658.

[42] A.C. Ford, E. Ching, P. Moayyedi, Meta-analysis: yield of diagnostic tests for coeliac disease in dyspepsia, Aliment. Pharmacol. Ther. 30 (2009) 28–36.

[43] S. Bodé, E. Gudmand-Høyer, Symptoms and haematologic features in consecutive adult coeliac patients, Scand. J. Gastroenterol. 31 (1996) 54–60.

[44] J.W. Harper, S.F. Holleran, R. Ramakrishnan, G. Bhagat, P.H.R. Green, Anemia in celiac disease is multifactorial in etiology, Am. J. Hematol. 82 (2007) 996–1000.

[45] T.R. Halfdanarson, M.R. Litzow, J.A. Murray, Hematologic manifestations of celiac disease, Blood 109 (2007) 412–421.

[46] A. Dahele, S. Ghosh, Vitamin B12 deficiency in untreated celiac disease, Am. J. Gastroenterol. 96 (2001) 745–750.

[47] M. Haapalahti, P. Kulmala, T.J. Karttunen, L. Paajanen, K. Laurila, M. Mäki, et al., Nutritional status in adolescents and young adults with screen-detected celiac disease, J. Pediatr. Gastroenterol. Nutr. 40 (2005) 566–570.

[48] M.R. Howard, A.J. Turnbull, P. Morley, P. Hollier, R. Webb, A. Clarke, A prospective study of the prevalence of undiagnosed coeliac disease in laboratory defined iron and folate deficiency, J. Clin. Pathol. 55 (2002) 754–757.

[49] P. Singh, S. Arora, G.K. Makharia, Presence of anemia in patients with celiac disease suggests more severe disease, Indian J. Gastroenterol. Off. J. Indian Soc. Gastroenterol. 33 (2014) 161–164.

[50] S. Mahadev, M. Laszkowska, J. Sundström, M. Björkholm, B. Lebwohl, P.H.R. Green, et al., Prevalence of Celiac Disease in Patients With Iron Deficiency Anemia—A Systematic Review With Meta-analysis, Gastroenterology 155 (2018) 374–382 e1.

[51] N. Afredj, S. Metatla, S.A. Faraoun, A. Nani, N. Guessab, M. Benhalima, et al., Association of Budd-Chiari syndrome and celiac disease, Gastroenterol. Clin. Biol. 34 (2010) 621–624.

[52] P. Singh, A. Agnihotri, G. Jindal, P.K. Sharma, M. Sharma, P. Das, et al., Celiac disease and chronic liver disease: is there a relationship? Indian J. Gastroenterol. Off. J. Indian Soc. Gastroenterol. 32 (2013) 404–408.

[53] R. Sonti, B. Lebwohl, S.K. Lewis, H. Abu Daya, H. Klavan, K. Aguilar, et al., Men with celiac disease are shorter than their peers in the general population, Eur. J. Gastroenterol. Hepatol. 25 (2013) 1033–1037.

[54] S. Nardecchia, R. Auricchio, V. Discepolo, R. Troncone, Extra-Intestinal Manifestations of Coeliac Disease in Children: Clinical Features and Mechanisms, Front. Pediatr. 7 (2019) 56.

[55] B. Boersma, R.H.J. Houwen, W.F. Blum, J. van Doorn, J.M. Wit, Catch-up growth and endocrine changes in childhood celiac disease. Endocrine changes during catch-up growth, Horm. Res. 58 (Suppl 1) (2002) 57–65.

[56] A. Comba, G. Çaltepe, Y.ü.c.e. Ö, E. Erena, A.G. Kalaycı, Effects of age of diagnosis and dietary compliance on growth parameters of patients with celiac disease, Arch. Argent. Pediatr. 116 (2018) 248–255.

[57] A. Luciano, M. Bolognani, A. Di Falco, C. Trabucchi, P. Bonetti, A. Castellarin, [Catch-up growth and final height in celiac disease], Pediatr. Medica. E. Chir. Med. Surg. Pediatr. 24 (2002) 9–12.

[58] A.D. Singh, P. Singh, N. Farooqui, T. Strand, V. Ahuja, G.K. Makharia, Prevalence of celiac disease in patients with short stature: A systematic review and Meta-analysis, J. Gastroenterol. Hepatol. (2020), doi:10.1111/jgh.15167.

[59] C. Meazza, S. Pagani, K. Laarej, F. Cantoni, P. Civallero, A. Boncimino, et al., Short stature in children with coeliac disease, Pediatr. Endocrinol. Rev. PER. 6 (2009) 457–463.

[60] R. Troncone, R. Kosova, Short stature and catch-up growth in celiac disease, J. Pediatr. Gastroenterol. Nutr. 51 (Suppl 3) (2010) S137–S138.

[61] P. Elfström, J. Sundström, J.F. Ludvigsson, Systematic review with meta-analysis: associations between coeliac disease and type 1 diabetes, Aliment. Pharmacol. Ther. 40 (2014) 1123–1132.

[62] P. Collin, K. Kaukinen, M. Välimäki, J. Salmi, Endocrinological disorders and celiac disease, Endocr. Rev. 23 (2002) 464–483.

[63] A. Roy, M. Laszkowska, J. Sundström, B. Lebwohl, P.H.R. Green, O. Kämpe, et al., Prevalence of Celiac Disease in Patients with Autoimmune Thyroid Disease: A Meta-Analysis, Thyroid. Off. J. Am. Thyroid. Assoc. 26 (2016) 880–890.

[64] G. Bellastella, M.I. Maiorino, P. Cirillo, M. Longo, V. Pernice, A. Costantino, et al., Remission of Pituitary Autoimmunity Induced by Gluten-Free Diet in Patients With Celiac Disease, J. Clin. Endocrinol. Metab. 105 (2020), doi:10.1210/clinem/dgz228.

[65] R. Mazure, H. Vazquez, D. Gonzalez, C. Mautalen, S. Pedreira, L. Boerr, et al., Bone mineral affection in asymptomatic adult patients with celiac disease, Am. J. Gastroenterol. 89 (1994) 2130–2134.

[66] T. Kemppainen, H. Kröger, E. Janatuinen, I. Arnala, V.M. Kosma, P. Pikkarainen, et al., Osteoporosis in adult patients with celiac disease, Bone 24 (1999) 249–255.

[67] R. Ganji, M. Moghbeli, R. Sadeghi, G. Bayat, A. Ganji, Prevalence of osteoporosis and osteopenia in men and premenopausal women with celiac disease: a systematic review, Nutr. J. 18 (2019) 9.

[68] M. Olmos, M. Antelo, H. Vazquez, E. Smecuol, E. Mauriño, J.C. Bai, Systematic review and meta-analysis of observational studies on the prevalence of fractures in coeliac disease, Dig. Liver. Dis. Off. J. Ital. Soc. Gastroenterol. Ital. Assoc. Study Liver 40 (2008) 46–53.

[69] M. Laszkowska, S. Mahadev, J. Sundström, B. Lebwohl, P.H.R. Green, K. Michaelsson, et al., Systematic review with meta-analysis: the prevalence of coeliac disease in patients with osteoporosis, Aliment. Pharmacol. Ther. 48 (2018) 590–597.

[70] E.M. Stein, H. Rogers, A. Leib, D.J. McMahon, P. Young, K. Nishiyama, et al., Abnormal Skeletal Strength and Microarchitecture in Women With Celiac Disease, J. Clin. Endocrinol. Metab. 100 (2015) 2347–2353.

[71] M.B. Zanchetta, V. Longobardi, F. Costa, G. Longarini, R.M. Mazure, M.L. Moreno, et al., Impaired Bone Microarchitecture Improves After One Year On Gluten-Free Diet: A Prospective Longitudinal HRpQCT Study in Women With Celiac Disease, J. Bone Miner. Res. Off. J. Am. Soc. Bone Miner. Res. 32 (2017) 135–142.

[72] A. Sainsbury, D.S. Sanders, A.C. Ford, Meta-analysis: coeliac disease and hypertransaminasaemia, Aliment. Pharmacol. Ther. 34 (2011) 33–40.

[73] J.F. Ludvigsson, P. Elfström, U. Broomé, A. Ekbom, S.M. Montgomery, Celiac disease and risk of liver disease: a general population-based study, Am. Gastroenterol. Assoc. 5 (2007) 63–69 e1.

[74] S. Lindgren, K. Sjöberg, S. Eriksson, Unsuspected coeliac disease in chronic 'cryptogenic' liver disease, Scand. J. Gastroenterol. 29 (1994) 661–664.

[75] K. Kaukinen, L. Halme, P. Collin, M. Färkkilä, M. Mäki, P. Vehmanen, et al., Celiac disease in patients with severe liver disease: gluten-free diet may reverse hepatic failure, Gastroenterology 122 (2002) 881–888.

[76] J. Wakim-Fleming, M.R. Pagadala, A.J. McCullough, R. Lopez, A.E. Bennett, D.S. Barnes, et al., Prevalence of celiac disease in cirrhosis and outcome of cirrhosis on a gluten free diet: a prospective study, J. Hepatol. 61 (2014) 558–563.

[77] A. Rubio-Tapia, J.A. Murray, The liver in celiac disease, Hepatol. Baltim. Md. 46 (2007) 1650–1658.

[78] U. Volta, L. De Franceschi, N. Molinaro, F. Cassani, L. Muratori, M. Lenzi, et al., Frequency and significance of anti-gliadin and anti-endomysial antibodies in autoimmune hepatitis, Dig. Dis. Sci. 43 (1998) 2190–2195.

[79] D. Villalta, D. Girolami, E. Bidoli, N. Bizzaro, M. Tampoia, M. Liguori, et al., High prevalence of celiac disease in autoimmune hepatitis detected by anti-tissue tranglutaminase autoantibodies, J. Clin. Lab. Anal. 19 (2005) 6–10.

[80] N.M. van Gerven, S.F. Bakker, Y.S. de Boer, B.I. Witte, H. Bontkes, C.M. van Nieuwkerk, et al., Seroprevalence of celiac disease in patients with autoimmune hepatitis, Eur. J. Gastroenterol. Hepatol. 26 (2014) 1104–1107.

[81] A. Lawson, J. West, G.P. Aithal, R.F.A. Logan, Autoimmune cholestatic liver disease in people with coeliac disease: a population-based study of their association, Aliment. Pharmacol. Ther. 21 (2005) 401–405.

[82] S. Caprai, P. Vajro, A. Ventura, M. Sciveres, G. Maggiore, Autoimmune Liver Disease Associated With Celiac Disease in Childhood: A Multicenter Study, Clin. Gastroenterol. Hepatol. 6 (2008) 803–806.

[83] M.B. Jacobsen, O. Fausa, K. Elgjo, E. Schrumpf, Hepatic lesions in adult coeliac disease, Scand. J. Gastroenterol. 25 (1990) 656–662.

[84] I. Singh, A. Agnihotri, A. Sharma, A.K. Verma, P. Das, B. Thakur, et al., Patients with celiac disease may have normal weight or may even be overweight, Indian J. Gastroenterol. Off. J. Indian Soc. Gastroenterol. 35 (2016) 20–24.

[85] R. Tortora, P. Capone, G. De Stefano, N. Imperatore, N. Gerbino, S. Donetto, et al., Metabolic syndrome in patients with coeliac disease on a gluten-free diet, Aliment. Pharmacol. Ther. 41 (2015) 352–359.

[86] A. Agarwal, A. Singh, W. Mehtab, V. Gupta, A. Chauhan, M.S. Rajput, et al., Patients with celiac disease are at high risk of developing metabolic syndrome and fatty liver, Intest. Res. (2020), doi:10.5217/ir.2019.00136.

[87] A. Diamanti, T. Capriati, M.S. Basso, F. Panetta, V.M. Di Ciommo Laurora, F. Bellucci, et al., Celiac disease and overweight in children: an update, Nutrients 6 (2014) 207–220.

[88] M. Papastamataki, I. Papassotiriou, A. Bartzeliotou, A. Vazeou, E. Roma, G.P. Chrousos, et al., Incretins, amylin and other gut-brain axis hormones in children with coeliac disease, Eur. J. Clin. Invest. 44 (2014) 74–82.

[89] W.M. Weinstein, J.R. Brow, F. Parker, C.E. Rubin, The small intestinal mucosa in dermatitis herpetiformis. II. Relationship of the small intestinal lesion to gluten, Gastroenterology 60 (1971) 362–369.

[90] E. Mansikka, K. Hervonen, K. Kaukinen, P. Collin, H. Huhtala, T. Reunala, et al., Prognosis of Dermatitis Herpetiformis Patients with and without Villous Atrophy at Diagnosis, Nutrients 10 (2018), doi:10.3390/nu10050641.

[91] P.H.R. Green, S.N. Stavropoulos, S.G. Panagi, S.L. Goldstein, D.J. Mcmahon, H. Absan, et al., Characteristics of adult celiac disease in the USA: results of a national survey, Am. J. Gastroenterol. 96 (2001) 126 –1.

[92] R. Eliakim, D.M. Sherer, Celiac disease: fertility and pregnancy, Gynecol. Obstet. Invest. 51 (2001) 3–7.

[93] A. Gasbarrini, E.S. Torre, C. Trivellini, S. De Carolis, A. Caruso, G. Gasbarrini, Recurrent spontaneous abortion and intrauterine fetal growth retardation as symptoms of coeliac disease, Lancet. Lond. Engl. 356 (2000) 399–400.

[94] J.S. Morris, A.B. Adjukiewicz, A.E. Read, Coeliac infertility: an indication for dietary gluten restriction? Lancet. Lond. Engl. 1 (1970) 213–214.

[95] D. Zugna, L. Richiardi, O. Akre, O. Stephansson, J.F. Ludvigsson, A nationwide population-based study to determine whether coeliac disease is associated with infertility, Gut. 59 (2010) 1471–1475.

[96] P. Singh, S. Arora, S. Lal, T.A. Strand, G.K. Makharia, Celiac Disease in Women With Infertility: A Meta-Analysis, J. Clin. Gastroenterol. 50 (2016) 33–39.

[97] K. Bürk, M.-L. Farecki, G. Lamprecht, G. Roth, P. Decker, M. Weller, et al., Neurological symptoms in patients with biopsy proven celiac disease, Mov. Disord. Off, J. Mov. Disord. Soc. 24 (2009) 2358–2362.

[98] M. Hadjivassiliou, P. Aeschlimann, D.S. Sanders, M. Mäki, K. Kaukinen, R.A. Grünewald, et al., Transglutaminase 6 antibodies in the diagnosis of gluten ataxia, Neurology 80 (2013) 1740–1745.

[99] M. Hadjivassiliou, A.P. Duker, D.S. Sanders, Gluten-related neurologic dysfunction, Handb. Clin. Neurol. 120 (2014) 607–619.

[100] L. Pastore, A. Carroccio, D. Compilato, V. Panzarella, R. Serpico, L. Lo Muzio, Oral manifestations of celiac disease, J. Clin. Gastroenterol. 42 (2008) 224–232.

[101] L. Aine, M. Mäki, P. Collin, O. Keyriläinen, Dental enamel defects in celiac disease, J. Oral. Pathol. Med. Off. Publ. Int. Assoc. Oral. Pathol. Am. Acad. Oral. Pathol. 19 (1990) 241–245.

[102] D. Souto-Souza, da Consolação, M.E. Soares, V.S. Rezende, P.C. de Lacerda Dantas, E.L. Galvão, S.G.M. Falci, Association between developmental defects of enamel and celiac disease: A meta-analysis, Arch. Oral. Biol. 87 (2018) 180–190.

[103] A.F.B. Oliveira, A.M.B. Chaves, A. Rosenblatt, The influence of enamel defects on the development of early childhood caries in a population with low socioeconomic status: a longitudinal study, Caries. Res. 40 (2006) 296–302.

[104] A.K.L. Tsang, The Special Needs of Preterm Children - An Oral Health Perspective, Dent. Clin. North Am. 60 (2016) 737–756.

[105] S.O. Akintoye, M.S. Greenberg, Recurrent aphthous stomatitis, Dent. Clin. North Am. 58 (2014) 281–297.

[106] J. Cheng, T. Malahias, P. Brar, M.T. Minaya, P.H.R. Green, The association between celiac disease, dental enamel defects, and aphthous ulcers in a United States cohort, J. Clin. Gastroenterol. 44 (2010) 191–194.

[107] M. Nieri, E. Tofani, E. Defraia, V. Giuntini, L. Franchi, Enamel defects and aphthous stomatitis in celiac and healthy subjects: Systematic review and meta-analysis of controlled studies, J. Dent. 65 (2017) 1–10.

[108] G. Cervino, L. Fiorillo, L. Laino, A.S. Herford, F. Lauritano, G.L. Giudice, et al., Oral Health Impact Profile in Celiac Patients: Analysis of Recent Findings in a Literature Review, Gastroenterol. Res. Pract. 2018 (2018) 7848735.

[109] A. Rubio–Tapia, R.A. Kyle, E.L. Kaplan, D.R. Johnson, W. Page, F. Erdtmann, T.L. Brantner, W.R. Kim, T.K. Phelps, B.D. Lahr, A.R. Zinsmeister, Increased prevalence and mortality in undiagnosed celiac disease, Gastroenterology 137 (1) (2009) 88–93 Jul 1.

[110] G. Corrao, G.R. Corazza, V. Bagnardi, G. Brusco, C. Ciacci, M. Cottone, C.S. Guidetti, P. Usai, P. Cesari, M.A. Pelli, S. Loperfido, Mortality in patients with coeliac disease and their relatives: a cohort study, The Lancet 358 (9279) (2001) 356–361 Aug 4.

[111] J.F. Ludvigsson, Mortality and malignancy in celiac disease, Gastrointestinal Endoscopy Clinics 22 (4) (2012) 705–722 Oct 1.

[112] B. Lebwohl, P.H. Green, J. Söderling, B. Roelstraete, J.F. Ludvigsson, Association between celiac disease and mortality risk in a Swedish population, JAMA 323 (13) (2020) 1277–1285 Apr 7.

[113] C. Catassi, E. Fabiani, G. Corrao, et al., Risk of non-Hodgkin lymphoma in celiac disease, JAMA 287 (11) (2002) 1413–1419.

[114] J.F. Ludvigsson, K. Michaelsson, A. Ekbom, S.M. Montgomery, Coeliac disease and the risk of fractures—a general population-based cohort study, Aliment. Pharmacol. Ther. 25 (3) (2007) 273–285.

[115] P. Elfström, J. Sundström, J.F. Ludvigsson, Systematic review with meta-analysis: associations between coeliac disease and type 1 diabetes, Aliment. Pharmacol. Ther. 40 (10) (2014) 1123–1132.

[116] J.F. Ludvigsson, B. Lindelöf, F. Zingone, C. Ciacci, Psoriasis in a nationwide cohort study of patients with celiac disease, J. Invest. Dermatol. 131 (10) (2011) 2010–2016.

CHAPTER 2

Pediatric coeliac disease

Kalle Kurppa[a], Daniel Agardh[b]
[a]Tampere University and Tampere University Hospital, Tampere, Finland
[b]Department of Clinical Sciences, Lund University, Malmö, Sweden

2.1 Introduction: Distinct features of pediatric coeliac disease

Coeliac disease was for long considered as a chronic intestinal disorder mainly affecting infants and toddlers. The first international clinical guidelines were therefore laid by paediatric gastroenterology organizations [1,2]. Over time, the diagnostic procedures have been improved and novel disease-specific autoantibodies discovered. The evolution of manageable immunoassays for clinical practice has facilitated the screening of large populations. As such, the conception of the disease have evolved from a rare paediatric malabsorption syndrome to a common disease presenting with diverse manifestations and affecting individuals at all ages [3]. As a consequence, the diagnostic criteria have also been adapted by the European Society for Paediatric Gastroenterology and Hepatology (ESPGHAN) from the long-standing histological definition toward less non-invasive approach relying more on serological markers [4,5].

Although the aforesaid paradigm shift is shared with all patients regardless of age, the perception of coeliac disease differs markedly between children and adults in terms of diagnosis, treatment, and follow-up (Table 2.1) [3,6–8]. Paediatric patients were previously considered to present mostly with gastrointestinal symptoms and failure to thrive, whereas less characteristic forms, including various extra-intestinal manifestations, were frequently observed as the first and only clinical finding in adults [9–11]. Through numerous performed screenings in children with no previous coeliac disease suspicion, it is now well established that the majority of them actually also present with atypical and subclinical forms [12,13]. There is thus a risk for long-term complications if these children are not diagnosed and treated early, which calls for effective case-finding and screening strategies.

There are also aspects differing between children and adults in regards to the adherence to the gluten-free diet and long-term follow-up of coeliac disease (Table 2.1). Patients diagnosed in early childhood may not later recall their presenting symptoms and burden of the disease at diagnosis, which may predispose them to poor compliance later during turbulent adolescence [14,15]. This emphasizes the importance of successful transition from paediatric to adult follow-up care [16]. Furthermore, in contrast to adults, continuous monitoring of growth and development is an integral part of the management in children and adolescents with coeliac disease. The aim of this chapter is to give the

Coeliac Disease and Gluten-Related Disorders.
DOI: https://doi.org/10.1016/B978-0-12-821571-5.00002-7

Table 2.1 Features of coeliac disease that differ between children and adults.

Feature	Children	Adults
Genetics	High-risk haplotypes overrepresented	Moderate-risk haplotypes prevail
Presentation	Often clinically and histologically severe	Varies from asymptomatic to RCD
Diagnosis	Non-biopsy approach possible	Histology is the gold standard
Treatment	Under parents control until adolescence	Self-responsibility
Follow-up	Organized by healthcare, re-biopsy rarely performed	Self-responsibility, often re-biopsy
Complications	Poor growth and reduced bone accrual	Osteoporosis, infertility, malignancies
Prognosis	Excellent in case of early diagnosis	Complications possible even on a GFD

Abbreviations: HLA, human leucocyte antigen; GFD, gluten-free diet; RCD, refractory celiac disease

current state of knowledge about the diagnosis, treatment, and monitoring of paediatric coeliac disease.

2.2 Clinical presentation

The archetypical presentation of celiac disease is diarrhea, malabsorption, and poor growth/failure to thrive in infancy (Fig. 2.1) [17].

Although gastrointestinal symptoms may still dominate as the presenting signs in paediatric patients, particularly in low-income countries, the overall clinical picture is becoming far more heterogeneous (Table 2.2). Based on experience form population-based screenings, it is now becoming widely accepted that many if not even the majority of children only have mild symptoms or no apparent symptoms [11,18–20]. With increasing awareness and more active screening of the condition, age at diagnosis has in turn increased from early childhood to middle school age [11]. However, patients presenting with severe forms of malabsorption syndromes, including the rare cases of coeliac crisis, are found almost exclusively in children younger than two years of age [21–24]. This phenotype, often referred as classical coeliac disease, is at least in partly explained by high genetic predisposition [25].

In addition to symptoms described in Table 2.2, possible additional oral findings are, for example, aphthous stomatitis, angular cheilitis, atrophic glossitis, and dry mouth [26–28] Also, specific forms of gluten-sensitive epilepsy have been described in certain limited geographical areas [29–31]. Rickets is nowadays a rare complication of an untreated coeliac disease in children [17,32,33], although poor bone accrual and osteomalacia may still be present, predisposing to increased risk of osteoporosis and low-energy fractures later in adolencense and adulthood [34–38].

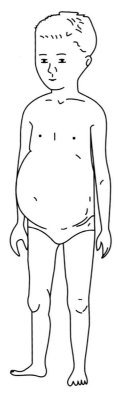

Figure 2.1 Nowadays in high-income countries fairly infrequent case of "classical" coeliac disease, designating a patient presenting in early infancy with severe protein-losing enteropathy, poor growth, distended abdomen, and muscle wasting. Figure drawn by Siiri Lahti.

Table 2.2

Symptom or sign	Prevalence, %	Comments
Gastrointestinal symptoms		
Diarrhea/fatty stools	12–49	"Classical" presentation of coeliac disease
Vomiting/nausea	3–17	Appears to be quite rare nowadays
Abdominal pain	14–77	Difficult to recognize in infants
Constipation	5–41	Probably markedly underrecognized symptom
Extra-intestinal symptoms		
Poor growth	11–48	Reduced adult height possible if unrecognized
Enamel defects	1–6	An early sign that may not respond to GFD
Anaemia	7–74	Quite common and sometimes the sole presenting sign
Neurologic/psychiatric	0–24	E.g. headache, developmental delay, depression
Elevated transaminases	9–40	Usually mild and easily reversible in children
Delayed puberty	ND	Exact prevalence unclear
Arthralgia or arthritis	0–11	Difficult to differentiate e.g. from growth pains
Skin symptoms	0–15	Dermatitis herpetiformis is rare in childhood

Abbreviations: GFD, gluten-free diet; ND, no data

Anemia and increased hepatic transaminases are frequently encountered laboratory findings, although sings of liver damage are usually milder compared with adults [39–44]. Common deranged hematologic findings found in paediatric coeliac disease patients are iron deficiency and decreased levels of other micronutrients, albeit deficiencies of vitamin B12 and folate are quite rare in the pediatric age group [44–48]. These laboratory abnormalities do not necessarily associate with any apparent signs of malabsorption.

Children with coeliac disease may also present with psychological manifestations and various constitutional symptoms, such as tiredness, lack of energy, feel of discomfort, irritability, and underachievement [21,49,50]. Other symptoms frequently described in adults, such as gluten ataxia, significant neuropathy, and intestinal malignancies, are exceptional in patients diagnosed in childhood [29,51,52].

Since the clinical picture is so multifaceted, it is more or less impossible to differentiate coeliac disease clinically from other common conditions such as inflammatory bowel disease and functional gastrointestinal disorders [53,54]. In fact, as coeliac disease is so frequent in the population, the symptoms leading to the original suspicion and screening of the disease might have been just a coincidence or caused by coexisting disorder [55–58]. The wide panorama of the clinical presentations support a low-threshold case-finding or, in the future, even large-scale serological screening in order to facilitate the detection of the majority of affected individuals (see Section 2.4) [59–63].

2.3 Diagnosis

The diagnostic criteria of coeliac disease have continuously been revised during the past decades [1,2,4,5,64]. To date, the noninvasive approach by using serology instead of intestinal biopsy to confirm the diagnosis is more uniformly implemented in pediatric than adult patients. The current adult guidelines still rely strongly on the demonstration of duodenal pathology, with some exceptions which have accepted the serological approach for the diagnosis also in a part of the adult patients [65–67]. Age-related discrepancies in the diagnostic guidelines may thus be a potential source of misconception about the diagnosis between paediatric and adult gastroenterologists when the child is referred to the adult care. Of note, the criteria by European Society for the Study of Coeliac Disease accepts the serology-based diagnosis as laid by ESPGHAN [66].

It is recommended that the initial serological testing in a case of coeliac disease suspicion should be performed by assessing IgA class tissue transglutaminase antibodies (tTGA) while the child is still on a gluten-containing diet. It is worth measuring serum total IgA once to exclude IgA deficiency (see Fig. 2.2). Other conditions should be considered in symptomatic children with negative tTGA [68]. Possible reasons for false-negative tTGA include extra-intestinal manifestations of coeliac disease, short duration of the disease in infants, reduced gluten consumption (not uncommon in family members or self-initiated gluten-free diet) and immunosuppression [5,71]. In such cases, an extended

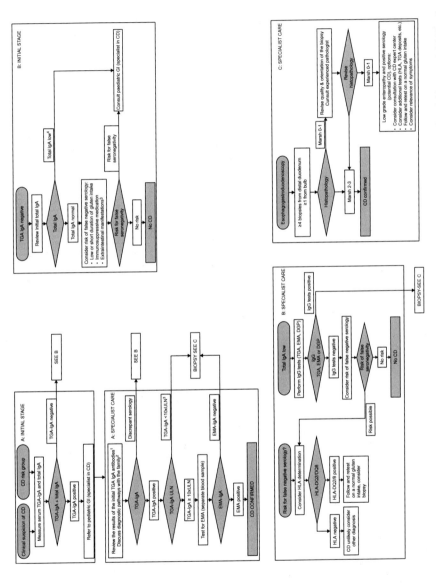

Figure 2.2 The ESPGHAN 2020 criteria for coeliac disease. (1) Only validated assays with calibration curve should be used; (2) The family should be informed that the diagnosis confirms the need for a life-long gluten-free diet; (3) In case of tTGA borderline levels of positivity, sufficient gluten intake and re-testing should be considered; (4) Low for age or <0.2 g/l above the age of three years; (5) E.G. dermatitis herpetiformis. Modified from Husby et al. J Pediatr Gastroenterol Nutr 2020;70:141-56 with permission [5].

gluten challenge and re-testing is indicated [4,72]. In addition to these diagnostic pitfalls, the physician should be aware of variations in the diagnostic performances between different commercial tTGA kits [69,70]. Endomysium autoantibodies (EmA) and antibodies to native or deamidated gliadin peptides are other available biomarkers of coeliac disease, but they are currently not recommended for the initial testing. In addition, although coeliac disease is strongly associated with HLA-DR3-DQ2 and DR4-DQ8, these haplotypes are frequent occurring in the population and should not be used as a diagnostic test [73,74]. However, the strong negative predictive value of not having coeliac disease in non-HLA-DQ2/DQ8 carriers may instead be used for excluding coeliac disease in dubious cases [5].

According to ESPGHAN, diagnosis of coeliac disease in children can be established on basis of either serological or histological criteria (Fig. 2.2). If the serological approach is applied, a minimum level of tTGA ten times the upper limit of normal in validated immunoassays for tTGA, in an optimal setting confirmed by a positive EmA test, is required for the diagnosis [5,75]. However, assessment of EmA may not be available in all centers, and whether it could be omitted in the future in patients fulfilling the tTGA criteria remains to be determined.

Using the histological approach, ESPGHAN states that an intestinal biopsy showing Marsh II or more severe lesion is required for the diagnosis (Fig. 2.2). Interpretation of histological findings showing less advanced damage is difficult [83–85]. In unclear cases, it is important to ensure that sufficient number of samples (≥4 and ≥1 from the anatomical duodenal bulb) were taken to exclude patchy duodenal lesions [76,79]. Moreover, sufficient quality and correct orientation of the biopsy cuttings are critical for reliable histological analysis [79–81]. In theory, possibly associated comorbidities such as eosinophilic esophagitis might be missed when omitting the endoscopy [82], but risk for this is considered to be low in children [5,75,76]. A gastroduodenoscopy is nevertheless recommended in case of alarm symptoms suggesting other upper gastrointestinal disease. The intestinal biopsy still remains mandatory for the diagnosis of coeliac disease in children according the 2005 published US guidelines, although current implementation of these criteria in clinical routine is unclear [64]. Rare cases of pediatric dermatitis herpetiformis, dermatological form of coeliac disease, can be diagnosed by skin biopsy (see Chapter 8).

Diagnostic challenges in clinical practice occur particularly when there are discordances between the results of serology and histology. Table 2.3 shows conditions that may need additional diagnostic investigations to differentiate them from paediatric coeliac disease [68].

Another clinical challenge is if to treat and follow children having normal or only minor mucosal findings (Marsh 0-I) despite persistently elevated tTGA levels [71], a condition coined potential coeliac disease. The management of these children remains debated, but there is some evidence that they benefit from an early-initiated gluten-free

Table 2.3 Conditions that may mimic celiac disease in children.

Immunologic and allergic diseases	Infections and other conditions
Autoimmune enteropathy	Cryptosporidiosis, giardiasis
Severe cow's milk allergy	H. Pylori infection
Common variable immunodeficiency and other severe immunedeficiencies	Viral gastroenteritis, HIV
Eosinophilic gastroenteritis	Non-steroidal anti-inflammatory drugs
Inflammatory bowel disease	Malnutrition, severe vitamin deficiencies

Abbreviations: HIV, human immunodeficiency virus

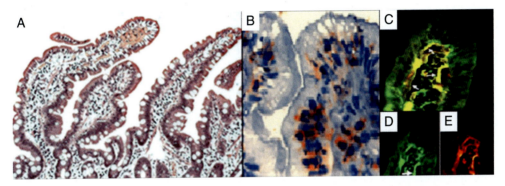

Figure 2.3 A 11-year-old tissue transglutaminase (tTG) antibody positive girl suffered from diarrhea and bloating but had non-diagnostic duodenal histology (A). Specific immunohistochemistric investigations from frozen sample revealed increased mucosal inflammation (B) and disease-specific co-localization (C) of IgA (D) and tTG (E). Gluten-free diet resulted positive clinical and serological response. Modified with permission from Kurppa et al. J Pediatr 2010;157:373-80 [86].

diet or develop manifest coeliac disease later in life if continuing gluten consumption [86–89], Under these circumstances, consultation of a center specialized in coeliac disease is recommended for discussion on extended analysis utilizing more sophisticated diagnostic methodologies (Fig. 2.3) [85,90,91].

2.4 Screening of children for coeliac disease

Without active screening, the majority of children affected by a coeliac disease will likely remain undiagnosed and thus prone to develop secondary complications later in life. Screening of children with conditions associated with an increased risk for coeliac disease is becoming increasingly implemented [92–95]. The best-characterized of these at-risk groups are presented in Table 2.4. Beside these, many other immune-mediated diseases have also been suspected to be overrepresented in coeliac disease patients, at least in adulthood, but the evidence with these is less conclusive [96,97].

Table 2.4 The reported prevalence of celiac disease in at-risk subjects [98,100–108].

Risk group	Study cohort	Prevalence, %	Country
First-degree relatives	4,508 children and adults	4.5	United States
	14,225 children and adults	5.6	India
Type 1 diabetes	4,322 children	6.8	Italy
	1,151 children	9.1	Sweden
Thyroidal diseases	302 children	2.3	United States
	952 adults	10.2	United States
IgA deficiency	126 children	8.7	Italy
	34 children and adults	6.0	Brazil
Addison's disease	109 children	2.7	Italy
	925 adults	0.3	United States
Trisomy 21	105 children	3.8	United States
	72 children	5.6	Brazil
Turner syndrome	87 children	4.6	Sweden
	389 children and adults	6.4	Italy

Table 2.5 Recommendations for celiac disease screening in international guidelines [5,64,67,109,110].

Organization	Age groups	Screening recommendation
NICE	All	T1D, AIT, family risk
BSG	Adults	T1D, irritable bowel syndrome, Down syndrome, family risk
ACG	All	Symptomatic T1D, family risk
ESPGHAN	Children	T1D, AIT, liver diseases, syndromes[1], IgA deficiency, family risk
NASPGHAN	Children	T1D, AIT, liver diseases, syndromes[1], family risk

[1] Trisomy 21, Turner, Williams. Abbreviations: ACG, American College of Gastroenterology; BSG, British Society of Gastroenterology; AIT, autoimmune thyroidal disease; ESPGHAN and NASPGHAN, European and North American Society for Pediatric Gastroenterology Hepatology and Nutrition; NICE, The National Institute for Health and Care Excellence; T1D, type 1 diabetes.

First-degree relatives of coeliac disease patients comprise the largest risk group, their likelihood to develop the disease being approximately 5–10% [98,99]. Another important group is individuals with a previously diagnosed other autoimmune condition [3]. The current international guidelines for coeliac disease are not uniform as regards on which risk groups should be screened (Table 2.5).

The common nominator for being affected by a double diagnosis is attributed to shared genetics, although environmental factors might also play a role [111]. It has been suggested that HLA genotyping to identify individuals at risk for coeliac disease could be applied as a first step for determining who should be screened, followed by serological screening among those with HLA-DQ2 and/or DQ8 as a second step [112]. By this approach, a substantial proportion of the children might be saved from unnecessary testing, since less than half the population are carriers of these two haplotypes [113–116]. Such a stepwise approach may be useful in some but not all of the aforementioned risk groups [117,118].

It is still unclear when the optimal age for the first screening is and how often the negative testing should be repeated [119]. Coeliac disease can affect individuals at any age, although data from longitudinal birth cohorts indicate that the frequency of new tTGA seroconversions is low after puberty [120]. Onset of the disease is likely significantly dependent on genetics factors; it is known that HLA-DQ2/DQ2 homozygous children have higher risk of early onset as compared with HLA-DQ8 carriers [112,120]. The results anyhow indicate that at-risk children should be tested at several time points [122]. One option could be a two-step strategy, comprising e.g. a first screening before school age and a second before teen age in asymptomatic individuals. However, longitudinal data from prospective studies are warranted before this approach can be applied as a general screening strategy.

Whether or not to perform general mass screening is a matter of debate. Variations in the incidence between countries should be considered, as screening is not sensible in low-prevalence populations [13,123]. Screening-detected children also more often have potential coeliac disease and, as mentioned, diagnostic approach to these subjects remains disputed [89,124]. There is, however, evidence that patients detected in screenings attain excellent dietary adherence and good quality of life on a gluten-free diet [11,12,18, 124–126]. Moreover, even apparently asymptomatic children with coeliac disease children may have signs of reduced bone density and suffer from anemia and poor growth [12,18]. More research on the possible benefits and optimal age, frequency and cost-effectiveness of mass screening are needed.

2.5 Treatment and monitoring of paediatric coeliac disease

To date there is still no cure for coeliac disease. Albeit several potential pharmaceutical approaches have been tested, none has proven to be more efficient and tolerable than the gluten-free diet. Accordingly, the only available treatment is a strict lifelong gluten-free diet [4,127]. After the diagnosis has been established, the child with guardian should be referred to a dietitian for a consultation. In practice, the dietary treatment means strict exclusion of all gluten-containing foods. The food products are defined as gluten-free when containing less than 20 milligrams of gluten per kilogram. Use of gluten-free oats as a complement to the diet is still controversial in some parts of the world, but based on current evidence considered well tolerated and safe option in the majority of pediatric celiac disease patients [128–130].

Although a qualified dietitian for initial dietary guidance is strongly recommended [131]. it is not always feasible in all healthcare systems. In practice, maintaining a gluten-free diet is mostly a concern of the caregivers until the responsibility is gradually transferred to the child in early adolescence. This is typically the period when problems with dietary compliance occur due to concurrent other life changes and peer pressure [15,132]. Accepting a restricted diet might be particularly difficult in screen-detected asymptomatic patients and those who experience no or only negligible symptoms after gluten exposure [133,134].

Although actual evidence is sparse, long-term monitoring of coeliac disease is considered essential for ensuring a good compliance to the gluten-free diet and early detection of possible complications and co-morbidities [135–138]. Optimal frequency and type of the follow-up has not been clearly defined and depend partly on access to health care professionals. The general recommendation is annual visits to a paediatrician and/or dietitian with special interest in coeliac disease [139]. This schedule might nevertheless be tailored according to the local circumstances and family preferences. There are indications that e-health visits could be a practical and reliable alternative for a part of the clinical visits [140].

During the follow-up, special attention should be paid to the normal growth and pubertal development after the recovery of initial clinical and laboratory findings. The majority of baseline abnormalities, including for example iron deficiency anemia, secondary lactose intolerance and poor growth, usually improve markedly within the first year on a gluten-free diet [141–143]. Vitamin and micronutrient supplements might be considered in selected cases. Parameters that should be routinely monitored have not been clearly defined, but wide-scale laboratory surveillance and bone densitometry are usually not needed in children on clinical remission [144–146]. Further investigations e.g. for hormone disturbances could be considered if there is poor growth and/or pubertal delay despite a strict gluten-free diet. It is also good to realize that a too restrictive diet may also predispose to nutritional imbalances, anorexia and, paradoxically, even to obesity.

The dietary compliance should be regularly checked for all patients and particularly in non-responders. Measurement of the tTGA helps assessing the compliance, but sensitivity for minor lapses can be suboptimal and, on the other hand, sometimes seroconversion to normal levels take even years [147]. In children with fluctuating or persistently elevated tTGA, intentional or inadvertent dietary transgressions should be suspected and a comprehensive dietary evaluation and even a duodenal biopsy controlling for mucosal healing should be considered [146]. Monitoring IgA deficient patients is challenging as IgG-class antibodies may remain elevated for a long time despite of gluten exclusion. The recently introduced urine and fecal gluten immunogenic peptides could more accurate markers for the compliance than serology, although more evidence in children is needed [148,149]. If the symptoms persist even with a strict gluten-free diet, presence of coexisting gastrointestinal or autoimmune comorbidity should be excluded. Severe complications, including refractory coeliac disease and intestinal malignancies, are extremely rare in children [150].

Seamless transition of adolescents with coeliac disease from paediatric to adult care is essential for successful long-term treatment. Due to the aforesaid increased risk for compliance problems in adolescence, sufficient knowledge about the importance of a strict gluten exclusion should be ensured and additional healthcare visits organized if needed. Furthermore, the possible health concerns experienced by the adolescent should be enquired, preferably with systematic questionnaires, in order to find patients needing psychological counseling and to avoid unnecessary restrictions in the daily life [151,152].

2.6 Toward primary prevention of coeliac disease?

A steep increase in the incidence of coeliac disease that occurred in Sweden in the mid-1980's and a subsequent decline after changes in the national feeding practices led to the hypothesis about a window of opportunity for inducing tolerance to gluten during weaning [153,154]. This theory was tested in a randomized multicenter trial in which at-risk infants were given either 100 mg gluten or placebo daily between 4 and 6 months of age. However, no significant difference in prevalence of coeliac disease between the groups at 3 years of age was observed [155]. Concomitantly, an Italian group observed the age of gluten introduction (6 vs. 12 months) or breastfeeding to have no effect on the risk of the disease at the age of 5 years [156]. There is anyhow growing evidence that environmental triggers, particularly certain viruses, and the amount of gluten intake in infancy, could modulate the risk of coeliac disease among at-risk individuals [157–161]. Results from the first prospective clinical trials are still pending whether modifying these factors could actually prevent or least delay later development of coeliac disease in at-risk children.

2.7 Summary

The view on coeliac disease in children has changed considerably over time from a rare malabsorption syndrome to a condition presenting with plethora of symptoms affecting individuals at all ages. The challenge to identifying coeliac disease also in children with subtle or atypical symptoms emphasize the role of low-treshold case-finding and active screening of at-risk individuals. In recent years, an early diagnosis of the condition has been made easier by the possibility to omit the duodenal biopsy in selected children, although there currently is no worldwide consensus in this. Hopefully, in the course of time, the less invasive approach could be applied globally and possible even more broadly to the majority of tTGA positive children. It is also important for the physicians to acknowledge the special challenges encountered by adolescent patients and enhance their skills for coping through this vulnerable age. Successful transition to adult care is of particular importance in these circumstances. Important goals of the current scientific research are effective prevention of the secondary complications and perhaps even prevention of coeliac disease.

References

[1] G.W. Meeuwisse, Round Table discussion. Diagnostic criteria for celiac disease, Acta. Paediatr. Scand. 59 (1970) 461.

[2] J.A. Walker-Smith, S. Guandalini, J. Schmitz, D.H. Shmerling, J.K. Visakorpi, Revised criteria for diagnosis of celiac disease, Arch. Dis. Child. 65 (1990) 909–911.

[3] K. Lindfors, C. Ciacci, K. Kurppa, et al., Coeliac disease, Nat. Rev. Dis. Primers 5 (2019) 3.

[4] S. Husby, S. Koletzko, I.R. Korponay-Szabó, et al., European Society for Pediatric Gastroenterology, Hepatology, and Nutrition Guidelines for the diagnosis of coeliac disease, J. Pediatr. Gastroenterol. Nutr. 54 (2012) 136–160.

[5] S. Husby, S. Koletzko, I.R. Korponay-Szabó, et al., European Society Paediatric Gastroenterology, Hepatology and Nutrition Guidelines for diagnosing coeliac disease 2020, J. Pediatr. Gastroenterol. Nutr. 70 (2020) 141–156.

[6] P.H. Green, S. Krishnareddy, B. Lebwohl, Clinical Manifestations of celiac disease, Dig. Dis. 33 (2015) 137-4.

[7] A.J. Lucendo, A. García-Manzanares, A. Arias, et al., Coeliac disease in the 21st century: no longer "kids' stuff", Gastroenterology Res. 4 (2011) 268–276.

[8] R. Ciccocioppo, P. Kruzliak, G. Cangemi, et al., The spectrum of differences between childhood and adulthood celiac disease, Nutrients 7 (2015) 8733–8751.

[9] S. Vivas, J. Ruiz de Morales, M. Fernandez, et al., Age-related clinical, serological, and histopathological features of celiac disease, Am. J. Gastroenterol. 10 (2008) 2360–2365.

[10] L. Rodrigo-Sáez, D. Fuentes-Álvarez, I. Pérez-Martínez, et al., Differences between pediatric and adult celiac disease, Rev. Esp. Enferm. Dig. 103 (2011) 238–244.

[11] L. Kivelä, K. Kaukinen, M.L. Lähdeaho, et al., Presentation of celiac disease in Finnish children is no longer changing: a 50-year perspective, J. Pediatr. 167 (2015) 1109–1115.

[12] L. Kivelä, K. Kaukinen, H. Huhtala, M.L. Lähdeaho, M. Mäki, K. Kurppa, At-risk screened children with celiac disease are comparable in disease severity and dietary adherence to those found because of clinical suspicion: a large cohort study, J. Pediatr. 183 (2017) 115–121.

[13] L. Kivelä, K. Kurppa, Screening for coeliac disease in children, Acta. Paediatr. 107 (2018) 1879–1887.

[14] L. Kivelä, S. Hekkala, H. Huhtala, K. Kaukinen, K. Kurppa, Lack of long-term follow-up after paediatric-adult transition in coeliac disease is not associated with complications, ongoing symptoms or dietary adherence, United European Gastroenterol. J. 8 (2020) 157–166.

[15] K. Kurppa, O. Lauronen, P. Collin, et al., Factors associated with dietary adherence in celiac disease: a nationwide study, Digestion. 86 (2012) 309–314.

[16] J. Ludvigsson, A. Agreus, C. Carolina Ciacci, et al., Transition from childhood to adulthood in coeliac disease: the Prague consensus report, Gut. 65 (2016) 1242–1251.

[17] J.K. Visakorpi, P. Kuitunen, P. Pelkonen, Intestinal malabsorption: a clinical study of 22 children over 2 years of age, Acta. Paediatr. Scand. 59 (1970) 273–280.

[18] I.R. Korponay-Szabó, K. Szabados, J. Pusztai, et al., Population screening for coeliac disease in primary care by district nurses using a rapid antibody test: diagnostic accuracy and feasibility study, BMJ 335 (2007) 1244–1247.

[19] S. Björck, C. Brundin, M. Karlsson, D. Agardh, Reduced bone mineral density in children with screening-detected celiac disease, J. Pediatr. Gastroenterol. Nutr. 65 (2017) 526–532.

[20] M. Verkasalo, O. Raitakari, J. Viikari, J. Marniemi, E. Savilahti, Undiagnosed silent coeliac disease: a risk for underachievement, Scand. J. Gastroenterol. 40 (2005) 1407–1412.

[21] E. Savilahti, K.L. Kolho, M. Westerholm-Ormio, M. Verkasalo, Clinics of coeliac disease in children in the 2000s, Acta Paediatr. 99 (2010) 1026–1030.

[22] P. Tanpowpong, S. Broder-Fingert, A. Katz, C. Camargo, Clin Age-related patterns in clinical presentations and gluten-related issues among children and adolescents with celiac disease, Transl. Gastroenterol. 3 (2012) e9.

[23] M. Ravikumara, D. Tuthill, H. Jenkins, The changing clinical presentation of coeliac disease, Arch. Dis. Child. 91 (2006) 969–971.

[24] E. Liu, K. Wolter-Warmerdam, J. Marmolejo, D. Daniels, G. Prince, F. Hickey, Routine screening for celiac disease in children with down syndrome improves case finding, J. Pediatr. Gastroenterol. Nutr. (2020), doi:10.1097/MPG.0000000000002742.

[25] P. Zubillaga, M. Vidales, I. Zubillaga, et al., HLA-DQA1 and HLA-DQB1 genetic markers and clinical presentation in celiac disease, J. Pediatr. Gastroenterol. Nutr. 34 (2002) 548–554.

[26] G. Campisi, C. Di Liberto, A. Carroccio, et al., Coeliac disease: oral ulcer prevalence, assessment of risk and association with gluten-free diet in children, Dig. Liver Dis. 40 (2008) 104–107.

[27] P. Bucci, F. Carile, A. Sangianantoni, F. D'Angiò, A. Santarelli, L. Lo Muzio, Oral aphthous ulcers and dental enamel defects in children with coeliac disease, Acta. Paediatr. 95 (2006) 203–207.

[28] G. Campisi, C. Di Liberto, G. Iacono, et al., Oral pathology in untreated coeliac disease, Aliment. Pharmacol. Ther. 26 (2007) 1529–1536.

[29] O. Khalafalla, I. Bushara, Neurologic presentation of celiac disease, Gastroenterology 128 (2005) S92–S97.

[30] E. Ferlazzo, S. Polidoro, G. Gobbi, et al., Epilepsy, cerebral calcifications, and gluten-related disorders: are anti-transglutaminase 6 antibodies the missing link? Seizure. 73 (2019) 17–20.

[31] J. Bai, A. Mota, E. Mauriño, et al., Class I and class II HLA antigens in a homogeneous Argentinian population with Whipple's disease: lack of association with HLA-B27, Am. J. Gastroenterol. 8 (1988) 992–994.

[32] B.A. Al-Sharafi, S.A. Al-Imad, A.M. Shamshair, D.H. Al-Faqeeh, Severe rickets in a young girl caused by celiac disease: the tragedy of delayed diagnosis: a case report, BMC. Res. Notes. 7 (2014) 701.

[33] A. Assiri, A. Saeed, A. AlSarkhy, M. El Mouzan, W. Matary, Celiac disease presenting as rickets in Saudi children, Ann. Saudi. Med. 33 (2013) 49–51.

[34] C. Hartman, B. Hino, A. Lerner, et al., Bone quantitative ultrasound and bone mineral density in children with celiac disease, J. Pediatr. Gastroenterol. Nutr. 39 (2004) 504–510.

[35] C. Tau, C. Mautalen, S. De Rosa, A. Roca, X. Valenzuela, Bone mineral density in children with celiac disease. effect of a gluten-free diet, Eur. J. Clin. Nutr. 60 (2006) 358–363.

[36] R. Heyman, P. Guggenbuhl, A. Corbel, et al., Effect of a gluten-free diet on bone mineral density in children with celiac disease, Gastroenterol. Clin. Biol. 33 (2009) 109–114.

[37] S. Blazina, N. Bratanic, A. Campa, R. Blagus, R. Orel, Bone mineral density and importance of strict gluten-free diet in children and adolescents with celiac disease, Bone. 47 (2010) 598–603.

[38] F. Tovoli, G. Negrini, V. Sansone, et al., Celiac disease diagnosed through screening programs in at-risk adults is not associated with worse adherence to the gluten-free diet and might protect from osteopenia/osteoporosis, Nutrients. 10 (2018) 1940.

[39] A. Di Biase, A. Colecchia, E. Scaioli, et al., Autoimmune liver diseases in a paediatric population with coeliac disease - a 10-year single-centre experience, Aliment. Pharmacol. Ther. 31 (2010) 253–260.

[40] C. Anania, E. De Luca, G. De Castro, C. Chiesa, L. Pacifico, Liver involvement in pediatric celiac disease, World. J. Gastroenterol. 21 (2015) 5813–5822.

[41] L. Äärelä, S. Nurminen, L. Kivelä, et al., Prevalence and associated factors of abnormal liver values in children with celiac disease, Dig. Liver Dis. 48 (2016) 1023–1029.

[42] H. Jericho, S. Guandalini, Extra-intestinal manifestation of celiac disease in children, Nutrients. 10 (2018) 755.

[43] M. Repo, T. Rajalahti, P. Hiltunen, et al., Diagnostic findings and long-term prognosis in children with anemia undergoing GI endoscopies, Gastrointest. Endosc. 91 (2020) 1272–1281.

[44] M. Repo, K. Lindfors, M. Mäki, et al., Anemia and iron deficiency in children with potential celiac disease, J. Pediatr. Gastroenterol. Nutr. 64 (2017) 56–62.

[45] J. Harper, S. Holleran, R. Ramakrishnan, G. Bhagat, P. Green, Anemia in celiac disease is multifactorial in etiology, Am. J. Hematol. 82 (2007) 996–1000.

[46] R. Martín-Masot, M. Nestares, J. Diaz-Castro, et al., Multifactorial etiology of anemia in celiac disease and effect of gluten-free diet: a comprehensive review, Nutrients. 11 (2019) 2557.

[47] G. Dinler, E. Atalay, A. Kalayci, Celiac disease in 87 children with typical and atypical symptoms in Black Sea region of Turkey, World. J. Pediatr. 5 (2009) 282–286.

[48] T. Nestares, R. Martín-Masot, A. Labella, et al., Is a gluten-free diet enough to maintain correct micronutrients status in young patients with celiac disease? Nutrients. 12 (2020) 844.

[49] P. Pynnönen, E. Isometsä, M. Verkasalo, et al., Gluten-free diet may alleviate depressive and behavioural symptoms in adolescents with coeliac disease: a prospective follow-up case-series study, BMC. Psychiatry. 5 (2005) 14.

[50] L. Smith, K. Kurppa, D. Agardh, Further support for psychological symptoms in pediatric celiac disease, Pediatrics. 144 (2019) e20191683.

[51] S. Nurminen, L. Kivelä, H. Huhtala, K. Kaukinen, K. Kurppa, Extraintestinal manifestations were common in children with coeliac disease and were more prevalent in patients with more severe clinical and histological presentation, Acta. Paediatr. 108 (2019) 681–687.

[52] N. Zelnik, A. Pacht, R. Obeid, A. Lerner, Range of neurologic disorders in patients with celiac disease, Pediatrics. 113 (2004) 1672–1676.

[53] K. Katz, S. Rashtak, B. Lahr, et al., Screening for celiac disease in a North American population: sequential serology and gastrointestinal symptoms, Am. J. Gastroenterol. 106 (2011) 1333–1339.

[54] A. Rosén, O. Sandström, A. Carlsson, L. Högberg, O. Olén, H. Stenlund, A. Ivarsson, Usefulness of symptoms to screen for celiac disease, Pediatrics. 133 (2014) 211–218.

[55] A. Irvine, W. Chey, A. Ford, Screening for celiac disease in irritable bowel syndrome: an updated systematic review and meta-analysis, Am. J. Gastroenterol. 112 (2017) 65–76.

[56] F. Cristofori, C. Fontana, A. Magistà, et al., Increased prevalence of celiac disease among pediatric patients with irritable bowel syndrome: a 6-year prospective cohort study, JAMA. Pediatr. 168 (2014) 555–556.

[57] V. Pascual, R. Dieli-Crimi, N. López-Palacios, A. Bodas, L. Medrano, C. Núñez, Inflammatory bowel disease and celiac disease: overlaps and differences, World. J. Gastroenterol. 20 (2014) 4846–4856.

[58] A.K. Kamboj, A.S. Oxentenko, Clinical and histologic mimickers of celiac disease, Clin. Transl. Gastroenterol. 8 (2017) e114.

[59] M. Jansen, S. Beth, D. van den Heuvel, et al., Ethnic Differences in coeliac disease autoimmunity in childhood: The Generation R study, Arch. Dis. Child. 102 (2017) 529–534.

[60] D. Agardh, H.S. Lee, K. Kurppa, et al., Clinical features of celiac disease: a prospective birth cohort, Pediatrics. 135 (2015) 627–634.

[61] A. Tsouka, F. Mahmud, M. Marcon, Celiac disease alone and associated with type 1 diabetes mellitus, J. Pediatr. Gastroenterol. Nutr. 61 (2015) 297–302.

[62] D. Hansen, B. Brock-Jacobsen, E. Lund, et al., Clinical benefit of a gluten-free diet in type 1 diabetic children with screening-detected celiac disease: a population-based screening study with 2 years' follow-up, Diabetes. Care. 29 (2006) 2452–2456.

[63] A. Popp, L. Kivelä, V. Fuchs, K. Kurppa, Diagnosing celiac disease: towards wide-scale screening and serology-based criteria? Gastroenterol. Res. Pract. (2019) 2916024.

[64] I. Hill, M. Dirks, G. Liptak, et al., Guideline for the diagnosis and treatment of celiac disease in children: recommendations of the North American Society for Pediatric Gastroenterology, Hepatology and Nutrition, J. Pediatr. Gastroenterol. Nutr. 40 (2005) 1–19.

[65] V. Fuchs, K. Kurppa, H. Huhtala, et al., Serology-based criteria for adult coeliac disease have excellent accuracy across the range of pre-test probabilities, Aliment. Pharmacol. Ther. 49 (2019) 277–284.

[66] A. Al-Toma, U. Volta, R. Auricchio, et al., European Society for the Study of Coeliac Disease (ESsCD) guideline for coeliac disease and other gluten-related disorders, United. European. Gastroenterol. J. 7 (2019) 583–613.

[67] A. Rubio-Tapia, I. Hill, C. Kelly, A. Calderwood, J. Murray, ACG Clinical Guidelines: diagnosis and management of celiac disease, Am. J. Gastroenterol. 108 (2013) 656–676.

[68] I. Gustafsson, M. Repo, A. Popp, et al., Prevalence and diagnostic outcomes of children with duodenal lesions and negative celiac serology, Dig. Liver. Dis. 52 (2020) 289–295.

[69] A. Naiyer, L. Hernandez, E. Ciaccio, et al., Comparison of commercially available serologic kits for the detection of celiac disease, J. Clin. Gastroenterol. 43 (2009) 225–232.

[70] K. Giersiepen, M. Lelgemann, N. Stuhldreher, et al., Accuracy of diagnostic antibody tests for coeliac disease in children: summary of an evidence report, J. Pediatr. Gastroenterol. Nutr. 54 (2012) 229–241.

[71] C. Catassi, A. Fasano, Is this really celiac disease? Pitfalls in diagnosis, Curr. Gastroenterol. Rep. 10 (2008) 466–472.

[72] D. Leffler, D. Schuppan, K. Pallav, et al., Kinetics of the histological, serological and symptomatic responses to gluten challenge in adults with coeliac disease, Gut. 62 (2013) 996–1004.

[73] L. Sollid, G. Markussen, J. Ek, H. Gjerde, F. Vartdal, E. Thorsby, Evidence for a primary association of celiac disease to a particular HLA-DQ alpha/beta heterodimer, J. Exp. Med. 169 (1989) 345–350.

[74] L. Sollid, B. Jabri, Triggers and drivers of autoimmunity: lessons from coeliac disease, Nat. Rev. Immunol. 13 (2013) 294–302.

[75] K. Werkstetter, I.R. Korponay-Szabó, A. Popp, et al., Accuracy in diagnosis of celiac disease without biopsies in clinical practice, Gastroenterology 153 (2017) 924–935.

[76] S. Hommeida, M. Alsawas, M. Murad, et al., The association between celiac disease and eosinophilic esophagitis: mayo experience and meta-analysis of the literature, J. Pediatr. Gastroenterol. Nutr. 65 (2017) 58–63.

[77] S. Kröger, K. Kurppa, M. Repo, et al., Severity of villous atrophy at diagnosis in childhood does not predict long-term outcomes in celiac disease, J. Pediatr. Gastroenterol. Nutr. (2020), doi:10.1097/MPG.0000000000002675.

[78] A. Ravelli, V. Villanacci, C. Monfredini, S. Martinazzi, V. Grassi, S. Manenti, How patchy is patchy villous atrophy? Distribution pattern of histological lesions in the duodenum of children with celiac disease, Am. J. Gastroenterol. 105 (2010) 2103–2110.

[79] A. Ravelli, V. Villanacci, Tricks of the trade: how to avoid histological pitfalls in celiac disease, Pathol. Res. Pract. 208 (2012) 197–202.

[80] J. Taavela, O. Koskinen, H. Huhtala, et al., Validation of morphometric analyses of small-intestinal biopsy readouts in celiac disease, PLoS. One. 8 (2013) e76163.

[81] G. Corazza, V. Villanacci, Coeliac Disease, J. Clin. Pathol. 58 (2005) 573–574.

[82] M. Kurien, J. Ludvigsson, D. Sanders, A no biopsy strategy for adult patients with suspected coeliac disease: making the world gluten-free, Gut. 64 (2015) 1003–1004.

[83] G. Corazza, V. Villanacci, C. Zambelli, et al., Comparison of the interobserver reproducibility with different histologic criteria used in celiac disease, Clin. Gastroenterol. Hepatol. 5 (2007) 838–843.

[84] C. Arguelles-Grande, C. Tennyson, S. Lewis, P. Green, G. Bhagat, Variability in small bowel histopathology reporting between different pathology practice settings: impact on the diagnosis of coeliac disease, J. Clin. Pathol. 65 (2012) 242–247.

[85] A. Popp, T. Arvola, J. Taavela, et al., Nonbiopsy approach for celiac disease is accurate when using exact duodenal histomorphometry: prospective study in 2 countries, J. Clin. Gastroenterol. (2020), doi:10.1097/MCG.0000000000001349.

[86] K. Kurppa, M. Ashorn, S. Iltanen, et al., Celiac Disease without villous atrophy in children: a prospective study, J. Pediatr. 157 (2010) 373–380.

[87] O. Koskinen, P. Collin, I.R. Korponay-Szabo, et al., Gluten-dependent small bowel mucosal transglutaminase 2-specific iga deposits in overt and mild enteropathy coeliac disease, J. Pediatr. Gastroenterol. Nutr. 47 (2008) 436–442.

[88] R. Auricchio, A. Tosco, E. Piccolo, et al., Potential celiac children: 9-year follow-up on a gluten-containing diet, Am. J. Gastroenterol. 109 (2014) 913–921.

[89] R. Auricchio, R. Mandile, M. Del Vecchio, et al., Progression of celiac disease in children with antibodies against tissue transglutaminase and normal duodenal architecture, Gastroenterology 157 (2019) 413–420.

[90] T. Salmi, P. Collin, T. Reunala, M. Mäki, K. Kaukinen, Diagnostic methods beyond conventional histology in coeliac disease diagnosis, Dig. Liver. Dis. 42 (2010) 28–32.

[91] C. Trovato, M. Montuori, F. Valitutti, B. Leter, S. Cucchiara, S. Oliva, The challenge of treatment in potential celiac disease, Gastroenterol. Res. Pract. (2019) 8974751.

[92] A. Laitinen, D. Agardh, L. Kivelä, et al., Coeliac patients detected during type 1 diabetes surveillance had similar issues to those diagnosed on a clinical basis, Acta. Paediatr. 106 (2017) 639–646.

[93] S. Kinos, K. Kurppa, A. Ukkola, et al., Burden of illness in screen-detected children with celiac disease and their families, J. Pediatr. Gastroenterol. Nutr. 55 (2012) 412–416.

[94] E. Fröhlich-Reiterer, S. Kaspers, S. Hofer, et al., Anthropometry, metabolic control, and follow-up in children and adolescents with type 1 diabetes mellitus and biopsy-proven celiac disease, J. Pediatr. 158 (2011) 589–593.

[95] J. Simmons, G. Klingensmith, K. McFann, et al., Celiac autoimmunity in children with type 1 diabetes: a two-year follow-up, J. Pediatr. 158 (2011) 276–281.

[96] E. Lauret, L. Rodrigo, Celiac disease and autoimmune-associated conditions, Biomed. Res. Int. 2013 (2013) 127589.

[97] L. Conti, E. Lahner, G. Galli, G. Esposito, M. Carabotti, B. Annibale, Risk factors associated with the occurrence of autoimmune diseases in adult coeliac patients, Gastroenterol. Res. Pract. 2018 (2018) 3049286.

[98] P. Singh, S. Arora, S. Lal, T. Strand, G. Makharia, Risk of celiac disease in the first- and second-degree relatives of patients with celiac disease: a systematic review and meta-analysis, Am. J. Gastroenterol. 110 (2015) 1539–1544.

[99] K. Kurppa, J. Salminiemi, A. Ukkola, et al., Utility of the new ESPGHAN criteria for the diagnosis of celiac disease in at-risk groups, J. Pediatr. Gastroenterol. Nutr. 54 (2012) 387–389.

[100] A. Fasano, I. Berti, T. Gerarduzzi, et al., Prevalence of celiac disease in at-risk and not-at-risk groups in the United States: a large multicenter study, Arch. Intern. Med. 163 (2003) 286–292.

[101] A. Ivarsson, A. Carlsson, A. Bredberg, et al., Prevalence of coeliac disease in turner syndrome, Acta. Paediatr. 88 (1999) 933–936.

[102] M. Bonamico, A. Pasquino, P. Mariani, et al., Prevalence and clinical picture of celiac disease in Turner syndrome, J. Clin. Endocrinol. Metab. 87 (2002) 5495–5498.

[103] F. Cerutti, G. Bruno, F. Chiarelli, et al., Younger age at onset and sex predict celiac disease in children and adolescents with type 1 diabetes: an Italian multicenter study, Diabetes. Care. 27 (2004) 1294–1298.

[104] M. Bybrant, E. Örtqvist, S. Lantz, L. Grahnquist, High prevalence of celiac disease in Swedish children and adolescents with type 1 diabetes and the relation to the Swedish epidemic of celiac disease: a cohort study, Scand. J. Gastroenterol. 49 (2014) 52–58.

[105] N. Sattar, F. Lazare, M. Kacer, et al., Celiac disease in children, adolescents, and young adults with autoimmune thyroid disease, J. Pediatr. 158 (2011) 272–275.

[106] K. Fahl, C. Silva, A. Pastorino, M. Carneiro-Sampaio, C. Jacob, Autoimmune diseases and autoantibodies in pediatric patients and their first-degree relatives with immunoglobulin a deficiency, Rev. Bras. Reumatol. 55 (2015) 197–202.

[107] C. Betterle, F. Lazzarotto, A. Spadaccino, et al., Celiac disease in North Italian patients with autoimmune Addison's disease, Eur. J. Endocrinol. 154 (2006) 275–279.

[108] R. Nisihara, L. Kotze, S. Utiyama, et al., Celiac disease in children and adolescents with Down syndrome, J. Pediatr. (Rio J) 81 (2005) 373–376.

[109] L. Downey, R. Houten, S. Murch, D. Longson, Recognition, assessment, and management of coeliac disease: summary of updated NICE guidance, BMJ 351 (2015) 4513.

[110] J. Ludvigsson, J. Bai, F. Biagi, et al., Diagnosis and management of adult coeliac disease: guidelines from the British society of gastroenterology, Gut. 63 (2014) 1210–1228.

[111] K. Kurppa, A. Laitinen, D. Agardh, Coeliac disease in children with type 1 diabetes, Lancet. Child. Adolesc. Health. 2 (2018) 133–143.

[112] E. Liu, H.S. Lee, C. Aronsson, et al., Risk of pediatric celiac disease according to HLA haplotype and country, N. Engl. J. Med. 371 (2014) 42–49.

[113] S. Björck, C. Brundin, E. Lörinc, K. Lynch, D. Agardh, Screening detects a high proportion of celiac disease in young HLA-genotyped children, J. Pediatr. Gastroenterol. Nutr. 50 (2010) 49–53.

[114] J. Wouters, M. Weijerman, M. van Furth, et al., Prospective human leukocyte antigen, endomysium immunoglobulin A antibodies, and transglutaminase antibodies testing for celiac disease in children with Down syndrome, J. Pediatr. 154 (2009) 239–242.

[115] K. Kaukinen, J. Partanen, M. Mäki, P. Collin, HLA-DQ typing in the diagnosis of celiac disease, Am. J. Gastroenterol. 97 (2002) 695–699.

[116] N. Brown, S. Guandalini, C. Semrad, S. Kupfer, A clinician's guide to celiac disease HLA genetics, Am. J. Gastroenterol. 114 (2019) 1587–1592.

[117] E. Binder, T. Rohrer, D. Denzer, et al., Screening for coeliac disease in 1624 mainly asymptomatic children with type 1 diabetes: is genotyping for coeliac-specific human leucocyte antigen the right approach? Arch. Dis. Child. 104 (2019) 354–359.

[118] E. Binder, M. Martina Loinger, A. Mühlbacher, et al., Genotyping of coeliac-specific human leucocyte antigen in children with type 1 diabetes: does this screening method make sense? Arch. Dis. Child. 102 (2017) 603–606.

[119] R. Choung, S. Khaleghi, A. Cartee, et al., Community-based study of celiac disease autoimmunity progression in adults, Gastroenterology 158 (2020) 151–159.

[120] W. Hagopian, H.S. Lee, E. Liu, et al., Co-occurrence of type 1 diabetes and celiac disease autoimmunity, Pediatrics. 140 (2017) e20171305.

[121] E. Liu, F. Dong, A. Barón, et al., High incidence of celiac disease in a long-term study of adolescents with susceptibility genotypes, Gastroenterology 152 (2017) 1329–1336.

[122] S. Björck, K. Lynch, C. Brundin, D. Agardh, Repeated screening can be restricted to at-genetic-risk birth cohorts, J. Pediatr. Gastroenterol. Nutr. 62 (2016) 271–275.

[123] J.F. Ludvigsson 1, T.R. Card 2, K. Kaukinen 3, et al., Screening for celiac disease in the general population and in high-risk groups, United. European Gastroenterol. J. (3) (2015) 106–120.

[124] K. Kurppa, P. Collin, M. Viljamaa, et al., Diagnosing mild enteropathy celiac disease: a randomized, controlled clinical study, Gastroenterology 136 (2009) 816–823.

[125] L. Kivelä, A. Popp, T. Arvola, H. Huhtala, K. Kaukinen, K. Kurppa, Long-term health and treatment outcomes in adult coeliac disease patients diagnosed by screening in childhood, United. European Gastroenterol. J. 6 (2018) 1022–1031.

[126] A. Tommasini, T. Not, V. Kiren, V. Baldas, et al., Mass screening for coeliac disease using antihuman transglutaminase antibody assay, Arch. Dis. Child. 89 (2004) 512–515.

[127] J.A. Tye-Din, H.J. Galipeau, D. Agardh, Celiac disease: a review of current concepts in pathogenesis, prevention, and novel therapies, Front. Pediatr. 6 (2018) 350.

[128] M. Pinto-Sánchez, N. Causada-Calo, P. Premysl Bercik, et al., Safety of adding oats to a gluten-free diet for patients with celiac disease: systematic review and meta-analysis of clinical and observational studies, Gastroenterology 153 (2017) 395–409.

[129] E. Lionetti, S. Gatti, T. Galeazzi, et al., Safety of oats in children with celiac disease: a double-blind, randomized, placebo-controlled trial, J. Pediatr. 194 (2018) 116–122.

[130] K. Aaltonen, P. Laurikka, H. Huhtala, M. Mäki, K. Kaukinen, K. Kurppa, The long-term consumption of oats in celiac disease patients is safe: a large cross-sectional study, Nutrients 9 (2017) 611.

[131] L. Downey, R. Houten, S. Murch, D. Longson, Guideline Development Group. Recognition, assessment, and management of coeliac disease: summary of updated NICE guidance, BMJ 351 (2015) h4513.

[132] J. Arnone, V. Fitzsimons, Adolescents with celiac disease: a literature review of the impact developmental tasks have on adherence with a gluten-free diet, Gastroenterol. Nurs. 35 (2012) 248–254.

[133] E. Fabiani, M. Taccari, I. Rätsch, S. Di Giuseppe, G. Coppa, C. Catassi, Compliance with gluten-free diet in adolescents with screening-detected celiac disease: a 5-year follow-up study, J. Pediatr. 136 (2000) 841–843.

[134] E. Koppen, J. Schweizer, C. Csizmadia, et al., Long-term health and quality-of-life consequences of mass screening for childhood celiac disease: a 10-year follow-up study, Pediatrics 123 (2009) 582–588.

[135] L. Barnea, Y. Mozer-Glassberg, I. Hojsak, C. Hartman, R. Shamir, Pediatric celiac disease patients who are lost to follow-up have a poorly controlled disease, Digestion 90 (2014) 248–253.

[136] O. Jadresin, Z. Misak, S. Kolacek, Z. Sonicki, V. Zizić, Compliance with gluten-free diet in children with coeliac disease, J. Pediatr. Gastroenterol. Nutr. 47 (2008) 344–348.

[137] M. Haines, R. Anderson, P. Gibson, Systematic review: the evidence base for long-term management of coeliac disease, Aliment. Pharmacol. Ther. 28 (2008) 1042–1066.

[138] F. Valitutti, C.M. Trovato, M. Montuori, et al., Pediatric celiac disease: follow-up in the spotlight, Adv. Nutr. 8 (2017) 356–361.

[139] K. Johansson, E. Malmberg Hård Af Segerstad, H. Mårtensson, D. Agardh, Dietitian visits were a safe and cost-effective form of follow-up care for children with celiac disease, Acta. Paediatr. 108 (2019) 676–678.

[140] S. Vriezinga, A. Borghorst, E. van den Akker-van Marle, et al., E-healthcare for celiac disease-a multicenter randomized controlled trial, J. Pediatr. 195 (2018) 154–160.

[141] C. Kirsaclioglu, Z. Kuloglu, A. Tanca, et al., Bone mineral density and growth in children with coeliac disease on a gluten free-diet, Turk. J. Med. Sci. 46 (2016) 1816–1821.

[142] R. Kuchay, B. Thapa, A. Mahmood, M. Anwar, S. Mahmood, Lactase genetic polymorphisms and coeliac disease in children: a cohort study, Ann. Hum. Biol. 42 (2015) 101–104.

[143] T. Rajalahti, M. Repo, L. Kivelä, et al., Anemia in pediatric celiac disease: association with clinical and histological features and response to gluten-free diet, J. Pediatr. Gastroenterol. Nutr. 64 (2017) 1–6.

[144] J. Burger, J. van der Laan, T. Jansen, et al., Low yield for routine laboratory checks in follow-up of coeliac disease, J. Gastrointestin. Liver Dis. 27 (2018) 233–239.

[145] C. Webb, A. Myléus, F. Norström, et al., High adherence to a gluten-free diet in adolescents with screening-detected celiac disease, J. Pediatr. Gastroenterol. Nutr. 60 (2015) 54–59.

[146] S. Koletzko, R. Auricchio, J. Dolinsek, et al., No need for routine endoscopy in children with celiac disease on a gluten-free diet, J. Pediatr. Gastroenterol. Nutr. 65 (2017) 267–269.

[147] K. Kaukinen, S. Sulkanen, M. Mäki, P. Collin, IgA-class transglutaminase antibodies in evaluating the efficacy of gluten-free diet in coeliac disease, Eur. J. Gastroenterol. Hepatol. 14 (2002) 311–315.

[148] K. Gerasimidis, K. Zafeiropoulou, M. Mackinder, et al., Comparison of clinical methods with the faecal gluten immunogenic peptide to assess gluten intake in coeliac disease, J. Pediatr. Gastroenterol. Nutr. 67 (2018) 356–360.

[149] M.L. Moreno, Á. Cebolla, A. Muñoz-Suano, et al., Detection of gluten immunogenic peptides in the urine of patients with coeliac disease reveals transgressions in the gluten-free diet and incomplete mucosal healing, Gut. 66 (2017) 250–257.

[150] A. Mubarak, J.H. Oudshoorn, C.M. Kneepkens, et al., A child with refractory coeliac disease, J. Pediatr. Gastroenterol. Nutr. 53 (2011) 216–218.

[151] H. Leinonen, L. Kivelä, M.L. Lähdeaho, H. Huhtala, K. Kaukinen, K. Kurppa, Daily life restrictions are common and associated with health concerns and dietary challenges in adult celiac disease patients diagnosed in childhood, Nutrients 11 (2019) 1718.

[152] R. Wolf, B. Lebwohl, A. Lee, et al., Hypervigilance to a gluten-free diet and decreased quality of life in teenagers and adults with celiac disease, Dig. Dis. Sci. 63 (2018) 1438–1448.

[153] A. Ivarsson, L. Persson, L. Nyström, et al., Epidemic of coeliac disease in Swedish children, Acta. Paediatr. 89 (2000) 165–171.

[154] A. Ivarsson, The Swedish epidemic of coeliac disease explored using an epidemiological approach - some lessons to be learnt, Best Pract. Res. Clin. Gastroenterol. 19 (2005) 425–440.

[155] S. Vriezinga, R. Auricchio, E. Bravi, et al., Randomized feeding intervention in infants at high risk for celiac disease, N. Engl. J. Med. 371 (2014) 1304–1315.

[156] E. Lionetti, S. Castellaneta, R. Francavilla, et al., Introduction of gluten, HLA status, and the risk of celiac disease in children, N. Engl. J. Med. 371 (2014) 1295–1303.

[157] K. Lindfors, J. Lin, H.S. Lee, et al., Metagenomics of the faecal virome indicate a cumulative effect of enterovirus and gluten amount on the risk of coeliac disease autoimmunity in genetically at risk children: the TEDDY study, Gut. (2019) gutjnl-2019-319809.

[158] C. Kahrs, K. Chuda, G. Tapia, et al., Enterovirus as trigger of coeliac disease: nested case-control study within prospective birth cohort, BMJ 364 (2019) l231.

[159] R. Bouziat, R. Hinterleitner, J. Brown, et al., Reovirus infection triggers inflammatory responses to dietary antigens and development of celiac disease, Science 356 (2017) 44–50.

[160] C. Aronsson, H.S. Lee, E. Hård Af Segerstad, et al., Association of gluten intake during the first 5 years of life with incidence of celiac disease autoimmunity and celiac disease among children at increased risk, JAMA 322 (2019) 514–523.

[161] K. Mårild, F. Dong, N. Lund-Blix, et al., Gluten Intake and Risk of Celiac Disease: Long-Term Follow-up of an At-Risk Birth Cohort, Am. J. Gastroenterol. 114 (2019) 1307–1314.

CHAPTER 3

Pathogenesis of coeliac disease – a disorder driven by gluten-specific CD4+ T cells

Shiva Dahal-Koirala[a,b], Louise Fremgaard Risnes[a,c], Ludvig M. Sollid[a,b,c]
[a]K.G. Jebsen Coeliac Disease Research Centre, Institute of Clinical Medicine, University of Oslo, Oslo, Norway
[b]Department of Immunology, University of Oslo, Oslo, Norway
[c]Department of Immunology, Oslo University Hospital, Oslo, Norway

Coeliac disease (CeD) is a common and acquired disorder that is caused by an immune response to cereal gluten proteins. The lesion of the disease located in the upper small bowel is characterized by villous blunting and infiltration of lymphocytes both in the epithelium and the lamina propria. On adopting a gluten-free diet, the great majority of patients experience normalization of their gut pathology and relief of most clinical symptoms. The disease reappears on re-exposure to gluten. Gluten is thus a driver of the condition [1]. Initial diagnostic schemes for CeD involved elimination/provocation diets that is typical of food intolerances. This is no more so. The patients with active disease have highly disease-specific antibodies to the autoantigen transglutaminase (TG2) as well as to deamidated gluten peptides (DGP). The diagnosis can be made with high certainty if high amounts of IgA anti-TG2 is present in the serum. For paediatric disease the diagnosis can be made solely by detection of autoantibodies [2]. Thus, as perceived by clinicians, CeD is a condition that over the last 50 years have made a move from the camp of food intolerances to the camp of autoimmune diseases. Even though CeD can be considered an autoimmune disease, still the adaptive immune response mounted by the gut immune system to the foreign antigen gluten is the central part. The specialized immune system of the gut has a distinct organization that facilitates effective immune responses to a plethora of antigens including gluten proteins.

3.1 Inductive and effector sites within the gut immune system

The adaptive immune system in the intestine is comprised of separate inductive and effector sites [3]. The Peyer's patches, isolated lymphoid follicles and the mesenteric lymph nodes are the primary inductive sites whereas the lumen lining epithelium and the lamina propria are the typical effector sites (Fig. 3.1). The inductive sites contain organized lymphoid tissue with T cells, B cells and antigen-presenting cells (APCs). The

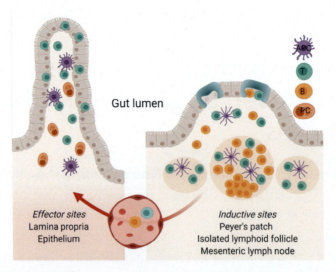

Figure 3.1 *Inductive and effector sites in the intestine.* The intestinal immune system is composed of separate inductive sites (Peyer's patches, isolated lymphoid follicles and mesenteric lymph nodes) and effector sites (lumen lining epithelium and lamina propria). Following induction of an antigen specific adaptive immune response to a given antigen, antigen-specific T and B cells travel through blood and populate the effector sites as effector memory T cells and plasma cells. Created with BioRender.com.

priming of naïve T cells and B cells happens here. APCs, such as dendritic cells, have an important role in capturing antigen and bringing it to the immune structures of the priming sites where the antigens are presented to T cells. This role of dendritic cells is certainly important for mesenteric lymph nodes and likely also for Peyer's patches, yet there is evidence that B cells in Peyer's patches can capture antigen from the lumen without involvement of classical dendritic cells [4]. B cells are very important APCs for T cells, and this is particularly so at the gut inductive immune sites. Cognate interaction of T cells and B cells leads to activation of both cell types with clonal expansion and generation of effector T cells and B cells (i.e. plasma cells). Interaction between T cells and B cells is a prerequisite to mount an efficient adaptive immune response. The T cells and B cells that have become activated by antigen in organized lymphoid structures, either as naïve cells in primary response or as memory cells in a recall response, will enter the blood stream and populate effector sites along large segments of the gut (Fig. 3.1). The B cells travel in blood as plasmablasts and settle in the lamina propria where some of the cells may live for decades as plasma cells [5]. The knowledge we have today on the involvement of T cells, B cells, plasma cells and dendritic cells in CeD is so far based on the studies of cells isolated from effector sites and from peripheral blood and not from inductive sites. Much can probably also be learned from studies of cells at inductive sites, but an undertaking of such studies in humans will be restricted by what is ethically acceptable.

3.2 Induction of adaptive immune response to gluten

3.2.1 Involvement HLA II molecules

Genes are important for development of CeD. The concordance rate in monozygotic twins is 75% [6] and about 10% of first-degree relatives are affected [7]. The single most important genetic factor is HLA, and HLA-DQ encoding genes are responsible for the effect of HLA [8]. Over 90% of the individuals with CeD express a variant of HLA-DQ2 (*DQA1* 05* and *DQB1* 02*) known as HLA-DQ2.5 while the remaining minority express another HLA-DQ2 variant (*DQA1* 02:01* and *DQB1* 02:02*) known as HLA-DQ2.2 or HLA-DQ8 (*DQA1* 03* and *DQB1* 03:02*) [9–11]. Very few CeD subjects who do not express any of these three HLA-DQ variants exist. Most of these subjects seem to express HLA-DQ7.5 (*DQA1* 05:05* and *DQB1* 03:01*) [11]. The risk of CeD is considerably increased in individuals who are homozygous for DQ2.5 [12], DQ2.2 [12] or DQ8 [9]. Genome-wide association studies (GWAS) have demonstrated that HLA class II genes account for about 40% of genetic variation in CeD [8]. Although GWAS has identified several other non-HLA genes as risk genes in CeD, the impacts of each these genes are much smaller than that accounted for by HLA alone.

A seminal finding was the observation that gluten-reactive T cells generated by stimulating the biopsies of CeD subjects with gluten digest are selectively restricted by HLA-DQ2.5 and not by any other HLA molecule expressed by the patients [13]. Similar experiments performed on the biopsies from CeD subjects expressing HLA-DQ8 [10] and HLA-DQ2.2 [14] later demonstrated that in such subjects the gluten-reactive T cells are restricted by these HLA-DQ allotypes. These pieces of evidence indicate that the CeD-associated HLA molecules exert their role in the pathogenesis by an exclusive ability to present gluten antigen to CD4+ T cells. These CeD-associated HLA-DQ allotypes commonly prefer binding of peptides with negatively charged residues. The negatively charged residues are introduced in gluten peptides by the process of deamidation which is mediated by the enzyme transglutaminase 2 (TG2). The molecular basis for the preference of binding negatively charged peptides, by HLA-DQ2.5, HLA-DQ2.2, and HLA-DQ8 has become clear by several X-ray crystallographic studies [15–19]. CeD is thus a disease for which the mechanism underlying the HLA association is unusually clear [20].

3.2.2 Transglutaminase 2 in CeD

TG2 has important roles in CeD pathogenesis. Not only is TG2 mediating deamidation of gluten peptides, but the enzyme is also the target for CeD specific autoantibodies. Notwithstanding the essential role of TG2 in the pathogenesis of CeD, it is not known where the pathogenic TG2 is localized. A recent study on immunoglobulin knock-in mice expressing an anti-TG2 B-cell receptor demonstrated that TG2-specific B cells are present in the periphery of these mice with no signs of B-cell tolerance induction [21].

This observation suggests that TG2-specific B cells have not encountered TG2 during their development or during their patrolling of peripheral tissues. It was also observed in this mouse model that autoreactive B cells were able to produce antibodies once they were provided T-cell help. Following the notion that under normal physiological conditions there is minimal expression of extracellular TG2, the long-standing question of where pathogenic TG2 is derived from got new traction. A solution to this conundrum was possibly provided by the suggestion that the pathogenic TG2 in CeD is gut luminal enzyme released from shed enterocytes [22]. This luminal TG2 would be shielded from B cells during their development, and thus could explain the existence of ignorant B cells specific for TG2. The concentration of gluten peptides would he high in the lumen favoring their interaction as substrates with the enzyme TG2. Gluten-TG2 complexes could be formed, which as we discuss later, likely are essential for the formation of anti-TG2 antibodies in CeD subjects consuming gluten. The primary induction site for gut luminal antigens is considered to Peyer's patches, and such luminal TG2-gluten peptide complexes could be effective to prime adaptive immune responses. In addition, to the generation of TG2-gluten peptide complexes, luminal TG2 could also be implicated in the generation of deamidated gluten peptides, which upon translocation of the epithelium, could be presented to gluten-specific effector memory CD4+ T cells residing in the lamina propria (Fig. 3.2).

3.2.3 Gluten: The driver of T-cell response

Gluten is the driver of CeD as the disease can be turned off and on by exposure to this environmental agent. Strictly speaking, the term "gluten" is the cohesive mass of proteins that remains after washing wheat flour with water [23]. However, nowadays, gluten has become a common term for storage proteins in wheat, barley, rye and oats that are known to be associated with CeD [24]. While gluten proteins derived from wheat, barley and rye are universally excluded from diets for treatment of CeD, the exclusion of oats remains debatable [25,26]. The wheat gluten proteins are categorized into gliadins (α, γ and ω subtypes) and glutenins (LMW and HMW subtypes). The gluten proteins in barley and rye are called hordeins (B, γ, C and D subtypes) and secalins (γ, ω and HMW subtypes), respectively [23].

The fragments of proteins and peptides that are recognized by T cells are termed T-cell epitopes. Most of the known T-cell epitopes in CeD are derived from the gliadin or glutenin moieties of wheat [24,27–30]. The T-cell epitopes of hordeins, secalins and avenins are highly homologous to those found in wheat [31–33]. It is striking that despite the vast size of the gluten proteome, only a limited number of peptide sequences give rise to T-cell epitopes. This suggests that there is a very stringent epitope selection process. The selection of the peptide sequences to become gluten T-cell epitopes is known to be influenced by proteolytic stability of the individual gluten proteins,

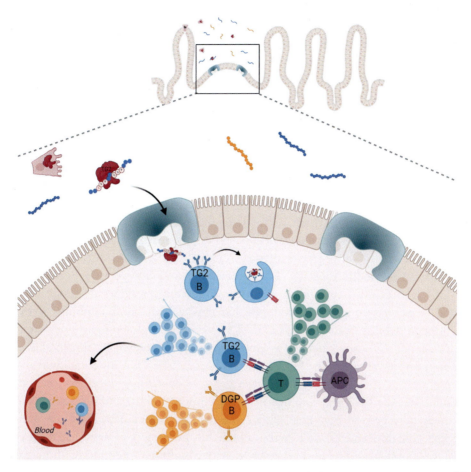

Figure 3.2 *Inductive immune sites implicated in the generation of adaptive immune responses of CeD.* The adaptive immune responses typical of CeD (T cells:gluten-specific and B cell:anti-TG2 and anti-gluten) are initiated at inductive immune sites of the gut immune system. Here, B cells and T cells interact in antigen-specific fashion, and a productive interaction leads to clonal amplification of both T cells and B cells. Generated effector T and B cells travel via blood and populate gut tissue at distant sites. Possibly the pathogenic TG2 in celiac disease is luminal enzyme derived from the cytosol of shed enterocytes. Luminal TG2 may use gluten peptides, which are present in relatively high concentration in the lumen due to their resistance to proteolysis, as substrates. Intermediates of enzyme-substrate will be formed and as there are fairly stable, they can bind to the B-cell receptor of TG2-specific B cells and be internalized. Upon internalization of the TG2:gluten peptide complex, a deamidated peptide will be released from the enzyme's active site. This deamidated peptide may bind to HLA-DQ and be presented to gluten-specific CD4+ T cells. The gluten-specific CD4+ T cells can in this way provide help to TG2-specific B cells. The gluten-specific T cells can also provide T-cell help to B cells specific for deamidated gluten peptides. Antibodies to TG2 and DGP thus report on productive T-cell and B-cell interactions which can explain why they are such good diagnostic markers for CeD. Created with BioRender.com.

selectivity of the TG2 enzyme in mediating deamidation and binding preference of HLA molecules [34].

Gluten peptides must survive the gut proteases to be a target for adaptive immune responses (Fig. 3.3). Gluten is rich in proline residues. Since proline-rich polypeptides are poorly digested by the gut proteases, long peptide fragments of gluten survive the proteolytic digestion. These long proteolytically stable polypeptide fragments typically harbor several CeD-relevant T-cell epitopes [35]. Several DQ2.5-restricted T-cell epitopes in gliadin proteins are located in regions rich in proline [36] and the most frequently recognized T-cell epitopes are more resistant to enzymatic degradation than those that are recognized more infrequently [37], indicating the importance of proteolytic stability in selection of gluten T-cell epitopes.

TG2-mediated deamidation is another crucial factor for selection of gluten T-cell epitopes. Gluten proteins in their native states have few negatively charged residues. It has been demonstrated that negatively charges are necessary for stable binding of gluten peptides to binding groove of the CeD-associated HLA-DQ molecules. The post-translational modification by TG2 converts gluten polypeptides into excellent HLA-DQ binders. TG2 specifically deamidates glutamine residues in gluten peptides resulting in generation of T-cell epitopes that display high binding affinity as well as better kinetic stability [27,38,39]. The TG2 targets mainly glutamine residues residing in QXP sequence motifs where Q is glutamine residue targeted by the enzyme, X is any amino acid and P is proline [40,41]. The importance of TG2-mediated deamidation is exemplified by the observations that the peptides containing T-cell epitopes are all good substrates for TG2 [42], and that from the complex mixture of peptides generated by proteolytic digestion of whole gluten, the peptides selected by TG2 are typically those that contain known T-cell epitopes [43]. All the three CeD-associated allotypes, HLA-DQ2.5, HLA-DQ2.2 and HLA-DQ8, prefer binding of peptides with negatively charged anchor residues. However, the preference for positioning of the negatively charged glutamate residue differs (Fig. 3.4). HLA-DQ2.5 is known to prefer negatively charged residues at the anchor positions P4, P6, and P7 [44–46] while deamidation at P1 and P9 is preferred by HLA-DQ8 [47]. For HLA-DQ2.2, similar to HLA-DQ2.5, negatively charged residues are preferred at P4, P6 and P7, but notably this HLA molecule has an additional requirement for serine (or threonine) at P3 [14,19,48]. In total, 27 HLA-DQ2.5-restricted, 5 HLA-DQ8-restricted and 3 HLA-DQ2.2-restricted epitopes have been identified so far [49].

3.2.4 Gluten trafficking from lumen to lamina propria

The mechanism by which gluten polypeptides in gut lumen penetrate the relatively impermeable surface of the intestinal epithelial layer to reach the lamina propria is a highly debated topic. To this end, paracellular and transcellular mechanisms have been proposed.

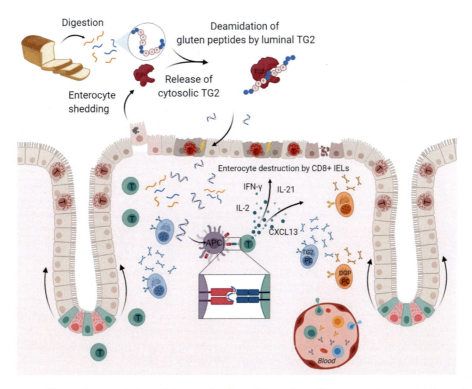

Figure 3.3 *The coeliac gut lesion of the proximal small intestine.* Gluten proteins are rich in proline residues, and for this reason long peptide fragments of gluten survive digestion. Gluten peptides are poor in negatively charged residues, yet the negatively charged amino acids are critical for binding of gluten peptides to the CeD-associated HLA molecules HLA-DQ2.5, HLA-DQ2.2, or HLA-DQ8. Introduction of the negatively charged glutamate into gluten peptides is thus essential for the CD4+ T cell response to gluten. The creation of gluten peptides with negatively charged glutamate resides is mediated in a process termed deamidation by the enzyme transglutaminase 2 (TG2). Where the pathogenic TG2 is localized in CeD is currently not known. One possibility is that it is derived from shed enterocytes which will release cytosolic TG2 on degradation in the gut lumen. Alternatively, deamidated gluten peptides are created by TG2 expressed in the lamina propria. At any rate, gluten peptides must cross the epithelium to the lamina propria to stimulate effector memory CD4+ T cells that reside in the lamina propria. On recognition of deamidated gluten peptides, the effector memory CD4+ T cells become activated. They produce cytokines/chemokines including IL2, IL21 and CXCL13 that may act on many other cells in the local environment and ultimately this activation of CD4+ T cells lead to formation of the CeD lesion crypt cell hyperplasia, increased enterocyte production, and turnover and villous blunting. One cell type which likely is influenced by the CD4+ gluten-specific T cells are CD8+ intraepithelial lymphocytes (CD8+ IELs). These cells, when becoming activated, can kill and destroy enterocytes. The CeD is also characterized by plasmacytosis. Many of the plasma cellss populating the lamina propria of the CeD lesion are specific for TG2 (10%) or deamidated gluten peptides (DGP). Like effector CD4+ T cells, the plasma cells are effector cells which have been generated at inductive immune sites (Peyer's patches, isolated lymphoid follicles or mesenteric lymph nodes) and which have traveled through blood to reach the lamina propria. The coeliac gut lesion in the proximal small intestine is thus populated by many effector cells of the adaptive immune system. Created with BioRender.com.

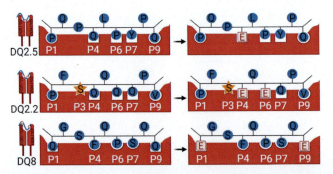

Figure 3.4 *Schematic representation of peptide binding preference of the CeD-associated HLA-molecules.* The CeD-associated HLA molecules (HLA-DQ2.5, HLA-DQ2.2 and HLA-DQ8) prefer peptides with negatively charged anchor residues. However, the native gluten peptides are poor in negatively charged residues and rich in proline (P) and glutamine (Q). Transglutaminase 2 (TG2) can deamidate the glutamine residues in a sequence-dependent manner into glutamate. Subsequently, the TG2-mediated deamidation converts the native gluten peptide into a good ligands for the CeD-predisposing HLA molecules. Further, each of these three different HLA molecules have specific preference as HLA-DQ2.5 and HLA-DQ2.2 prefer peptides with glutamate at positions P4, P6 and P7, while HLA-DQ8 prefers peptides with glutamate residue at P1 and/or P9. In addition, HLA-DQ2.2 prefers peptides with serine (S) or threonine (T) at P3 position. This representative figure shows the native and deamidated variants of the DQ2.5-glia-α2, DQ2.2-glut-L1 and DQ8-glia-α1 epitopes. Created with BioRender.com. Adapted by permission from Springer: Immunogenetics, The roles of MHC class II genes and post-translational modification in celiac disease [20]. https://doi.org/10.1007/s00251-017-0985-7.

The tight junctions in the epithelium restrict the movement of solutes through the space between the cells. As a consequence, most of the solutes are absorbed transcellularly. However, in cases of several infectious and inflammatory gastrointestinal conditions, this barrier protection is compromised resulting in uncontrolled flux of solutes between lumen and lamina propria [50–53]. Therefore, gluten exposure and these epithelial barrier disruptive conditions could synergize to allow access to the gluten peptides into the lamina propria via the paracellular route. Transcellular transport of gluten peptides has been proposed to occur using enterocytic vesicles carrying the peptide–HLA class II complex in untreated CeD subjects [54]. These vesicles are known to be capable of crossing the basal membrane and as a consequence, intact gluten peptides are trafficked into the lamina propria.

3.2.5 Antigen presenting cells in the lamina propria presenting gluten

The gluten peptides that get across the epithelium can be recognized by effector–memory CD4+ T cells sitting in the lamina propria. For this recognition to happen, the peptides must be picked up and presented by antigen presenting cells expressing HLA-DQ molecules. Dendritic cells and macrophages are the two main immune cells expressing the highest level of HLA-DQ in the normal duodenal mucosa [55]. An involvement in

CeD pathogenesis of CD11c+ dendritic cells with activated phenotype was suggested by the observation that such cells accumulate in the lamina propria of the gut lesion and the cells efficiently activate gluten-specific T cells. Further, a rapid accumulation of HLA-DQ expressing CD14+ CD11c+ dendritic cells in the gut mucosa was observed after a 3-day gluten challenge [56]. This gluten-induced accumulation of dendritic cell subset was found to occur before the histological changes in the intestine and before the increase in number of intraepithelial lymphocytes (IELs). Taken together, this evidence may suggest that this specific dendritic cell subset have the capacity to present gluten peptides to the CD4+ T cells in the lamina propria.

A hallmark of CeD is the presence of increased numbers of plasma cells in the lamina propria [57,58]. Remarkably, using an antibody that is specific for an HLA-DQ:gluten peptide complex, it was observed that plasma cells were the main population of the lamina propria cells of untreated CeD subjects that stained with the antibody [59]. It was also observed by antibody staining that some plasma cells express HLA-DQ as well as CD86 – a co-stimulatory molecule involved in T-cell activation. Bulk RNA-sequencing of lamina propria plasma cells populations confirmed some degree of expression of HLA class II molecules by plasma cells and also gave an indication that gut plasma cells have functions that go beyond immunoglobulin secretion, such as cytokine and chemokine secretion [60]. While is yet has to be demonstrated that lamina propria plasma cell indeed can present antigen to CD4+ T cells, the facts that they express surface IgA or IgM allowing sensing of antigen and that they may have immunomodulatory functions, suggest that plasma cells indirectly, and perhaps directly, can affect the function of gluten-specific CD4+ T cells.

3.3 Gluten-specific CD4+ T cells

Gluten-specific CD4+ T cells have been known as crucial contributors in the CeD pathogenesis for more than 25 years. Nevertheless, their roles are still being extensively investigated. The first landmark studies on the discovery and characterization of these cells showed that gluten-specific T-cell lines and clones restricted to HLA-DQ2.5 as well as HLA-DQ8 could be generated by stimulating the gut biopsies with gluten antigen [10,13]. In hindsight, we know that the success of these studies was due to the fact that the antigen used was pepsin/trypsin digests of gluten which contained artificially deamidated gluten peptides. Had digests of gluten been used that do not contain artificially deamidated gluten (i.e. trypsin digests), no T-cell responses to gluten in the *in vitro* assays would have been recorded and the discoveries would not have been made [20]. Other milestones events were the first sequence identification of gluten epitopes restricted by HLA-DQ2.5 and HLA-DQ8 [29,30]. Another seminal observation was the finding that deamidation of gluten is essential for T-cell recognition [25]. Soon thereafter this observation, it was discovered that the crucial deamidation of gluten is mediated by TG2 [47,61]. Further, while gluten-reactive T cells raised from gut biopsies were found to be uniquely restricted by either HLA-DQ2.5 or DQ8, this was not the case when raising gluten-reactive T cells from peripheral blood [62], and moreover the deamidation effect of TG2 was only apparent for HLA-DQ and not for the HLA-DR and HLA-DP restricted gluten-specific T cells [61]. These observations collectively suggest that priming to gluten without involvement of TG2 can occur and that T cells generated by such priming are unlikely to be involved in CeD pathogenesis. We now know that the frequency of gluten-specific and HLA-DQ-restricted CD4+ T cells in peripheral blood of both untreated and treated CeD is in the range 0.001–0.01% which explains the initial struggles to generate HLA-DQ2 or HLA-DQ8 restricted gluten-specific T cells from blood samples. In 2000 it was demonstrated that on day 6 after a 3-day gluten challenge the frequency of these HLA-DQ-restricted and gluten-specific T cells transiently surge in peripheral blood [63]. This approach of performing gluten-challenge has now become a standard procedure to study T-cell response in peripheral blood [32,64–70]. Another element that revolutionized the study of T-cell responses in CeD is the use of HLA-DQ:gluten tetramers – reagents that allow the identification and isolation of CD4+ T cells specific to individual gluten epitopes.

3.3.1 HLA-DQ:gluten tetramers

Visualization of antigen-specific T cells is more difficult than to visualize antigen-specific B cells as the TCR recognize a complex of HLA and peptide and as the affinity between TCR and peptide-bound HLA molecules is much weaker than for the interaction between the B-cell receptor (i.e. antibody) and antigen. The development of multimer/tetramer technologies in the late 1990's changed the field of antigen-specific T-cell research dramatically. Increased avidity and more stable interactions could be obtained by multimerization of peptide-HLA complexes on streptavidin and hence allowing for specific staining of antigen-specific T cells [71,72]. Due to the known HLA-DQ restriction and a variety of T-cell epitopes in CeD, it was possible to generate soluble, recombinant monomeric HLA-DQ2 or HLA-DQ8 molecules coupled to gluten T-cell epitopes of interest [73]. Thus, such HLA-DQ:gluten tetramers, made up of monomeric HLA-DQ molecules tetramerized on fluorochrome-conjugated streptavidin, allow for specific staining reagent of gluten-specific CD4+ T cells (Fig. 3.5).

In 2007 it was demonstrated that the HLA-DQ2.5:gluten tetramers could be used to directly detect gluten-specific T cells in the peripheral blood on day 6 after a 3-day gluten challenge, similar to IFN-γ ELISPOT-based studies [65]. Some of the striking findings were that the tetramer-detected CD4+ T cells showed mainly a memory phenotype and gut-homing capability by expression of integrin β7. At that time, these cells were neither detectable in blood of untreated coeliacs nor in treated coeliacs on a gluten-free diet. It took another six years to show that gluten-specific T cells could be directly detected by HLA-DQ:gluten tetramers in lamina propria in both untreated and treated CeD

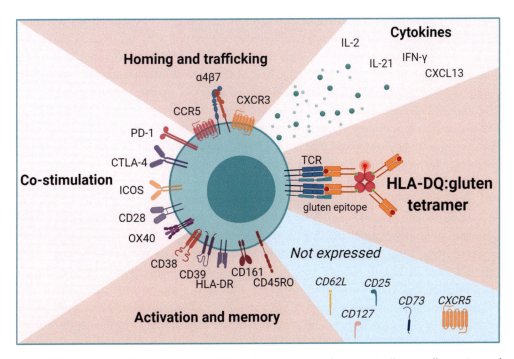

Figure 3.5 *Phenotype of gluten-specific CD4+ T cells.* Gluten-specific CD4+ T cells in small intestine and blood can be detected and characterized by the means of HLA-DQ:gluten tetramers. In blood these cells express gut-homing marker integrin $\alpha4\beta7$ and have an effector memory phenotype with low expression of L-selectin (CD62L). Extensive characterization has further revealed that gluten-specific CD4+ T cells have a narrow and distinct phenotype expressing a range of activation markers (CD38, CD39, HLA-DR and CD161) and co-stimulatory and checkpoint molecules. PD-1 and CTLA-4 are expressed among inhibitory checkpoint molecules and CD28, ICOS and OX40 among the stimulatory checkpoint molecules. Of chemokine receptors, gluten-specific CD4+ T cells express CXCR3 and CCR5 which is important for trafficking of the T cells, but not CXCR5 which is needed for entering B-cell follicles. *Ex vivo* analysis of key phenotype surface markers (CD25, CD127, CD73) implicated for regulatory T cells are not expressed in resting cells. Key cytokines/chemokines expressed by gluten-specific CD4+ T cells are IL-2, IL-21, IFN-γ and CXCL13. Created with BioRender.com.

patients [74]. By performing direct cloning of CD4+ T cells and HLA-DQ2.5:gluten tetramer staining, this study showed that around 1% of all lamina propria CD4+ T cells were gluten–specific T cells. As stated above, the frequency in peripheral blood of these cells is so low that it is impossible to detect them directly unless a procedure with magnetic bead enrichment is employed. Using this magnetic bead enrichment technique, the tetramer staining gains considerable improvement in sensitivity of detection and reduction of background staining [75,76]. With magnetic bead enrichment, the tetramer-binding cells become enriched a 100-fold, and gluten–specific CD4+ T cells in the blood can be enumerated in as few as 1 memory CD4+ T cell per million CD4+ T cells [77].

HLA-DQ:gluten tetramers have been recently explored as a blood-based diagnostic test blood, and they have proven to precisely classify treated CeD subjects from HLA-matched healthy subjects and those with non-coeliac gluten sensitivity [78]. In addition to being a potential diagnostic test, regardless of the test subjects are gluten consuming or not, HLA-DQ:gluten tetramers have indeed opened up the possibility to study gluten-specific T cells at various stages of the disease. The fast-moving developments in next-generation sequencing technologies during the last decade have facilitated a direct characterization of cells at single-cell level. The clonal relationship between T cells can be determined by sequencing their TCR genes. T cells with identical nucleotide sequences of both TCRα and TCRβ chain can be regarded as T-cell clones belonging to the same clonotype. By combining HLA-DQ:gluten tetramers and high-throughput TCR sequencing, it was recently demonstrated that gluten-specific T-cell clonotypes found at the site of inflammation are also present in circulation making up the same T-cell repertoire [66]. Most strikingly, in the same study persistence of the same clonotypes were observed in blood and gut biopsy samples from the same patient taken decades apart. Further, by comparing samples before and during gluten challenge, it was also revealed that the recall response was dominated by pre-existing clonotypes rather than recruitment of new clonotypes. These important findings strongly emphasize the crucial role of gluten-specific T cells in CeD and an explanation to why CeD is a chronic condition. However, it also supports the notion that unintended exposure to gluten despite a gluten-free diet may contribute in maintaining the existence of gluten-specific T cells.

3.3.2 TCR recognition of gluten

Generally it is observed that gluten-specific T cells recognize deamidated gluten peptides better than native gluten peptides [27,79]. However, in crystal structures that have been resolved, residues of the T-cell receptors (TCRs) make no direct interactions with glutamate residues of the peptides [15,19,80–82]. The study of TCRs recognizing the immunodominant HLA-DQ2.5- and HLA-DQ8-restricted gluten epitopes have reported the usage of biased usage of variable (V)-gene segments with or without conserved CDR3 motifs [15,16,64,79,82–84]. The DQ2.5-glia-α2-specific TCRs exhibit preferential expression of *TRAV26-1/TRBV7-2* with a conserved CDR3β arginine motif [64,79,81,84]. Crystal structures of *TRAV26-1/TRBV7-2* TCR:HLA-DQ2.5:DQ2.5-glia-α2 complexes showed that the conserved non-germline-encoded arginine in CDR3β serves as a lynchpin in the interaction with the DQ2.5-glia-α2 [81]. Similarly, the TCRs specific for DQ2.5-glia-α1a display biased expression of *TRAV4* and *TRBV20-1* or *TRBV29-1* [16,81]. A crystal structure of *TRAV4/TRBV20-1* TCR:HLA-DQ2.5:DQ2.5-glia-α1a complex revealed that the *TRAV4* bias is mostly attributed to the interactions between the germline-encoded TCR residues and the HLA-DQ2.5

molecule [81]. The TCRs specific to the ω-gliadins also exhibit biased V-gene usage as DQ2.5-glia-ω2-specific TCRs demonstrate biased usage of *TRAV4* and *TRBV4* gene segments [64] and DQ2.5-glia-ω1-specific TCRs display biased expression of *TRAV4* and *TRBV20-1* or *TRBV29-1* [16]. Similarly, the T cells specific to DQ2.5-hor-3, the immunodominant gluten epitope derived from barley also display preferential expression of *TRAV26-1* and *TRBV20-1* or *TRBV29-1* [83],85. The majority of DQ8-glia-α1-specific T cells express *TRBV9/TRBV26-2* and *TRAV8-3/TRBV6-1*, with a conserved CDR3β arginine motif. Crystal structures revealed that these CDR3-derived arginine residues are central for recognition of the DQ8-glia-α1. These biased TCRs are found in several donors indicating that public TCR bias underpins the T-cell response in CeD [19].

The T-cell epitopes presented by the HLA-DQ2.2 molecules have an additional feature in addition to the deamidation – the presence of a serine residue in the position P3 of the peptide [14]. The structural study has shown that the single polymorphic residue in HLA-DQ2.2 (Phe22α) effects the hydrogen bond network within the peptide-binding groove of HLA-DQ2.2 resulting in selective binding to gluten epitopes possessing a serine at the P3 position of the peptide [19].

3.3.3 Molecular mimicry between gluten T-cell epitopes and bacterial peptides

Several epidemiological studies have reported association between gastrointestinal infections caused by viruses and bacteria with CeD [86–97]. However, very little is known about how these infections contribute to CeD. Recently, it was shown that some microbial peptides derived from *Pseudomonas aeruginosa* with high sequence similarity with gluten epitopes could activate gluten-reactive HLA-DQ2.5-restricted T cells derived from the CeD subjects [98]. Further, gluten-reactive T cells in the blood of CeD subjects challenged with gluten were also reactive to these microbial peptides. By comparing the X-ray crystal structures of bacterial peptide bound to HLA-DQ2.5 in complex with a gluten-reactive TCR and gluten peptide bound to HLA-DQ2.5 in complex with the same gluten-reactive TCR, it was found that the TCR detects the features being similar in the two peptides. Hence, the TCR was cross-reactive to gluten and microbial epitope as a consequence of molecular mimicry. This indicates that by virtue of molecular mimicry any peptide that mimics gluten peptide, when presented on HLA-DQ2.5 has the ability to induce T cells that are cross-reactive to gluten or activate pre-existing gluten-reactive T cells that are cross-reactive to the microbial peptides. This phenomenon could potentially contribute to onset or maintenance of the disease in genetically predisposed individuals [87].

A model to explain the induction of CeD by molecular mimicry has some conceptual weaknesses. Firstly, none of the bacterial mimitopes identified for any given gluten T-cell epitope elicited activation of all T-cell clones specific for the gluten T-cell epitopes in question. This is not unexpected knowing the TCR specific for the same HLA-DQ:gluten peptides can differ in their fine specificity. Secondly, the large number of distinct gluten T-cell epitopes will require a number of different bacterial mimitopes to explain the diversity of the gluten-specific T-cell response of CeD subjects. Further, this model of bacterial mimitopes does not explain the formation of antibodies specific to TG2 and DGP in CeD. It is less difficult to envisage a role of bacterial mimitopes for the maintenance of an gluten-specific T-cell responsiveness in CeD subjects who stay on a strict gluten gluten-free diet or for causing symptoms in CeD subjects who do not respond to the gluten-free diet. The existence of the latter scenario should be testable by looking for bacteria that can generate gluten mimitopes in such diet non-responsive patients.

3.4 Functional role of gluten-specific CD4+ T cells

3.4.1 Phenotype of gluten-specific CD4+ T cells

Until recently, the phenotypic *ex vivo* characterization of gluten-specific T cells in blood was limited to multicolor flow cytometric assessment of relatively few surface markers in combination with specific tetramer staining. Recently, by combining HLA-DQ:gluten tetramer staining with mass cytometry (CyTOF) it became feasible to extensively characterize the phenotype of gluten-specific T cells [99]. The study demonstrated that gluten-specific CD4+ T cells both in the blood and gut of subjects with untreated CeD as well as in treated CeD subjects challenged with gluten exhibit a remarkably distinct and narrow phenotype [99] (Fig. 3.5). The cells express several checkpoint molecules with upregulation of PD-1 and CTLA-4 among the inhibitory molecules, and OX40, ICOS and CD28 among the stimulatory molecules. Importantly, T cells with similar phenotype were also found to be elevated in patients with other autoimmune conditions, e.g. systemic sclerosis and systemic lupus erythematosus. RNA sequencing of gluten-specific T cells from gut biopsies of untreated CeD subjects revealed production of the cytokines/chemokines IL-21 and CXCL13 suggestive of interaction of the T cells with B cells. A similar phenotype of CD4+ T cells has also been described in rheumatoid arthritis where the T cells were defined as peripheral helper T cells due to their lack of CXCR5 expression [100]. Of note, expression of CXCR5 is considered a hallmark of conventional follicular helper T (Tfh) cells.

Regulatory T (Tregs) cells are subgroup of CD4+ T cells that have important roles in regulating and controlling immune responses. In mice, it is well-established that Tregs typically express FoxP3 and CD25, however, definite markers of Tregs in humans are still lacking. Tregs are most commonly characterized by the expression of transcription factor FoxP3 together with high cell surface expression of CD25 and low expression of CD127. Notably, however, FoxP3 can also be expressed in activated human CD4+ T cells [101–103]. Therefore, expression of FoxP3 does not necessarily report on regulatory function of human CD4+CD25+ T cells. Another type of regulatory T cells, type 1 regulatory (Tr1) cells, does not stably express FoxP3, but can be distinguished based on co-expression of CD49b and LAG-3 together with high expression of the cytokines IL-10 and TGF-β. Theoretically, if regulatory T cells use certain HLA class II allotypes for antigen recognition, it can be predicted that these same HLA class II allotypes should exert dominant protection against disease development. Notably, no protective effect of HLA class II allotypes exists in CeD, speaking against a role of regulatory T cells in the pathogenesis of this disorder [20]. Notwithstanding, there have been several studies describing gluten-specific regulatory T cells with roles in CeD pathogenesis. It was reported that gliadin-specific Tr1 cells both producing IL10 and interferon-γ (IFN-γ) are present in the small intestine of CeD subjects, and these cells can suppress proliferation of pathogenic Th0 cells [104]. More recently, a study reported that after oral challenge of CeD subjects most of the gluten-specific T cells expressed CD39 and FoxP3 along with the phenotype OX40+/ CD25+/CD127low, thus the cells were classified as Tregs [105]. Noteworthy, the gluten-specific T cells in the study were detected by expression of CD25 and OX40, 44 hours after stimulation of whole blood samples with gluten peptides. It seems counterintuitive that CeD subjects should have many such regulatory T cells reactive with gluten, but as these cells after a short *in vitro* expansion demonstrated reduced suppressive function compared with polyclonal Treg cells, the authors concluded that a dysfunction of these Tregs cells may contribute to CeD pathogenesis [105]. At dissonance with these observations, another study with *ex vivo* analysis of gluten-specific T cells isolated using HLA-DQ:gluten tetramers revealed that some of the HLA-tetramer-binding T cells were FoxP3+ and all of them were CD25- [99]. Interestingly, upon *in vitro* stimulation and culturing the same cells demonstrated increase in expression of CD25 and FoxP3, thereby indicating that the CD25 expression likely results from the *in vitro* stimulation knowing that CD25 is an early activation maker of T cells [106]. There is little doubt that both studies have interrogated gluten-specific T cells, but the studies give opposing views as to whether the T cells have a regulatory function or whether the cells merely exhibit an activated phenotype after antigen stimulation.

3.4.2 Cytokine production

Decades ago it was shown that the *in vitro* cultured gluten-specific T-cell clones generated from the gut biopsies of CeD subjects produced IFN-γ and variable amounts of tumor necrosis factor-α (TNF-α), transforming growth factor-β (TGF-β), IL-4, IL-5, IL-6, and IL-10 [107]. Similar results were observed in a recent *in vitro* cytokine secretion study where the gluten-specific CD4+ T-cell clones produced IFN-γ, IL-10, IL-4, granulocyte-macrophage colony-stimulating factor (GM-CSF), MIP-1α, MIP-1β, TNF-α, IL-5 and IL-8 [70]. By identifying the gluten-specific T cells by HLA-DQ:gluten tetramers, it was found that the gluten-specific CD4+ T cells in polyclonal T-cell line as well as peripheral blood produce IFN-γ, IL-21, but not IL-17 [108]. Similar observation was made in the lamina propria of children with CeD where there was an increased number of IL-21-producing CD4+ T cells that also co-produced IL-21 and IFN-γ, but not IL-17 [109]. Upon RNA-sequencing, IL-21 was found to be one of the main differentially expressed genes in gluten-specific CD4+ T cells isolated from subjects with CeD [99]. Systemic cytokine release was recently linked to the immune reactivation to gluten and gastrointestinal symptoms in CeD. Specifically, IL-2 together with IL-8 and IL-10 were shown to be elevated only hours after gluten ingestion [70]. Moreover, a quantitative hierarchy of serum cytokines/chemokines was established with significant elevations of IL-2, CCL20, IL-6, CXCL9, CXCL8 (IL-8), IFN-γ, IL-10, IL-22, IL-17A, TNF-α, CCL2 and amphiregulin [69]. In both studies, IL-2 was the most prominently elevated and the earliest cytokine appearing in blood, and its appearance preceded the onset of gastrointestinal symptoms. The early elevations of IL-2, IL-17A, IL-22 and IFN-γ points to the involvement of rapidly activated T cells. As regards symptoms in conjunction with gluten challenge, it has been speculated that FODMAPs and a nocebo effect on the study participants may impact the immediate responses. These important issues were recently addressed in a masked gluten challenge with minimal content of FODMAPs [110]. It was here demonstrated that nausea and vomiting were directly linked to gluten ingestion and correlated with the release of IL-2 after 4 hours. Although it has been known that the kinetics of IL-2 is rapid and transient [111], this recent significance of IL-2 has probably been overlooked in previous studies performed on T-cell clones due to the lack of sensitive assays. This was exemplified by the observation that when fresh blood enriched for gluten-reactive T cells were stimulated with gluten peptides, IL-2 was one of the most prominent cytokines [70]. Taken together, the observations from these *in vitro* and *ex vivo* cytokine profile studies support the notion that CD4+ T cells are crucial in orchestrating the immune response against gluten in CeD.

3.4.3 T cell and B cell interaction: Antibody production and T-cell clonal expansion

Several pieces of evidence point toward a role of gluten-specific T cells in generation of anti-TG2 and anti-DGP antibodies. A role for gluten-specific T cells for generation of antibodies to gluten is rather obvious. This is not so for the generation of antibodies to TG2. The clinical observations that these antibodies are only formed in subject who express HLA-DQ2 or HLA-DQ8 [112] and that their active production is reliant on gluten exposure [113] are notable and striking. A model based on formation of complexes between TG2 and gluten can explain these clinical observations [114]. In brief, the model suggests that TG2-specific B cells of an HLA-DQ2 or HLA-DQ8 expressing subject can bind and internalize such hapten-carrier-like TG2-gluten complexes. After antigen processing by the B cells, a deamidated gluten epitope can be presented by HLA-DQ molecules of the B cells to gluten-specific T cells which then can provide T-cell help to the B cells.

A recent study with transgenic mice, made knock-in for a CeD-patient-derived TG2-specific BCR that has equal reactivity to human and mouse TG2, provided support to this hapten-carrier-like model [21]. Most notably, by expressing the anti-TG2 BCR in mice deficient of TG2 (i.e. *Tgm2-/-*), no signs of B-cell tolerance were observed. The non-tolerized TG2-specific B cells could readily be activated to produce antibodies. In adoptive transfer experiment, IgG anti-TG2 antibodies were produced in mice (transgenically expressing HLA-DQ2.5) that were adoptively transferred with anti-TG2 B cells and gluten-specific T cells and stimulated with a fusion protein of TG2 and a gluten peptide harboring the relevant T-cell epitope. The model supports the concept that interaction of anti-TG2 B cells and gluten-specific CD4+ T cells is important for clonal expansion of the T cells. This may explain why the anti-TG2 antibodies despite little evidence that they are pathogenic as soluble immunoglobulins by themselves, serve as such extremely good proxies for the disease. Presence of anti-TG2 antibodies report on productive interaction between TG2-specific B cells and gluten-specific CD4+ T cells.

Both TG2-specific and DGP-specific plasma cells exhibit biased usage of VH and VL genes with a dominance of certain VH/VL pairs and with a low frequency of somatic hypermutations in both VH and VL genes [57,58,115–117]. The short duration of the anti-TG2 and anti-DGP responses after initiation of the gluten-free diet, the low frequencies of somatic hypermutations and the gluten-specific CD4+ T cells lacking CXCR5, may suggest that these are extrafollicular antibody responses [118].

Taken together, evidence suggests that the interaction of gluten-specific CD4+ T cells with TG2-specific and DGP-specific B cells appears to be an essential part of the CeD pathogenesis. In parallel with the generation of anti-TG2 or anti-DGP plasma cells that seed lamina propria along the gut axis, this cross-talk also results in clonal expansion of gluten-specific CD4+ T cells - the presence of which is the fundamental problem of CeD. It seems very likely that T-cell and B-cell interaction is required for the initiation of CeD. Whether it is required for maintenance of already established disease is less clear as CD4+ effector memory cells residing in the lamina propria may be long lived and can have gluten antigenic peptides presented for their recall response by other APCs than B cells or plasma cells. It is for this reason not a given that anti-B-cell therapy, such as anti-CD20 treatment, will be an effective therapy for the disease.

3.4.4 CD4+ T-cell help to intraepithelial lymphocytes (IELs)

The increased number of IELs is also another hallmark of CeD. The main IEL populations are CD8+ $\alpha\beta$ T cells and $\gamma\delta$ T cells. The involvement of these IEL subsets have been investigated for several decades. Available evidence suggests a direct involvement of CD8+ IELs in pathogenesis of CeD, whereas the role of $\gamma\delta$ T cells is still very much unclear.

It has been demonstrated that CD8+ IELs can express innate C-type lectin natural killer receptors (NKRs), which recognize stress-induced ligands and which otherwise typically are expressed by natural killer cells [119]. In a healthy state, human IELs express the inhibitory CD94/NKG2A receptor and low levels of activating CD94 and NKG2D receptors. In CeD, CD8+ IELs typically lose expression of the inhibitory CD94/NKG2A receptors and upregulate the activating NKG2D and CD94/NKG2C receptors. MICA and MICB, which are ligands for NKG2D, are upregulated by stressed enterocytes [120]. Similarly, the enterocyte expression of HLA-E, the ligand for CD94/NKG2, is also increased in active CeD. NKG2D upregulation is thought to be driven by IL-15 in intestinal epithelial cells. IL-15 is induced upon stress and inflammation and is known to act in a cell contact-dependent manner. Altogether, these findings support the notion that CD8 +IELs acquire the ability to destroy distressed enterocytes.

In a mouse model of CeD that was recently published, it was demonstrated that the activated phenotype of IELs and their killing of enterocytes causing villous atrophy, is dependent on CD4+ T cells [121]. The molecular mechanism behind this CD4+ T-cell mediated control is not known, but evidence from the mouse model suggests that IFN-γ and IL-21 are involved.

3.4.5 Mechanisms underlying villous blunting in CeD

Two models, not mutually exclusive, can be envisioned to explain the typical features of the CeD lesion with increased epithelial cell turnover epithelial crypt cell hyperplasia and villous blunting [122,123]. The first model involves activated CD8+ IELs that kill enterocytes. Crypt cell hyperplasia is then a compensatory process to increased loss of enterocytes. The other model involves CD4+ T cells which either indirectly or directly affect enterocyte regeneration. MacDonald and co-workers suggested that activated CD4+ T cells produce cytokines that act on resident gut myofibroblasts to produce keratinocyte growth factor and matrix metalloproteinases [124,125]. The keratinocyte growth factor stimulates enterocytes to proliferate, and the matrix metalloproteinases reshapes the tissue scaffold onto which the enterocyte lining rests, thereby explaining the increased epithelial turnover and villous blunting. A mechanism of direct effect was described by Xavier and co-workers demonstrating in mice that activated CD4+ T cells in the intestine interact with stem cells to modulate their renewal and differentiation [126]. Of note, in both models CD4+ T cells are involved thus speaking to the central role these cells play in the pathogenesis of CeD.

3.5 Concluding remarks

Three decades of basic and clinical research have established CeD as a quintessential T-cell driven disorder. Unraveling of knowledge that gluten-specific CD4+ T cells are activated immediately upon gluten exposure, have distinct phenotype, exhibit usage of public TCRs and clonal persistence have highlighted their crucial role in CeD pathogenesis. Further, the collaboration between gluten-specific CD4+ T cells and TG2-specific/DGP-specific B cells appears to be a crucial step not only for generation of CeD-specific antibodies but also for activation and clonal expansion of gluten-specific T cells. Hence, in-depth studies on the T-B interaction in CeD will provide valuable knowledge on CeD pathogenesis. Further, the studies exploring the potential of using these disease-driving CD4+ T cells for diagnosis and treatment will be exciting avenues to embark upon in the future.

References

[1] L.M. Sollid, B. Jabri, Triggers and drivers of autoimmunity: lessons from coeliac disease, Nat. Rev. Immunol. 13 (4) (2013) 294–302 https://doi.org/10.1038/nri3407.

[2] S. Husby, S. Koletzko, I. Korponay-Szabo, K. Kurppa, M.L. Mearin, C. Ribes-Koninckx, R. Shamir, R. Troncone, R. Auricchio, G. Castillejo, R. Christensen, J. Dolinsek, P. Gillett, A. Hrobjartsson, T. Koltai, M. Maki, S.M. Nielsen, A. Popp, K. Stordal, M. Wessels, European Society Paediatric Gastroenterology, Hepatology and Nutrition Guidelines for Diagnosing Coeliac Disease 2020, J. Pediatr. Gastroenterol. Nutr. 70 (1) (2020) 141–156 https://doi.org/10.1097/mpg.0000000000002497.

[3] A.M. Mowat, W.W. Agace, Regional specialization within the intestinal immune system, Nat. Rev. Immunol. 14 (10) (2014) 667–685 https://doi.org/10.1038/nri3738.

[4] R.J. Komban, A. Strömberg, A. Biram, J. Cervin, C. Lebrero-Fernández, N. Mabbott, U. Yrlid, Z. Shulman, M. Bemark, N. Lycke, Activated Peyer's patch B cells sample antigen directly from M cells in the subepithelial dome, Nat. Commun. 10 (1) (2019) 2423 –2423 https://doi.org/10.1038/s41467-019-10144-w.

[5] O.J.B. Landsverk, O. Snir, R.B. Casado, L. Richter, J.E. Mold, P. Réu, R. Horneland, V. Paulsen, S. Yaqub, E.M. Aandahl, O.M. Øyen, H.S. Thorarensen, M. Salehpour, G. Possnert, J. Frisén, L.M. Sollid, E.S. Baekkevold, F.L Jahnsen, Antibody-secreting plasma cells persist for decades in human intestine, J. Exp. Med. 214 (2) (2017) 309–317 https://doi.org/10.1084/jem.20161590.

[6] L. Greco, R. Romino, I. Coto, N. Di Cosmo, S. Percopo, M. Maglio, F. Paparo, V. Gasperi, M.G. Limongelli, R. Cotichini, C. D'Agate, N. Tinto, L. Sacchetti, R. Tosi, M.A. Stazi, The first large population based twin study of coeliac disease, Gut 50 (5) (2002) 624–628 https://doi.org/10.1136/gut.50.5.624.

[7] M. Bonamico, M. Ferri, P. Mariani, R. Nenna, E. Thanasi, R.P. Luparia, A. Picarelli, F.M. Magliocca, B. Mora, M.T. Bardella, A. Verrienti, B. Fiore, S. Uccini, F. Megiorni, M.C. Mazzilli, C. Tiberti, Serologic and genetic markers of celiac disease: a sequential study in the screening of first degree relatives, J. Pediatr. Gastroenterol. Nutr. 42 (2) (2006) 150–154 https://doi.org/10.1097/01.mpg.0000189337.08139.83.

[8] S. Withoff, Y. Li, I. Jonkers, C. Wijmenga, Understanding celiac disease by genomics, Trends Genet. 32 (5) (2016) 295–308 https://doi.org/10.1016/j.tig.2016.02.003.

[9] K. Karell, A.S. Louka, S.J. Moodie, H. Ascher, F. Clot, L. Greco, P.J. Ciclitira, L.M. Sollid, J. Partanen, HLA types in celiac disease patients not carrying the DQA1*05-DQB1*02 (DQ2) heterodimer: results from the European Genetics Cluster on Celiac Disease, Hum. Immunol. 64 (4) (2003) 469–477 https://doi.org/10.1016/S0198-8859(03)00027-2.

[10] K.E. Lundin, H. Scott, O. Fausa, E. Thorsby, L.M. Sollid, T cells from the small intestinal mucosa of a DR4,DQ7/DR4,DQ8 celiac disease patient preferentially recognize gliadin when presented by DQ8, Hum. Immunol. 41 (4) (1994) 285–291 https://doi.org/10.1016/0198-8859(94)90047-7.

[11] L.M. Sollid, G. Markussen, J. Ek, H. Gjerde, F. Vartdal, E. Thorsby, Evidence for a primary association of celiac disease to a particular HLA-DQ α/β heterodimer, J. Exp. Med. 169 (1) (1989) 345–350 https://doi.org/10.1084/jem.169.1.345.

[12] R. Ploski, J. Ek, E. Thorsby, L.M. Sollid, On the HLA-DQ(α1*0501, β1*0201)-associated susceptibility in celiac disease: a possible gene dosage effect of DQB1*0201, Tissue Antigens 41 (4) (1993) 173–177 https://doi.org/10.1111/j.1399-0039.1993.tb01998.x.

[13] K.E. Lundin, H. Scott, T. Hansen, G. Paulsen, T.S. Halstensen, O. Fausa, E. Thorsby, L.M. Sollid, Gliadin-specific, HLA-DQ(α1*0501, β1*0201) restricted T cells isolated from the small intestinal mucosa of celiac disease patients, J. Exp. Med. 178 (1) (1993) 187–196 https://doi.org/10.1084/jem.178.1.187.

[14] M. Bodd, C.Y. Kim, K.E. Lundin, L.M. Sollid, T-cell response to gluten in patients with HLA-DQ2.2 reveals requirement of peptide-MHC stability in celiac disease, Gastroenterology 142 (3) (2012) 552–561 https://doi.org/10.1053/j.gastro.2011.11.021.

[15] S.E. Broughton, J. Petersen, A. Theodossis, S.W. Scally, K.L. Loh, A. Thompson, B.J. van, Y. Kooy-Winkelaar, K.N. Henderson, T. Beddoe, J.A. Tye-Din, S.I. Mannering, A.W. Purcell, J. McCluskey, R.P. Anderson, F. Koning, H.H. Reid, J. Rossjohn, Biased T cell receptor usage directed against human leukocyte antigen DQ8-restricted gliadin peptides is associated with celiac disease, Immunity 37 (4) (2012) 611–621 https://doi.org/10.1016/j.immuni.2012.07.013.

[16] S. Dahal-Koirala, L. Ciacchi, J. Petersen, L.F. Risnes, R.S. Neumann, A. Christophersen, K.E.A. Lundin, H.H. Reid, S.W. Qiao, J. Rossjohn, L.M Sollid, Discriminative T-cell receptor recognition of highly homologous HLA-DQ2-bound gluten epitopes, J. Biol. Chem. 294 (3) (2019) 941–952 https://doi.org/10.1074/jbc.RA118.005736.

[17] C.Y. Kim, H. Quarsten, E. Bergseng, C. Khosla, L.M. Sollid, Structural basis for HLA-DQ2-mediated presentation of gluten epitopes in celiac disease, Proc. Natl. Acad. Sci. USA 101 (12) (2004) 4175–4179 https://doi.org/10.1073/pnas.0306885101.

[18] K.H. Lee, K.W. Wucherpfennig, D.C. Wiley, Structure of a human insulin peptide-HLA-DQ8 complex and susceptibility to type 1 diabetes, Nat. Immunol. 2 (6) (2001) 501–507 https://doi.org/10.1038/88694.

[19] Y.T. Ting, S. Dahal-Koirala, H.S.K. Kim, S.-W. Qiao, R.S. Neumann, K.E.A. Lundin, J. Petersen, H.H. Reid, L.M. Sollid, J Rossjohn, A molecular basis for the T cell response in HLA-DQ2.2 mediated celiac disease, 117, 6th ed., Proc. Natl. Acad. Sci. USA, 2020, pp. 3063–3073, https://doi.org/10.1073/pnas.1914308117.

[20] L.M. Sollid, The roles of MHC class II genes and posttranslational modification in celiac disease, Immunogenetics 69 (12) (2017) https://doi.org/10.1007/s00251-017-0985-7.

[21] M.F. du Pré, J. Blazevski, A.E. Dewan, J. Stamnaes, C. Kanduri, G.K. Sandve, M.K. Johannesen, C.B. Lindstad, K. Hnida, L. Fugger, G. Melino, S.-W. Qiao, L.M. Sollid, B cell tolerance and antibody production to the celiac disease autoantigen transglutaminase 2, J. Exp. Med. 217 (2) (2020) e20190860 https://doi.org/10.1084/jem.20190860.

[22] R. Iversen, S.F. Amundsen, L. Kleppa, M. Fleur du Pre, J. Stamnaes, L.M. Sollid, Evidence that pathogenic transglutaminase 2 in celiac disease derives from enterocytes, Gastroenterology 159 (2) (2020) 788–790 https://doi.org/10.1053/j.gastro.2020.04.018.

[23] M.N. Marsh, P.R. Shewry, A.S. Tatham, D.D. Kasarda, Cereal proteins and coeliac disease, Coeliac Disease, Blackwell Scientific Publications, 1992, pp. 305–348.

[24] L.M. Sollid, S.W. Qiao, R.P. Anderson, C. Gianfrani, F. Koning, Nomenclature and listing of celiac disease relevant gluten T-cell epitopes restricted by HLA-DQ molecules, Immunogenetics 64 (6) (2012) 455–460 https://doi.org/10.1007/s00251-012-0599-z.

[25] H. Arentz-Hansen, B. Fleckenstein, O. Molberg, H. Scott, F. Koning, G. Jung, P. Roepstorff, K.E. Lundin, L.M. Sollid, The molecular basis for oat intolerance in patients with celiac disease, PLoS Med. 1 (1) (2004) e1 https://doi.org/10.1371/journal.pmed.0010001.

[26] M.Y. Hardy, J.A. Tye-Din, J.A. Stewart, F. Schmitz, N.L. Dudek, I. Hanchapola, A.W. Purcell, R.P. Anderson, Ingestion of oats and barley in patients with celiac disease mobilizes cross-reactive T cells activated by avenin peptides and immuno-dominant hordein peptides, J. Autoimmun. 56 (2015) 56–65 https://doi.org/10.1016/j.jaut.2014.10.003.

[27] H. Arentz-Hansen, R. Korner, O. Molberg, H. Quarsten, W. Vader, Y.M. Kooy, K.E. Lundin, F. Koning, P. Roepstorff, L.M. Sollid, S.N. McAdam, The intestinal T cell response to α-gliadin in adult celiac disease is focused on a single deamidated glutamine targeted by tissue transglutaminase, J. Exp. Med. 191 (4) (2000) 603–612 https://doi.org/10.1084/jem.191.4.603.

[28] L.W. Vader, A. de Ru, Y. van der Wal, Y.M. Kooy, W. Benckhuijsen, M.L. Mearin, J.W. Drijfhout, P. van Veelen, F. Koning, Specificity of tissue transglutaminase explains cereal toxicity in celiac disease, J. Exp. Med. 195 (5) (2002) 643–649 https://doi.org/10.1084/jem.20012028.

[29] Y. van de Wal, Y.M. Kooy, P.A. van Veelen, S.A. Pena, L.M. Mearin, O. Molberg, K.E. Lundin, L.M. Sollid, T. Mutis, W.E. Benckhuijsen, J.W. Drijfhout, F. Koning, Small intestinal T cells of celiac disease patients recognize a natural pepsin fragment of gliadin, Proc. Nat. Acad. Sci. USA 95 (17) (1998) 10050–10054 https://doi.org/10.1073/pnas.95.17.10050.

[30] H. Sjostrom, K.E. Lundin, O. Molberg, R. Korner, S.N. McAdam, D. Anthonsen, H. Quarsten, O. Noren, P. Roepstorff, E. Thorsby, L.M. Sollid, Identification of a gliadin T-cell epitope in coeliac disease: general importance of gliadin deamidation for intestinal T-cell recognition, Scand. J. Immunol. 48 (2) (1998) 111–115 https://doi.org/10.1046/j.1365-3083.1998.00397.x.

[31] H. Arentz-Hansen, B. Fleckenstein, Ø. Molberg, H. Scott, F. Koning, G. Jung, P. Roepstorff, K.E.A. Lundin, L.M Sollid, The molecular basis for oat intolerance in patients with celiac disease, PLoS Med. 1 (1) (2004) e1–e1 https://doi.org/10.1371/journal.pmed.0010001.

[32] J.A. Tye-Din, J.A. Stewart, J.A. Dromey, T. Beissbarth, D.A. van Heel, A. Tatham, K. Henderson, S.I. Mannering, C. Gianfrani, D.P. Jewell, A.V. Hill, J. McCluskey, J. Rossjohn, R.P. Anderson, Comprehensive, quantitative mapping of T cell epitopes in gluten in celiac disease, Sci. Transl. Med. 2 (41) (2010) https://doi.org/10.1126/scitranslmed.3001012.

[33] L.W. Vader, D.T. Stepniak, E.M. Bunnik, Y.M. Kooy, W. de Haan, J.W. Drijfhout, P.A. Van Veelen, F. Koning, Characterization of cereal toxicity for celiac disease patients based on protein homology in grains, Gastroenterology 125 (4) (2003) 1105–1113 https://doi.org/10.1016/S0016-5085(03)01204-6.

[34] L.M. Sollid, Coeliac disease: dissecting a complex inflammatory disorder, Nat. Rev. Immunol. 2 (9) (2002) 647–655 https://doi.org/10.1038/nri885.

[35] L. Shan, O. Molberg, I. Parrot, F. Hausch, F. Filiz, G.M. Gray, L.M. Sollid, C. Khosla, Structural basis for gluten intolerance in celiac sprue, Science 297 (5590) (2002) 2275–2279 https://doi.org/10.1126/science.1074129.

[36] H. Arentz-Hansen, S.N. McAdam, O. Molberg, B. Fleckenstein, K.E. Lundin, T.J. Jorgensen, G. Jung, P. Roepstorff, L.M. Sollid, Celiac lesion T cells recognize epitopes that cluster in regions of gliadins rich in proline residues, Gastroenterology 123 (3) (2002) 803–809 https://doi.org/10.1053/gast.2002.35381.

[37] S. Dorum, M. Bodd, L.E. Fallang, E. Bergseng, A. Christophersen, M.K. Johannesen, S.W. Qiao, J. Stamnaes, G.A. de Souza, L.M. Sollid, HLA-DQ molecules as affinity matrix for identification of gluten T cell epitopes, J. Immunol. 193 (9) (2014) 4497–4506 https://doi.org/10.4049/jimmunol.1301466.

[38] H. Quarsten, O. Molberg, L. Fugger, S.N. McAdam, L.M. Sollid, HLA binding and T cell recognition of a tissue transglutaminase-modified gliadin epitope, Eur. J. Immunol. 29 (8) (1999) 2506–2514 https://doi.org/10.1002/(SICI)1521-4141(199908)29:08/0742506:AID-IMMU2506/0763.0.CO;2-9.

[39] J. Xia, L.M. Sollid, C. Khosla, Equilibrium and kinetic analysis of the unusual binding behavior of a highly immunogenic gluten peptide to HLA-DQ2, Biochemistry 44 (11) (2005) 4442–4449 https://doi.org/10.1021/bi047747c.

[40] B. Fleckenstein, Ø. Molberg, S.-W. Qiao, D.G. Schmid, F. von der Mülbe, K. Elgstøen, G. Jung, L.M. Sollid, Gliadin T cell epitope selection by tissue transglutaminase in celiac disease: Role of enzyme specificity and pH influence on the transamidation versus deamidation reactions, J. Biol. Chem. 277 (37) (2002) 34109–34116 https://doi.org/10.1074/jbc.M204521200.

[41] W. Vader, Y. Kooy, P. Van Veelen, A. De Ru, D. Harris, W. Benckhuijsen, S. Pena, L. Mearin, J.W. Drijfhout, F. Koning, The gluten response in children with celiac disease is directed toward multiple gliadin and glutenin peptides, Gastroenterology 122 (7) (2002) 1729–1737 https://doi.org/10.1053/gast.2002.33606.

[42] S. Dorum, S.W. Qiao, L.M. Sollid, B. Fleckenstein, A quantitative analysis of transglutaminase 2-mediated deamidation of gluten peptides: Implications for the T-cell response in celiac disease, J. Proteome Res. 8 (4) (2009) 1748–1755 https://doi.org/10.1021/pr800960n.

[43] S. Dorum, M.O. Arntzen, S.W. Qiao, A. Holm, C.J. Koehler, B. Thiede, L.M. Sollid, B. Fleckenstein, The preferred substrates for transglutaminase 2 in a complex wheat gluten digest are peptide fragments harboring celiac disease T-cell epitopes, PLoS One 5 (11) (2010) e14056 https://doi.org/10.1371/journal.pone.0014056.

[44] B.H. Johansen, F. Vartdal, J.A. Eriksen, E. Thorsby, L.M. Sollid, Identification of a putative motif for binding of peptides to HLA-DQ2, Int. Immunol. 8 (2) (1996) 177–182 https://doi.org/10.1093/intimm/8.2.177.

[45] Y. van de Wal, Y.M. Kooy, J.W. Drijfhout, R. Amons, G.K. Papadopoulos, F. Koning, Unique peptide binding characteristics of the disease-associated DQ(α1*0501, β1*0201) vs the non-disease-associated DQ(α1*0501, β1*0202) molecule, Immunogenetics 46 (6) (1997) 484–492 https://doi.org/10.1007/s002510050309.

[46] F. Vartdal, B.H. Johansen, T. Friede, C.J. Thorpe, S. Stevanović, J.E. Eriksen, K. Sletten, E. Thorsby, H.G. Rammensee, L.M. Sollid, The peptide binding motif of the disease associated HLA-DQ(α1*0501, β1*0201) molecule, Eur. J. Immunol. 26 (11) (1996) 2764–2772 https://doi.org/10.1002/eji.1830261132.

[47] Y. van de Wal, Y. Kooy, P. van Veelen, S. Pena, L. Mearin, G. Papadopoulos, F. Koning , Selective deamidation by tissue transglutaminase strongly enhances gliadin-specific T cell reactivity, J. Immunol. 161 (4) (1998) 1585–1588.

[48] E. Bergseng, S. Dorum, M.O. Arntzen, M. Nielsen, S. Nygard, S. Buus, G.A. de Souza, L.M. Sollid, Different binding motifs of the celiac disease-associated HLA molecules DQ2.5, DQ2.2, and DQ7.5 revealed by relative quantitative proteomics of endogenous peptide repertoires, Immunogenetics 67 (2) (2015) 73–84 https://doi.org/10.1007/s00251-014-0819-9.

[49] L.M. Sollid, J.A. Tye-Din, S.-W. Qiao, R.P. Anderson, C. Gianfrani, F. Koning, Update 2020: nomenclature and listing of celiac disease–relevant gluten epitopes recognized by CD4$^+$ T cells, Immunogenetics 72 (1–2) (2020) 85–88 https://doi.org/10.1007/s00251-019-01141-w.

[50] F. Heller, P. Florian, C. Bojarski, J. Richter, M. Christ, B. Hillenbrand, J. Mankertz, A.H. Gitter, N. Bürgel, M. Fromm, M. Zeitz, I. Fuss, W. Strober, J.D. Schulzke, Interleukin-13 is the key effector Th2 cytokine in ulcerative colitis that affects epithelial tight junctions, apoptosis, and cell restitution, Gastroenterology 129 (2) (2005) 550–564 https://doi.org/10.1016/j.gastro.2005.05.002.

[51] A. Nusrat, C. von Eichel-Streiber, J.R. Turner, P. Verkade, J.L. Madara, C.A. Parkos, *Clostridium difficile* toxins disrupt epithelial barrier function by altering membrane microdomain localization of tight junction proteins, Infect. Immun. 69 (3) (2001) 1329 https://doi.org/10.1128/IAI.69.3.1329-1336.2001.

[52] H. Troeger, H.-J. Epple, T. Schneider, U. Wahnschaffe, R. Ullrich, G.-D. Burchard, T. Jelinek, M. Zeitz, M. Fromm, J.-D. Schulzke, Effect of chronic *Giardia lamblia* infection on epithelial transport and barrier function in human duodenum, Gut 56 (3) (2007) 328–335 https://doi.org/10.1136/gut.2006.100198.

[53] S. Zeissig, N. Bürgel, D. Günzel, J. Richter, J. Mankertz, U. Wahnschaffe, A.J. Kroesen, M. Zeitz, M. Fromm, J.D. Schulzke, Changes in expression and distribution of claudin 2, 5 and 8 lead to discontinuous tight junctions and barrier dysfunction in active Crohn's disease, Gut, 56 (1) (2007) 61–72 https://doi.org/10.1136/gut.2006.094375.

[54] K.P. Zimmer, H. Naim, P. Weber, H.J. Ellis, P.J. Ciclitira, Targeting of gliadin peptides, CD8, α/β-TCR, and γ/δ-TCR to Golgi complexes and vacuoles within celiac disease enterocytes, FASEB J. 12 (13) (1998) 1349–1357 https://doi.org/10.1096/fasebj.12.13.1349.

[55] M. Ráki, S. Tollefsen, Ø. Molberg, K.E.A. Lundin, L.M. Sollid, F.L Jahnsen, A unique dendritic cell subset accumulates in the celiac lesion and efficiently activates gluten-reactive T cells, Gastroenterology 131 (2) (2006) 428–438 https://doi.org/10.1053/j.gastro.2006.06.002.

[56] A.-C.R. Beitnes, M. Ráki, M. Brottveit, K.E.A. Lundin, F.L. Jahnsen, L.M Sollid, Rapid accumulation of CD14$^+$CD11c$^+$ dendritic cells in gut mucosa of celiac disease after *in vivo* gluten challenge, PLoS One 7 (3) (2012) e33556–e33556 https://doi.org/10.1371/journal.pone.0033556.

[57] R. Di Niro, L. Mesin, N.-Y. Zheng, J. Stamnaes, M. Morrissey, J.-H. Lee, M. Huang, R. Iversen, M.F. du Pré, S.-W. Qiao, K.E.A. Lundin, P.C. Wilson, L.M Sollid, High abundance of plasma cells secreting transglutaminase 2–specific IgA autoantibodies with limited somatic hypermutation in celiac disease intestinal lesions, Nat. Med. 18 (3) (2012) 441–445 https://doi.org/10.1038/nm.2656.

[58] Ø. Steinsbø, C.J.H. Dunand, M. Huang, L. Mesin, M. Salgado-Ferrer, K.E.A. Lundin, J. Jahnsen, P.C. Wilson, L.M Sollid, Restricted VH/VL usage and limited mutations in gluten-specific IgA of coeliac disease lesion plasma cells, Nat. Commun. 5 (1) (2014) 4041 https://doi.org/10.1038/ncomms5041.

[59] L.S. Høydahl, L. Richter, R. Frick, O. Snir, K.S. Gunnarsen, O.J.B. Landsverk, R. Iversen, J.R. Jeliazkov, J.J. Gray, E. Bergseng, S. Foss, S.-W. Qiao, K.E.A. Lundin, J. Jahnsen, F.L. Jahnsen, I. Sandlie, L.M. Sollid, G.Å Løset, Plasma cells are the most abundant gluten peptide MHC-expressing cells in inflamed intestinal tissues from patients with celiac disease, Gastroenterology 156 (5) (2019) 1428–1439 e10 https://doi.org/10.1053/j.gastro.2018.12.013.

[60] O. Snir, C. Kanduri, K.E.A. Lundin, G.K. Sandve, L.M Sollid, Transcriptional profiling of human intestinal plasma cells reveals effector functions beyond antibody production, United Eur. Gastroenterol. J. 7 (10) (2019) 1399–1407 https://doi.org/10.1177/2050640619862461.

[61] O. Molberg, S.N. McAdam, R. Korner, H. Quarsten, C. Kristiansen, L. Madsen, L. Fugger, H. Scott, O. Noren, P. Roepstorff, K.E. Lundin, H. Sjostrom, L.M. Sollid, Tissue transglutaminase selectively modifies gliadin peptides that are recognized by gut-derived T cells in celiac disease, Nat. Med. 4 (6) (1998) 713–717 https://doi.org/10.1038/nm0698-713.

[62] H.A. Gjertsen, L.M. Sollid, J. Ek, E. Thorsby, K.E. Lundin, T Cells from the peripheral blood of coeliac disease patients recognize gluten antigens when presented by HLA-DR, -DQ, or -DP molecules, Scand. J. Immunol. 39 (6) (1994) 567–574 https://doi.org/10.1111/j.1365-3083.1994.tb03414.x.

[63] R.P. Anderson, P. Degano, A.J. Godkin, D.P. Jewell, A.V. Hill, *In vivo* antigen challenge in celiac disease identifies a single transglutaminase-modified peptide as the dominant A-gliadin T-cell epitope, Nat. Med. 6 (3) (2000) 337–342 https://doi.org/10.1038/73200.

[64] S. Dahal-Koirala, L.F. Risnes, A. Christophersen, V.K. Sarna, K.E. Lundin, L.M. Sollid, S.W. Qiao, TCR sequencing of single cells reactive to DQ2.5-glia-α2 and DQ2.5-glia-ω2 reveals clonal expansion and epitope-specific V-gene usage, Mucosal. Immunol. 9 (3) (2016) 587–596 https://doi.org/10.1038/mi.2015.147.

[65] M. Raki, L.E. Fallang, M. Brottveit, E. Bergseng, H. Quarsten, K.E. Lundin, L.M. Sollid, Tetramer visualization of gut-homing gluten-specific T cells in the peripheral blood of celiac disease patients, Proc. Nat. Acad. Sci. USA 104 (8) (2007) 2831–2836 https://doi.org/10.1073/pnas.0608610104.

[66] L.F. Risnes, A. Christophersen, S. Dahal-Koirala, R.S. Neumann, G.K. Sandve, V.K. Sarna, K.E. Lundin, S.W. Qiao, L.M. Sollid, Disease-driving CD4$^+$ T cell clonotypes persist for decades in celiac disease, J. Clin. Invest. 128 (6) (2018) 2642–2650 https://doi.org/10.1172/JCI98819.

[67] V.K. Sarna, K.E.A. Lundin, L. Morkrid, S.W. Qiao, L.M. Sollid, A Christophersen, HLA-DQ-gluten tetramer blood test accurately identifies patients with and without celiac disease in absence of gluten consumption, Gastroenterology 154 (4) (2018) 886–896 e6https://doi.org/10.1053/j.gastro.2017.11.006.

[68] Jason A. Tye-Din, A.J.M. Daveson, H.C. Ee, G. Goel, J. MacDougall, S. Acaster, K.E. Goldstein, J.L. Dzuris, K.M. Neff, K.E. Truitt, R.P Anderson, Elevated serum interleukin-2 after gluten correlates with symptoms and is a potential diagnostic biomarker for coeliac disease, Aliment. Pharmacol. Ther. 50 (8) (2019) 901–910 https://doi.org/10.1111/apt.15477.

[69] G. Goel, A.J.M. Daveson, C.E. Hooi, J.A. Tye-Din, S. Wang, E. Szymczak, L.J. Williams, J.L. Dzuris, K.M. Neff, K.E. Truitt, R.P Anderson, Serum cytokines elevated during gluten-mediated cytokine release in coeliac disease, Clin. Exp. Immunol. 199 (1) (2020) 68–78 https://doi.org/10.1111/cei.13369.

[70] G. Goel, J.A. Tye-Din, S.-W. Qiao, A.K. Russell, T. Mayassi, C. Ciszewski, V.K. Sarna, S. Wang, K.E. Goldstein, J.L. Dzuris, L.J. Williams, R.J. Xavier, K.E.A. Lundin, B. Jabri, L.M. Sollid, R.P Anderson, Cytokine release and gastrointestinal symptoms after gluten challenge in celiac disease, Sci. Adv. 5 (8) (2019) eaaw7756 https://doi.org/10.1126/sciadv.aaw7756.

[71] J.D. Altman, P.A. Moss, P.J. Goulder, D.H. Barouch, M.G. McHeyzer-Williams, J.I. Bell, A.J. McMichael, M.M. Davis, Phenotypic analysis of antigen-specific T lymphocytes, Science 274 (5284) (1996) 94–96 https://doi.org/10.1126/science.274.5284.94.

[72] F. Crawford, H. Kozono, J. White, P. Marrack, J. Kappler, Detection of antigen-specific T cells with multivalent soluble class II MHC covalent peptide complexes, Immunity 8 (6) (1998) 675–682 https://doi.org/10.1016/s1074-7613(00)80572-5.

[73] H. Quarsten, S.N. McAdam, T. Jensen, H. Arentz-Hansen, O. Molberg, K.E. Lundin, L.M. Sollid, Staining of celiac disease-relevant T cells by peptide-DQ2 multimers, J. Immunol. 167 (9) (2001) 4861–4868 https://doi.org/10.4049/jimmunol.167.9.4861.

[74] M. Bodd, M. Raki, E. Bergseng, J. Jahnsen, K.E. Lundin, L.M. Sollid, Direct cloning and tetramer staining to measure the frequency of intestinal gluten-reactive T cells in celiac disease, Eur. J. Immunol. 43 (10) (2013) 2605–2612 https://doi.org/10.1002/eji.201343382.

[75] C.L. Day, N.P. Seth, M. Lucas, H. Appel, L. Gauthier, G.M. Lauer, G.K. Robbins, Z.M. Szczepiorkowski, D.R. Casson, R.T. Chung, S. Bell, G. Harcourt, B.D. Walker, P. Klenerman, K.W. Wucherpfennig, Ex vivo analysis of human memory CD4 T cells specific for hepatitis C virus using MHC class II tetramers, J. Clin. Invest. 112 (6) (2003) 831–842 https://doi.org/10.1172/JCI18509.

[76] J.J. Moon, H.H. Chu, M. Pepper, S.J. McSorley, S.C. Jameson, R.M. Kedl, M.K. Jenkins, Naive CD4$^+$ T cell frequency varies for different epitopes and predicts repertoire diversity and response magnitude, Immunity 27 (2) (2007) 203–213 https://doi.org/10.1016/j.immuni.2007.07.007.

[77] A. Christophersen, M. Raki, E. Bergseng, K.E. Lundin, J. Jahnsen, L.M. Sollid, S.W. Qiao, Tetramer-visualized gluten-specific CD4+ T cells in blood as a potential diagnostic marker for coeliac disease without oral gluten challenge, United European Gastroenterol. J. 2 (4) (2014) 268–278 https://doi.org/10.1177/2050640614540154.

[78] V.K. Sarna, G.I. Skodje, H.M. Reims, L.F. Risnes, S. Dahal-Koirala, L.M. Sollid, K.E.A Lundin, HLA-DQ:gluten tetramer test in blood gives better detection of coeliac patients than biopsy after 14-day gluten challenge, Gut 67 (9) (2018) 1606–1613 https://doi.org/10.1136/gutjnl-2017-314461.

[79] S.W. Qiao, M. Raki, K.S. Gunnarsen, G.A. Loset, K.E. Lundin, I. Sandlie, L.M. Sollid, Posttranslational modification of gluten shapes TCR usage in celiac disease, J. Immunol. 187 (6) (2011) 3064–3071 https://doi.org/10.4049/jimmunol.1101526.

[80] J. Petersen, Y. Kooy-Winkelaar, K.L. Loh, M. Tran, J. van Bergen, F. Koning, J. Rossjohn, H.H. Reid, Diverse T cell receptor gene usage in HLA-DQ8- associated celiac disease converges into a consensus binding solution, Structure 24 (10) (2016) 1643–1657 https://doi.org/10.1016/j.str.2016.07.010.

[81] J. Petersen, V. Montserrat, J.R. Mujico, K.L. Loh, D.X. Beringer, L.M. van, A. Thompson, M.L. Mearin, J. Schweizer, Y. Kooy-Winkelaar, B.J. van, J.W. Drijfhout, W.T. Kan, N.L. La Gruta, R.P. Anderson, H.H. Reid, F. Koning, J. Rossjohn, T-cell receptor recognition of HLA-DQ2-gliadin complexes associated with celiac disease, Nat. Struct. Mol. Biol. 21 (5) (2014) 480–488 https://doi.org/10.1038/nsmb.2817.

[82] J. Petersen, J. van Bergen, K.L. Loh, Y. Kooy-Winkelaar, D.X. Beringer, A. Thompson, S.F. Bakker, C.J. Mulder, K. Ladell, J.E. McLaren, D.A. Price, J. Rossjohn, H.H. Reid, F. Koning, Determinants of gliadin-specific T cell selection in celiac disease, J. Immunol. 194 (12) (2015) 6112–6122 https://doi.org/10.4049/jimmunol.1500161.

[83] S. Dahal-Koirala, R.S. Neumann, J. Jahnsen, K.E.A. Lundin, L.M Sollid, On the immune response to barley in celiac disease: Biased and public T-cell receptor usage to a barley unique and immunodominant gluten epitope, Eur. J. Immun. 50 (2) (2020) 256–269 https://doi.org/10.1002/eji.201948253.

[84] S.W. Qiao, A. Christophersen, K.E. Lundin, L.M. Sollid, Biased usage and preferred pairing of α- and β-chains of TCRs specific for an immunodominant gluten epitope in coeliac disease, Int. Immunol. 26 (1) (2014) 13–19 https://doi.org/10.1093/intimm/dxt037.

[85] M.Y. Hardy, A.K. Russell, C. Pizzey, C.M. Jones, K.A. Watson, N.L. La Gruta, D.J. Cameron, J.A. Tye-Din, Characterisation of clinical and immune reactivity to barley and rye ingestion in children with coeliac disease, Gut 69 (5) (2019) 830–840. https://doi.org/10.1136/gutjnl-2019-319093.

[86] S. Ashorn, H. Raukola, T. Välineva, M. Ashorn, B. Wei, J. Braun, I. Rantala, K. Kaukinen, T. Luukkaala, P. Collin, M. Mäki, S. Iltanen, Elevated serum anti-*Saccharomyces cerevisiae*, anti-I2 and anti-OmpW antibody levels in patients with suspicion of celiac disease, J. Clin. Immunol. 28 (5) (2008) 486–494 https://doi.org/10.1007/s10875-008-9200-9.

[87] A. Caminero, H.J. Galipeau, J.L. McCarville, C.W. Johnston, S.P. Bernier, A.K. Russell, J. Jury, A.R. Herran, J. Casqueiro, J.A. Tye-Din, M.G. Surette, N.A. Magarvey, D. Schuppan, E.F. Verdu, Duodenal bacteria from patients with celiac disease and healthy subjects distinctly affect gluten breakdown and immunogenicity, Gastroenterology 151 (4) (2016) 670–683 https://doi.org/10.1053/j.gastro.2016.06.041.

[88] A. Caminero, J.L. McCarville, H.J. Galipeau, C. Deraison, S.P. Bernier, M. Constante, C. Rolland, M. Meisel, J.A. Murray, X.B. Yu, A. Alaedini, B.K. Coombes, P. Bercik, C.M. Southward, W. Ruf, B. Jabri, F.G. Chirdo, J. Casqueiro, M.G. Surette, E.F. Verdu, Duodenal bacterial proteolytic activity determines sensitivity to dietary antigen through protease-activated receptor-2, Nat. Commun. 10 (1) (2019) 1198 –1198 https://doi.org/10.1038/s41467-019-09037-9 .

[89] V. D'Argenio, G. Casaburi, V. Precone, C. Pagliuca, R. Colicchio, D. Sarnataro, V. Discepolo, S.M. Kim, I. Russo, G. Del Vecchio Blanco, D.S. Horner, M. Chiara, G. Pesole, P. Salvatore, G. Monteleone, C. Ciacci, G.J. Caporaso, B. Jabrì, F. Salvatore, L Sacchetti, Metagenomics reveals dysbiosis and a potentially pathogenic *N. flavescens* strain in duodenum of adult celiac patients, Am. J. Gastroenterol. 111 (6) (2016) 879–890 https://doi.org/10.1038/ajg.2016.95.

[90] M.F. Kagnoff, Y.J. Paterson, P.J. Kumar, D.D. Kasarda, F.R. Carbone, D.J. Unsworth, R.K. Austin, Evidence for the role of a human intestinal adenovirus in the pathogenesis of coeliac disease, Gut 28 (8) (1987) 995–1001 https://doi.org/10.1136/gut.28.8.995.

[91] C.R. Kahrs, K. Chuda, G. Tapia, L.C. Stene, K. Mårild, T. Rasmussen, K.S. Rønningen, K.E.A. Lundin, L. Kramna, O. Cinek, K Størdal, Enterovirus as trigger of coeliac disease: nested case-control study within prospective birth cohort, Br. Med. J. 364 (2019) l231–l231 https://doi.org/10.1136/bmj.l231.

[92] M. Lawler, P. Humphries, C. O'Farrelly, H. Hoey, O. Sheils, M. Jeffers, D.S. O'Briain, D. Kelleher, Adenovirus 12 E1A gene detection by polymerase chain reaction in both the normal and coeliac duodenum, Gut 35 (9) (1994) 1226–1232 https://doi.org/10.1136/gut.35.9.1226.

[93] J. Mahon, G.E. Blair, G.M. Wood, B.B. Scott, M.S. Losowsky, P.D. Howdle, Is persistent adenovirus 12 infection involved in coeliac disease? A search for viral DNA using the polymerase chain reaction, Gut 32 (10) (1991) 1114–1116 https://doi.org/10.1136/gut.32.10.1114.

[94] M.S. Riddle, J.A. Murray, B.D. Cash, M. Pimentel, C.K. Porter, Pathogen-specific risk of celiac disease following bacterial causes of foodborne illness: A retrospective cohort study, Digest. Dis. Sci. 58 (11) (2013) 3242–3245 https://doi.org/10.1007/s10620-013-2733-7.

[95] E. Sánchez, E. Donat, C. Ribes-Koninckx, M.L. Fernández-Murga, Y. Sanz, Duodenal-mucosal bacteria associated with celiac disease in children, Appl. Environ. Microbiol. 79 (18) (2013) 5472–5479 https://doi.org/10.1128/aem.00869-13.

[96] L.C. Stene, M.C. Honeyman, E.J. Hoffenberg, J.E. Haas, R.J. Sokol, L. Emery, I. Taki, J.M. Norris, H.A. Erlich, G.S. Eisenbarth, M. Rewers, Rotavirus infection frequency and risk of celiac disease autoimmunity in early childhood: a longitudinal study, Am. J. Gastroenterol. 101 (10) (2006) 2333–2340 https://doi.org/10.1111/j.1572-0241.2006.00741.x.

[97] L. Viitasalo, L. Niemi, M. Ashorn, S. Ashorn, J. Braun, H. Huhtala, P. Collin, M. Mäki, K. Kaukinen, K. Kurppa, S. Iltanen, Early microbial markers of celiac disease, J. Clin. Gastroenterol. 48 (7) (2014) 620–624 https://doi.org/10.1097/MCG.0000000000000089.

[98] J. Petersen, L. Ciacchi, M.T. Tran, K.L. Loh, Y. Kooy-Winkelaar, N.P. Croft, M.Y. Hardy, Z. Chen, J. McCluskey, R.P. Anderson, A.W. Purcell, J.A. Tye-Din, F. Koning, H.H. Reid, J. Rossjohn, T cell receptor cross-reactivity between gliadin and bacterial peptides in celiac disease, Nat. Struct. Mol. Biol. 27 (1) (2020) 49–61 https://doi.org/10.1038/s41594-019-0353-4.

[99] A. Christophersen, E.G. Lund, O. Snir, E. Solà, C. Kanduri, S. Dahal-Koirala, S. Zühlke, Ø. Molberg, P.J. Utz, M. Rohani-Pichavant, J.F. Simard, C.L. Dekker, K.E.A. Lundin, L.M. Sollid, M.M Davis, Distinct phenotype of CD4$^+$ T cells driving celiac disease identified in multiple autoimmune conditions, Nat. Med. 25 (5) (2019) 734–737 https://doi.org/10.1038/s41591-019-0403-9.

[100] D.A. Rao, M.F. Gurish, J.L. Marshall, K. Slowikowski, C.Y. Fonseka, Y. Liu, L.T. Donlin, L.A. Henderson, K. Wei, F. Mizoguchi, N.C. Teslovich, M.E. Weinblatt, E.M. Massarotti, J.S. Coblyn, S.M. Helfgott, Y.C. Lee, D.J. Todd, V.P. Bykerk, S.M. Goodman, M.B. Brenner, Pathologically expanded peripheral T helper cell subset drives B cells in rheumatoid arthritis, Nature 542 (7639) (2017) 110–114 https://doi.org/10.1038/nature20810.

[101] M. Kmieciak, M. Gowda, L. Graham, K. Godder, H.D. Bear, F.M. Marincola, M.H. Manjili, Human T cells express CD25 and Foxp3 upon activation and exhibit effector/memory phenotypes without any regulatory/suppressor function, J. Transl. Med. 7 (1) (2009) 89 https://doi.org/10.1186/1479-5876-7-89.

[102] M.E. Morgan, J.H.M. van Bilsen, A.M. Bakker, B. Heemskerk, M.W. Schilham, F.C. Hartgers, B.G. Elferink, L. van der Zanden, R.R.P. de Vries, T.W.J. Huizinga, T.H.M. Ottenhoff, R.E.M. Toes, Expression of FOXP3 mRNA is not confined to CD4$^+$CD25$^+$ T regulatory cells in humans, Human Immunol. 66 (1) (2005) 13–20 https://doi.org/10.1016/j.humimm.2004.05.016.

[103] M.R. Walker, D.J. Kasprowicz, V.H. Gersuk, A. Benard, M. Van Landeghen, J.H. Buckner, S.F. Ziegler, Induction of FoxP3 and acquisition of T regulatory activity by stimulated human CD4$^+$CD25$^-$ T cells, J. Clin. Invest. 112 (9) (2003) 1437–1443 https://doi.org/10.1172/JCI19441.

[104] C. Gianfrani, M.K. Levings, C. Sartirana, G. Mazzarella, G. Barba, D. Zanzi, A. Camarca, G. Iaquinto, N. Giardullo, S. Auricchio, R. Troncone, M.-G. Roncarolo, Gliadin-specific type 1 regulatory T cells from the intestinal mucosa of treated celiac patients inhibit pathogenic T cells, J. Immunol. 177 (6) (2006) 4178–4186 (Baltimore, Md. : 1950) https://doi.org/10.4049/jimmunol.177.6.4178.

[105] L. Cook, C.M.L. Munier, N. Seddiki, D. van Bockel, N. Ontiveros, M.Y. Hardy, J.K. Gillies, M.K. Levings, H.H. Reid, J. Petersen, J. Rossjohn, R.P. Anderson, J.J. Zaunders, J.A. Tye-Din, A.D Kelleher, Circulating gluten-specific FOXP3$^+$CD39$^+$ regulatory T cells have impaired suppressive function in patients with celiac disease, J. Allergy Clin. Immunol. 140 (6) (2017) 1592–1603 e8 https://doi.org/10.1016/j.jaci.2017.02.015.

[106] M. Reddy, E. Eirikis, C. Davis, H.M. Davis, U. Prabhakar, Comparative analysis of lymphocyte activation marker expression and cytokine secretion profile in stimulated human peripheral blood mononuclear cell cultures: an in vitro model to monitor cellular immune function, J. Immunol. Methods 293 (1–2) (2004) 127–142 https://doi.org/10.1016/j.jim.2004.07.006.

[107] E.M. Nilsen, K.E. Lundin, P. Krajci, H. Scott, L.M. Sollid, P. Brandtzaeg, Gluten specific, HLA-DQ restricted T cells from coeliac mucosa produce cytokines with Th1 or Th0 profile dominated by interferon gamma, Gut 37 (6) (1995) 766–776 https://doi.org/10.1136/gut.37.6.766.

[108] M. Bodd, M. Raki, S. Tollefsen, L.E. Fallang, E. Bergseng, K.E. Lundin, L.M. Sollid, HLA-DQ2-restricted gluten-reactive T cells produce IL-21 but not IL-17 or IL-22, Mucosal. Immunol. 3 (6) (2010) 594–601 https://doi.org/10.1038/mi.2010.36.

[109] M.A. van Leeuwen, D.J. Lindenbergh-Kortleve, H.C. Raatgeep, L.F. de Ruiter, R.R. de Krijger, M. Groeneweg, J.C. Escher, J.N. Samsom, Increased production of interleukin-21, but not interleukin-17A, in the small intestine characterizes pediatric celiac disease, Mucosal. Immunol. 6 (6) (2013) 1202–1213 https://doi.org/10.1038/mi.2013.19.

[110] A.J.M. Daveson, J.A. Tye-Din, G. Goel, K.E. Goldstein, H.L. Hand, K.M. Neff, L.J. Williams, K.E. Truitt, R.P. Anderson, the R.C.S Group, Masked bolus gluten challenge low in FODMAPs implicates nausea and vomiting as key symptoms associated with immune activation in treated coeliac disease, Aliment. Pharmacol. Ther. 51 (2) (2020) 244–252 https://doi.org/10.1111/apt.15551.

[111] D.K. Sojka, D. Bruniquel, R.H. Schwartz, N.J. Singh, IL-2 secretion by CD4$^+$ T cells in vivo is rapid, transient, and influenced by TCR-specific competition, J. Immunol. 172 (10) (2004) 6136–6143 https://doi.org/10.4049/jimmunol.172.10.6136.

[112] S. Björck, C. Brundin, E. Lörinc, K.F. Lynch, D. Agardh, Screening detects a high proportion of celiac disease in young HLA-genotyped children, J. Ped. Gastroenterol. Nutr. 50 (1) (2010) 49–53 https://doi.org/10.1097/MPG.0b013e3181b477a6.

[113] D. Leffler, D. Schuppan, K. Pallav, R. Najarian, J.D. Goldsmith, J. Hansen, T. Kabbani, M. Dennis, C.P. Kelly, Kinetics of the histological, serological and symptomatic responses to gluten challenge in adults with celiac disease, Gut 62 (7) (2013) 996–1004 https://doi.org/10.1136/gutjnl-2012-302196.

[114] L.M. Sollid, O. Molberg, S. McAdam, K.E. Lundin, Autoantibodies in coeliac disease: tissue transglutaminase—guilt by association? Gut 41 (6) (1997) 851–852 https://doi.org/10.1136/gut.41.6.851.

[115] K. Hnida, J. Stamnaes, M.F. du Pré, S. Mysling, T.J.D. Jørgensen, L.M. Sollid, R Iversen, Epitope-dependent functional effects of celiac disease autoantibodies on transglutaminase 2, J. Biol. Chem. 291 (49) (2016) 25542–25552 https://doi.org/10.1074/jbc.M116.738161.

[116] R. Iversen, R. Di Niro, J. Stamnaes, K.E. Lundin, P.C. Wilson, L.M. Sollid, Transglutaminase 2–specific autoantibodies in celiac disease target clustered, N-terminal epitopes not displayed on the surface of cells, J. Immunol. 190 (12) (2013) 5981–5991 https://doi.org/10.4049/jimmunol.1300183.

[117] B. Roy, R.S. Neumann, O. Snir, R. Iversen, G.K. Sandve, K.E.A. Lundin, L.M Sollid, High-throughput single-cell analysis of B cell receptor usage among autoantigen-specific plasma cells in celiac disease, J. Immunol. 199 (2) (2017) 782 https://doi.org/10.4049/jimmunol.1700169.

[118] R. Iversen, L.M. Sollid, Autoimmunity provoked by foreign antigens, Science 368 (6487) (2020) 132–133 https://doi.org/10.1126/science.aay3037.

[119] T. Mayassi, B. Jabri, Human intraepithelial lymphocytes, Mucosal. Immunol. 11 (5) (2018) 1281–1289 https://doi.org/10.1038/s41385-018-0016-5.

[120] S. Hüe, J.-J. Mention, R.C. Monteiro, S. Zhang, C. Cellier, J. Schmitz, V. Verkarre, N. Fodil, S. Bahram, N. Cerf-Bensussan, S. Caillat-Zucman, A direct role for NKG2D/MICA interaction in villous atrophy during celiac disease, Immunity 21 (3) (2004) 367–377 https://doi.org/10.1016/j.immuni.2004.06.018.

[121] V. Abadie, S.M. Kim, T. Lejeune, B.A. Palanski, J.D. Ernest, O. Tastet, J. Voisine, V. Discepolo, E.V. Marietta, M.B.F. Hawash, C. Ciszewski, R. Bouziat, I. Horwath, M.A. Zurenski, I. Lawrence, A. Dumaine, V. Yotova, J.-C. Grenier, B. Jabri, IL-15, gluten and HLA-DQ8 drive tissue destruction in coeliac disease, Nature (2020) https://doi.org/10.1038/s41586-020-2003-8.

[122] A.J. Watson, N.A. Wright, Coeliac disease. Morphology and cell kinetics of the jejunal mucosa in untreated patients, Clin. Gastroenterol. 3 (1) (1974) 11–31.

[123] N.A. Wright, A.J. Watson, A.R. Morley, D.R. Appleton, J.M. Marks, Cell production rate in mucosa of untreated coeliac disease, Gut 13 (10) (1972) 846.

[124] T.T. MacDonald, M. Bajaj-Elliott, S.L. Pender, T cells orchestrate intestinal mucosal shape and integrity, Immunol. Today 20 (11) (1999) 505–510 https://doi.org/10.1016/s0167-5699(99)01536-4.

[125] T.T. MacDonald, J. Spencer, Evidence that activated mucosal T cells play a role in the pathogenesis of enteropathy in human small intestine, J. Exp. Med. 167 (4) (1988) 1341–1349 https://doi.org/10.1084/jem.167.4.1341.

[126] M. Biton, A.L. Haber, N. Rogel, G. Burgin, S. Beyaz, A. Schnell, O. Ashenberg, C.W. Su, C. Smillie, K. Shekhar, Z. Chen, C. Wu, J. Ordovas-Montanes, D. Alvarez, R.H. Herbst, M. Zhang, I. Tirosh, D. Dionne, L.T. Nguyen, R.J. Xavier, T helper cell cytokines modulate intestinal stem cell renewal and differentiation, Cell 175 (5) (2018) 1307–1320 e22 https://doi.org/10.1016/j.cell.2018.10.008.

CHAPTER 4

Seronegative villous atrophy

Annalisa Schiepatti[a], David S Sanders[b], Federico Biagi[a]
[a]Gastroenterology Unit, IRCCS Pavia, ICS Maugeri, University of Pavia, Italy
[b]Academic Department of Gastroenterology, Royal Hallamshire Hospital & University of Sheffield, UK

4.1 Introduction

Villous atrophy (VA), crypt hyperplasia and an increased intraepithelial lymphocyte count are the histological hallmarks of coeliac disease (CD) and its complications, a group of very rare premalignant and malignant conditions responsible for the higher mortality registered in CD than in the general population [1-5]. Although CD is surely the most common chronic enteropathy affecting nearly 1% of the Caucasian population, VA is not specific to CD and can be found in other enteropathies unrelated to gluten ingestion [3,6-16]. (Table 4.1) Diagnosis of conventional seropositive CD in adult patients is based on VA and positive IgA tissue transglutaminase (TTA) and/or endomysial antibodies (EmA) while on a normal gluten-containing diet (GCD) [1,2]. Although coeliac specific antibodies are characterized by high sensitivity and specificity [17,18], a minority of coeliac patients can present with VA and negative serology at diagnosis. This rare, yet poorly defined, form of CD is called seronegative CD (SNCD) and its diagnosis relies on the clinical and histological response to a gluten-free diet (GFD), after the exclusion of other non-coeliac enteropathies [3,16,19]. On the basis of 384 patients with SNVA described by five international centers [6-10,16], it emerges that SNCD is the most common cause of SNVA (Fig. 4.1). However, addressing the differential diagnosis of SNVA is still challenging for several reasons. Firstly, non-coeliac enteropathies are extremely heterogeneous and currently there are no specific international consensus guidelines addressing their differential diagnosis and management. Secondly, although certain principles of therapy for patients presenting with VA and severe malabsorption are similar regardless from the aetiology of VA (for example corticosteroids or total parenteral nutrition), there are also some unique therapeutic options that strictly depend on the underlying diagnosis. Consequently, misdiagnosis of SNVA may result in patients being placed unnecessarily on a lifelong GFD or even incorrectly classified as refractory CD and given immunosuppressive therapies [6-11,19,20]. Finally, it has recently been suggested that overall mortality in SNVA is higher than in conventional CD [14,15,20-22]. Therefore, correct identification and management of non-coeliac enteropathies is mandatory in order to avoid long-term morbidity and mortality. In this chapter we will summarize the current evidence for the investigation and clinical management of patients with SNVA.

Coeliac Disease and Gluten-Related Disorders.
DOI: https://doi.org/10.1016/B978-0-12-821571-5.00001-5

Table 4.1 Etiological classification of seronegative villous atrophy.

SERONEGATIVE VILLOUS ATROPHY	
Non-coeliac SNVA	**Seronegative coeliac disease**
Immuno-mediated	**True seronegative CD**
Autoimmune enteropathy	Early stage of the disease
Common variable immunodeficiency	Late stage of the disease
Crohn's disease	Refractory CD and/or EATL
Lymphoproliferative	Dermatitis herpetiformis
Type 2 EATL (MEITL)	First degree relatives of coeliac patients
CD4+ indolent lymphomas	**Conditions influencing serological response**
IPSID	GFD and/or immunosuppressants
Medication-related and iatrogenic	**Seronegative CD associated to common variable immunodeficiency**
Angiotensin type 2 receptor blockers	
Azathioprine	
Methotrexate	
Micophenolate	
mephenamic acid (±other NSAIDs?)	
Chemotherapy	
Radiotherapy	
Graft–versus–host disease	
Transplanted small bowel	
Infectious	
Giardiasis	
HIV enteropathy	
Tubercolosis	
Whipple's disease	
Tropical sprue	
Small intestinal bacterial overgrowth (?)	
Helicobacter Pylori (?)	
Inflammatory	
Eosinophilic gastro–enteritis	
Peptic duodenitis	
Collagenous sprue	
Idiopathic	
IVA 1–transient VA likely post–infective	
IVA 2– persistent non–lymphoproliferative VA	
IVA 3– persistent VA with lymphoproliferative features	

VA: villous atrophy, CD: coeliac disease; IVA: idiopathic villous atrophy, GFD: gluten-free diet, NSAIDS: non-steroidal anti-inflammatory drugs, EATL: enteropathy associated T-cell lymphoma, MEITL: monomorphic epitheliotropic T-cell lymphoma, IPSID: immune-proliferative small intestinal disease, (?) indicates conditions for which clear consensus on whether they can cause villous atrophy is still lacking.

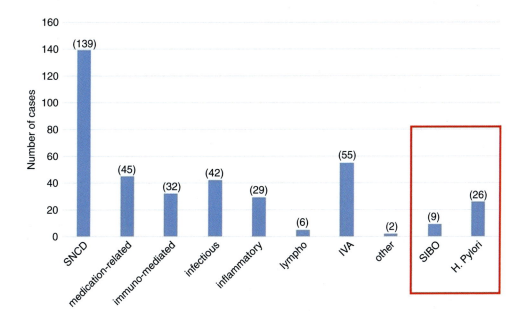

Figure 4.1 *Frequency of different causes of seronegative villous atrophy based on a total of 384 published cases by five major international centers [7-10,16].* The number of cases for each condition is provided in brackets. SNCD: seronegative coeliac disease; IVA: idiopathic villous atrophy; SIBO: small intestinal bacterial overgrowth; red line keep together conditions for which association with seronegative villous atrophy is still debatable (see text). The two patients belonging to the group "other" were affected by a form of villous atrophy due to sarcoidosis [8] and extensive gastric metaplasia [7]. **Permission process not started.** Select figure and clickicon in top toolbar to start permission process or indicate no permission is required.

4.2 Seronegative coeliac disease

Seronegative CD (SNCD) is a rare and poorly defined form of CD characterized by negative IgA/IgG TTA/EmA and VA showing clinical and histological response to a GFD [1-3,8-10,19]. In the last years our understanding on SNCD has been growing and it is likely that different forms of SNCD may exist. According to literature, forms of true SNCD may be found in an early phase of the disease with a lesser degree of VA, as well as in the late stage of disease (with possible refractory CD or lymphoma) and in dermatitis herpetiformis [23-27]. Remarkably, up to 30% of patients with biopsy proven dermatitis herpetiformis can have a negative serology at diagnosis [28]. Also patients who had already been started on a GFD or steroids/immunosuppressants prior to serological testing may present with IgA negative TTA/EmA at diagnosis [19,29]. Forms of SNCD may also rarely occur in first degree relatives of coeliac patients [30,31]. Patients with total IgA deficiency may be found to have positive class IgG TTA or IgG EmA. Taking into consideration that these patients generate an IgG-based serological response, this point may be in favor of a diagnosis of conventional seropositive CD instead of that of true SNCD [19]. Finally, SNCD can be found in association with common variable immune-deficiency (CVID), and in this context the histological response to a GFD is the main diagnostic criterion, as it will be discussed in the dedicated paragraph on CVID. The true prevalence of SNCD among coeliac patients still needs to be elucidated. While the first papers on SNCD reported a prevalence of SNCD of 10-20% of all coeliac patients [23-26,32], more recent studies show a lower prevalence between 2-6.5% [9,10,33]. It is still unclear if this difference may be ascribed to a true change in the prevalence of SNCD over the years or to evolving diagnostic criteria for this condition. Finally, the higher sensitivity of Ema and TTA may have led to an increase in the proportion of patients with seropositive CD, thus reducing the number of those affected by SNCD [19]. Conversely, SNCD is reported to be the most common cause of SNVA (Fig. 4.1) [6-10,16,19]. Although the main literature findings suggest that patients with SNCD are older at diagnosis and can more frequently present with severe malabsorption [8-10,19,26], there is a lack of clinical data about the long-term follow-up of this condition. Moreover, there are no specific markers for SNCD. Some interest was initially dedicated to intestinal deposits of IgA tTG2 antibodies that were found in the small bowel mucosa of seronegative coeliac patients, but not in other noncoeliac enteropathies [26]. However, their specificity has recently been questioned [34], so their diagnostic relevance in everyday clinical practice is limited. Future research should focus on the study of natural history of SNCD, identification of risk factors for poor outcomes and strategies for follow-up.

4.3 Immunomediated non-coeliac enteropathies

4.3.1 Autoimmune enteropathy

Autoimmune enteropathy (AE) is a very rare condition, affecting both children and adults, characterized by SNVA with severe malabsorption refractory to any dietary exclusion, serum enterocyte antibodies and/or associated autoimmune conditions [35-38]. While in the very first diagnostic criteria for AE proposed by Walker-Smith [35] absence of immunodeficiencies was mandatory for diagnosis, in the last years, cases of AE have also been described in association with rare primary immunodeficiencies [39,40]. Patients affected by AE are often misdiagnosed as refractory CD or SNCD [36,41,42]. Apart from VA that is mandatory for diagnosis [35-42], other histological features have been described in AE, but they are not entirely specific for this condition. These include a low count of intraepithelial lymphocytes with dense lymphoplasmacellular infiltrates of the lamina propria, neutrophilic cryptitis with or without crypt micro-abscesses, and increased cryptic apoptosis [12,42,43]. Enterocyte antibodies are the mainstay for serological diagnosis of AE [36,41,42]. Even if positive enterocyte antibodies were described in HIV

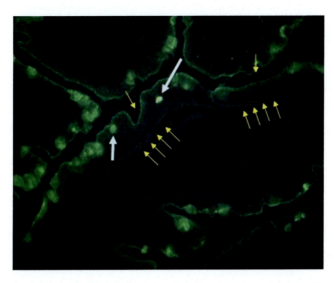

Figure 4.2 *Anti-enterocyte antibodies. Indirect immunofluorescence on monkey jejunum showing enterocyte antibodies (thin arrows).* Goblet cell antibodies are also shown (thick arrows). Goblet cell antibodies are totally non-specific and should not be taken into account for the diagnosis of autoimmune enteropathy [46,51]. **Permission process not started.** Select figure and clickicon in top toolbar to start permission process or indicate no permission is required.

patients suffering from chronic diarrhea, it is likely that these discrepancies occurred because of the lack of standardized criteria for their detection by means of indirect immunofluorescence [44]. However, on the basis of our clinical experience, we and others [45-48] suggest that positive enterocyte antibodies in indirect immunofluorescence on monkey jejunum slides should be identified according to the bright staining of enterocyte cytoplasm with negative nuclear structures [45]. Homogeneous staining of the enterocyte and enterocytes brighter than the lamina propria are further patterns that should be taken into account [46-48], as shown in Fig. 4.2. Finally, anti-goblet cell antibodies were proposed as a further marker for AE [42], they are totally nonspecific and their use should definitely be discouraged for the diagnosis of AE [49-51]. We support positive IgA and/or IgG enterocyte antibodies together with persistent VA unresponsive to a GFD as cornerstone criteria for the diagnosis of AE in adults. HLA typing is not relevant for diagnosis of AE [36,41,42]. Immunosuppressive therapy is necessary but severe complications can develop even after many years and despite good initial clinical and histological response [52,53]. Prognosis is poor in our experience. Only one of the five patients we diagnosed in Pavia is still alive almost 20 years after the diagnosis (patient 2 of [36]) while the other four all died (71 $\pm$ 63 months after diagnosis). Although two of them died of apparently unrelated causes (myocardial infarction and pneumonia), a third one died of haemophagocytic lymphohistiocytosis five years after our description

[54], and the last one died of intestinal lymphoma, likely as a complication of the enteropathy [53].

4.3.2 Enteropathy associated to common variable immunodeficiency

Common variable immunodeficiency (CVID) is the most frequent symptomatic form of primary humoral immunodeficiency, whose diagnosis is based on a marked decrease of IgG (at least 2 standard deviations-SD below the mean for age) and at least one of either IgM or IgA. The following diagnostic criteria must also be fulfilled: onset of immunodeficiency after the age of two years; absent isohemagglutinins and/or poor response to vaccines; exclusion of secondary causes of hypogammaglobulinemia such as malignancies, drugs, infections and genetic disorders [55,56]. Gastrointestinal involvement in CVID is frequent [57-66] and persistent diarrhea has been described in up to 20-60% of cases [63,67-70]. Gastrointestinal symptoms and lesions may be due to infectious/parasitic agents, particularly *Giardia Lamblia*. More recently, Norovirus infection has been also suggested to have a causative role in the severe CVID enteropathy [57-59,71]. Apart from intestinal lesions of infective aetiology, a wide range of noninfectious histopathological lesions have been described in CVID. They include increased intraepithelial lymphocytosis with normal duodenal architecture (46% [66]), VA undistinguishable from untreated CD (2%, 5% [21,66]), Crohn's like lesions, lymphoid nodular hyperplasia, and absence of plasma cells [57-66,72,73]. None of these histopathological lesions, however, is strictly specific for CVID. Their frequency in CVID, as well as their relationship with gastrointestinal symptoms and their relevance in the diagnostic process needs to be confirmed by the whole clinical and biochemical scenario. As far as therapy is concerned, iv immunoglobulins do not improve clinical symptoms and histological lesions, whereas corticosteroids are effective in improving symptoms and intestinal lesions [74,75]. Patients with CVID and VA are burdened by a very high mortality mainly due to onco-immunological disorders and infections [21,70,76,77], therefore a strict clinical and endoscopic follow-up is required in these patients.

Recognizing coeliac disease in patients with common variable immunodeficiency

In clinical practice, it can be very difficult to distinguish between CVID and SNCD, not only because of the similarity in the intestinal lesions, but also because of the possible coexistence of both CVID with VA and CD in the same patients [72-75]. In this specific setting, the histological recovery of VA after gluten withdrawal is the mainstay for confirming CD, and negative HLA DQ2 and DQ8 are very useful to rule out CD [74,75]. On the contrary, coeliac antibodies have no role in this diagnostic work-up. Class IgG EmA have been found in CVID patients who had previously received iv immunoglobulin replacement therapy and in whom CD was excluded, due to the absence of HLA DQ2 and DQ8 molecules [66,74,75]. Finally, certain histopathological features such as absence of plasma cells and the presence of a polymorphonuclear infiltrate and graft-versus-host disease like lesions [60,63,72-75] were described in patients affected by CVID and duodenal VA but not in untreated CD. However, their role is only supportive for diagnosis of CVID.

4.3.3 Iatrogenic forms of villous atrophy

Non-coeliac VA can develop as the consequence of the use of certain medications, which include angiotensin II type 2 receptors blockers (ARB2s) particularly olmesartan [78], azathioprine [79], methotrexate [80] and mycophenolate mofetil [81]. SNVA has been described also in patients treated with chemotherapy, radiotherapy, graft versus host disease and transplanted small intestine [82,83]. Although there are some evidences supporting mefenamic acid as cause of non-coeliac VA [84], the causative role of other NSAIDS needs to be confirmed [85]. In terms of frequency, iatrogenic forms of VA rank as the second most common form of SNVA due to a known cause, if we consider a summary of all the reports from international referral centers (see Fig. 4.1). Diagnosis of iatrogenic enteropathies is guided by a thorough pharmacological history after the exclusion of other causes of SNVA. Histological and clinical improvement after drug withdrawal confirm the diagnosis. Prognosis of iatrogenic forms of VA is usually good since enteropathy and malabsorption respond very quickly to drug withdrawal [78-81,84].

Enteropathy associated to use of angiotensin II type 2 receptors blockers. Olmesartan-associated enteropathy (OAE) was first reported by Rubio-Tapia in 2012 [78], and can be considered the prototype of drug-induced enteropathies. Cases related to other ARB2s such as irbesartan [86], valsartan [87], telmisartan [88] and possibly candesartan [8] were also described. Although this may suggest a class effect of ARB2s [88], a large French population study and a multi-database US study show that users of ARB2s other than olmesartan are not at increased risk of developing malabsorption [89,90]. Epidemiology and natural history of OAE are still poorly understood. Estimated prevalence may be within 1/4000 and 1/5000 olmesartan users [91], particularly in patients older than 40 years old and with no appreciable gender difference [78,86]. While in some papers the length of treatment seems to be a major determinant of the enteropathy [78,86,89,92], an American study found a higher rate of enteropathy in olmesartan initiators [90]. HLA typing has no diagnostic role in these patients [78,86,91,92]. Although antinuclear antibodies have been described by some authors in OAE [86,92], the lack of specific serological markers for OAE is still a major limitation to our understanding of the natural history of this condition. Apart from VA, granulocytic infiltration (both neutrophils and eosinophils), increased crypt apoptosis [93,94] and increased deposition of sub-epithelial collagen have also been observed on duodenal biopsies in OAE [93-95]. The diagnostic relevance of these histological findings should, however, be matched with the clinical data on a case-by-case basis. At present, only histological and clinical improvement after drug withdrawal confirm the diagnosis.

4.4 Enteropathies due to infectious causes

According to published data, infections represent the third major cause for SNVA (Fig. 4.1). This is the case for giardiasis [7,10,96], Whipple's disease [97], tuberculosis [8,98], HIV enteropathy [99], and tropical sprue [100,101]. However, while VA and malabsorption can prompt a clinical suspicion of giardiasis and tropical sprue, VA is not the primary manifestation in Whipple's disease, tuberculosis, and HIV enteropathy. An isolated finding of VA hardly ever leads to their diagnosis in the absence of a suggestive clinical picture. *Giardia Lamblia* infection should be suspected in all patients with malabsorption, particularly in those affected by a primary or acquired immunodeficiency [7-10,96]. Identification of cysts or trophozoites in the stool, stool specific Giardia antigens, PCR on duodenal biopsies aspirate or direct identification of the parasite by the pathologist on formalin-fixed paraffin embedded hematoxylin-eosin stained small-bowel specimens are reliable diagnostic methods [96,102,103]. Whipple's disease should be suspected predominantly in middle-aged Caucasians men with a long-lasting history of seronegative arthritis and fever, that anticipates the onset of a severe malabsorption syndrome [97,104]. Lymphoadenopathy, endocarditis and potentially life-threatening neurological involvement can complete the clinical picture. Duodenal biopsy showing a PAS+ diastase resistant macrophagic infiltration of the lamina propria are key to diagnosis. PCR for WD is highly specific but should be reserved to certain sterile sites, such as the central nervous system [97]. Diagnosis of tropical sprue and HIV enteropathy should be guided by significant history (residence of travels to tropical regions; positive HIV testing). Finally, it has been reported that also small intestinal bacteria overgrowth and *Helicobacter Pylori* may be associated to non-coeliac VA [8,105]. However, the strength of evidence is poor for these two conditions. In patients affected by SIBO a wide variety of histological lesions have been reported [12], but VA is not a key diagnostic element. Whereas mild VA seems to occur only in the most severe cases of SIBO [7,8,10], in most patients a normal villous architecture has been described [105,106]. Moreover, in patients with VA and absence of any predisposing conditions to SIBO a positive glucose H_2 breath test is quite likely to be a consequence of to the histological lesions themselves rather than their cause [107]. So, in patients suffering from a malabsorption syndrome and with predisposing factors to SIBO a glucose H_2 breath test should be performed and antibiotic treatment considered [122]. Few reports suggest peptic duodenitis with or without *Helicobacter Pylori* infection [6,8,108] as cause of SNVA, but for there is scarce evidence in favor of its causative role in SNVA.

4.5 Lymphoproliferative disorders causing villous atrophy

Primary lymphoproliferative disorders of the small-bowel are rare and only few of them can manifest with VA in the uninvolved nonneoplastic mucosa [109]. This is mainly the case for enteropathy associated T-cell lymphoma (type 1 and type 2), indolent T-cell lymphoproliferative disease of gastrointestinal tract, and immunoproliferative small intestinal disease (IPSID). Type 1 enteropathy associated T-cell lymphoma (EATL) is well-known to be a complication of CD, arising both as *de novo* lymphoma or as an evolution from type 2 refractory CD [109-114]. Typical histological pattern on immunohistochemistry consists of medium to large sized tumor $CD3^+,CD4^-,CD5^-$, $CD7^+,CD8^-,CD30^+,CD56^-,CD103^+,TCRbeta^+$ cells [109]. Clinical picture is usually dominated by severe malabsorption with fever, increased inflammatory markers, bleeding or abdominal pain. Abdominal CT scan, positron emission tomography, and capsule endoscopy should be part of the diagnostic workup in these patients [113]. Type 2 EATL, also known as monomorphic epitheliotropic T-cell lymphoma occurs mostly in Asian populations and it is characterized by monomorphic infiltrate of small to medium-sized $CD3^+,CD4^-,CD8^+,CD30^-,CD56^+,TCRbeta^+$ cells [109]. Although cases in Asia do not seem to be related to CD, a few European cases with a history of CD have been documented [109,115-117]. Indolent T-cell lymphoproliferative disease of gastrointestinal tract may represent a diagnostic challenge in the context of SNVA [109,118,119]. Although the $CD8^+$ forms have to be distinguished from aggressive lymphomas, a recent report suggests that $CD4^+$ T-cell lymphomas are often misdiagnosed as CD [119]. Neoplastic cells usually display a TCR-gamma monoclonal rearrangement. Identification of aberrant $CD103^+,CD3^-,CD8^-$ T-cell population by means of immunohistochemistry and flow cytometry excludes the diagnosis of type 2 refractory CD in these patients [119]. We also noticed that in two patients concomitant diagnoses of CD and AE were found [119]. Immunoproliferative small intestinal disease (IPSID) is a rare variant of MALT lymphoma with extensive plasmacytic differentiation, usually involving the duodenum or jejunum [109]. It usually affects young adults living in lower socioeconomic areas and in the Middle East, Far East and Africa and some infective causes have been postulated [120]. Histologically, it is characterized by a dense plasmacytic infiltrate of $CD20^+$ B cells and plasma cells expressing the alpha heavy chain but no light chains [109]. Given the clinico-epidemiological background and the histological features, differential diagnosis should be straightforward, but a full thickness intestinal biopsy can be necessary [117].

4.6 Villous atrophy due to inflammatory disorders

Collagenous sprue, Crohn's disease of the small bowel and eosinophilic gastroenteritis have been described as causes of SNVA [3,121,123]. Histopathological features together with specific tests such as faecal calprotectin, radiological imaging and past medical history are key to the correct diagnosis. Collagenous sprue was considered in the past a complication of CD or other enteropathies such as tropical sprue [112,121], however more recently, also cases potentially related to olmesartan use have been identified [95].

4.7 Idiopathic villous atrophy

In a relevant number of cases the aetiology of SNVA cannot be identified despite thorough investigations. These enteropathies are therefore defined as idiopathic villous atrophy (IVA) [3,6-9,11,20]. A recent international study on these forms of IVA has just shed some light on their clinical phenotype and natural history [20]. IVA patients can be sub-classified into three groups showing distinct clinical features and prognosis. IVA Group 1 is characterized by forms of transient self-limiting partial VA, likely due to an infectious agent and good prognosis (5-year survival 96%). IVA Group 2 is characterized by persistent nonclonal IVA with peculiar association with HLA DQB1*0301 and DQB1*06 and long-term survival (5-year survival was 100%). Finally, IVA group 3 is characterized by a cluster of enteropathies with lymphoproliferative features and poor outcome (5-years survival 27%). Hypoalbuminemia and age at diagnosis predicted mortality in IVA. Further studies should focus on the molecular mechanisms involved in these enteropathies and the identification of new options for treatment.

4.8 Methodological approach to differential diagnosis and management of seronegative villous atrophy

Currently, there is no standardized consensus about the nomenclature and the clinical management of SNVA [3,6-14,19]. We recently proposed an investigational work-up for the differential diagnosis of SNVA [3]. The first step is a careful review of duodenal biopsy slides by an expert gastrointestinal pathologist informed about the pharmacological history and clinical features of the patient. Confirmation of true presence of VA on correctly oriented duodenal specimens performed while on a gluten-containing diet is mandatory. Unfortunately, poor orientation of duodenal specimens is still a major diagnostic error, leading to overestimation of seronegative CD and unnecessary treatment [124-126]. It is important to ascertain modalities for coeliac antibodies testing, i.e. which "coeliac antibodies" were tested and whether the patient was on a gluten containing diet at time of serological testing. Class IgA EmA and TTA are the mainstay for serological diagnosis of CD in adult patients and in case of total IgA deficiency class IgG antibodies should be tested [1,2]. In the context of seronegative VA there may be a role for deamidated gliadin peptide antibodies [3]. Once SNVA is confirmed, all the possible aetiologies should be excluded before posing a diagnosis of SNCD or IVA [3,20].

Table 4.2 Diagnostic panel for serology negative villous atrophy.

Laboratory tests
HLA typing
Serum IgA, IgG, IgM
IgA and IgG Ema and TTA
Anti-enterocyte antibodies
HIV testing
Quantiferon
Stool tests
Giardia Iamblia and other parasites
Viruses
Helicobacter Pylori antigens
Fecal calprotectin
Duodenal biopsies
H&E and PAS staining
PCR for Giardia, tuberculosis and Whipple's disease
Small bowel aspirate
PCR for gamma-TCR clonality assessment
Flow cytometry for aberrant IELs
Other exams
Capsule endoscopy
Abdomen CT/PET
Colonoscopy

EmA: endomysial antibodies, TTA: tissue transglutaminase antibodies, H&E: hematoxylin and eosin staining, PAS: Periodic acid Shiff staining, IELs: intraepithelial lymphocytes, CT: computed tomography, PET: positron emission tomography.

Table 4.2 shows the diagnostic panel for SNVA and Fig. 4.3 shows a flowchart for the diagnosis of SNCD and IVA.

4.8.1 Principles of management of seronegative villous atrophy

Although histological response to a GFD is the mainstay for diagnosing SNCD, uncertainties still exist about the correct timing for a patient affected by SNVA to be started on a GFD. On the contrary, clinical response to a GFD should not be used as incontrovertible evidence to confirm SNCD, because it has been shown that SNVA patients who do not have a histological response may still have a clinical response at a level of up to 30% [3,4,12,14,127]. We think that it can be reasonable to take into account factors such as ethnicity, HLA typing, and severity of the clinical picture in order to decide whether or not to start a GFD. We suggest that in Caucasian patients carrying HLA-DQ2/DQ8 and in whom no alternative aetiology has been found, a GFD could be started. In patients with these characteristics and presenting with severe features of malabsorption and age> 50 years budesonide or other immunosuppressants may be considered to promote mucosal recovery. Alternatively, in patients with no cause found for VA, mild symptoms and absence of laboratory abnormalities at diagnosis (IVA 1), it may be reasonable to adopt

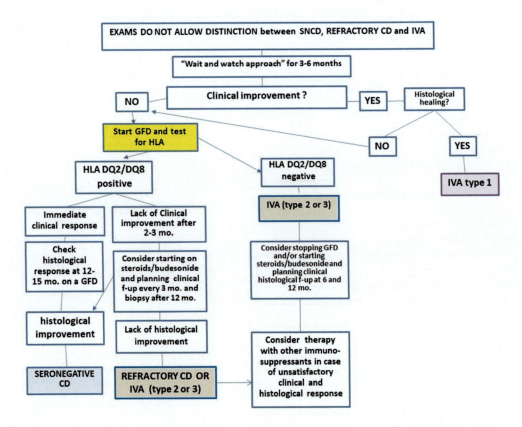

Figure 4.3 *Flowchart for diagnosis and management of seronegative coeliac disease and idiopathic villous atrophy.* CD: coeliac disease; SNCD: seronegative coeliac disease; GFD: gluten-free diet; IVA: idiopathic villous atrophy; **Permission process not started.** Select figure and clickicon in top toolbar to start permission process or indicate no permission is required.

a "watch and wait" approach and a histological follow–up, as this resulted in histological recovery of VA within twelve months from diagnosis [14,20].

4.9 Conclusions

Seronegative VA encompasses a wide range of rare and heterogeneous conditions for which international guidelines are still lacking and pitfalls in diagnosis are still common. Histological response to a GFD is the main criterion to distinguish between true SNCD and other forms of seronegative VA unrelated to gluten ingestion. A systematic and algo-rithmic approach may be useful to appropriately categorize these patients and treat them accordingly. Future research perspectives may include the development of standard con-sensus guidelines for diagnosis and a better definition of natural history of different forms of SNVA. Identification of molecular mechanisms beyond SNVA will also be precious as to develop target therapies, particularly for patients at higher risk of poor outcomes.

References

[1] J.F. Ludvigsson, J.C. Bai, F. Biagi, T.R. Card, C. Ciacci, P.J. Ciclitira, et al., BSG Coeliac Disease Guidelines Development Group; British Society of Gastroenterology. Diagnosis and management of adult celiac disease: guidelines from the British Society of Gastroenterology, Gut 63 (2014) 1210–1228.

[2] A. Rubio-Tapia, I.D. Hill, C.P. Kelly, A.H. Calderwood, J.A. Murray, American College of Gastroenterology. ACG clinical guidelines: diagnosis and management of celiac disease, Am J Gastroenterol 108 (2013) 656–676.

[3] A. Schiepatti, D.S. Sanders, M. Zuffada, O. Luinetti, A. Iraqi, F. Biagi, Overview in the clinical management of patients with seronegative villous atrophy, Eur J Gastroenterol Hepatol 31 (2019) 409–417.

[4] F. Biagi, G.R. Corazza, Mortality in celiac disease, Nat Rev Gastroenterol Hepatol 7 (2010) 158–162.

[5] F. Biagi, A. Schiepatti, G. Maiorano, G. Fraternale, S. Agazzi, F. Zingone, C. Ciacci, U. Volta, G. Caio, R. Tortora, C. Klersy, G.R. Corazza, Risk of complications in coeliac patients depends on age at diagnosis and type of clinical presentation, Dig Liver Dis 50 (2018) 549–552.

[6] K. Pallav, D.A. Leffler, S. Tariq, T. Kabbani, J. Hansen, A. Peer, et al., Non-coeliac enteropathy: the differential diagnosis of villous atrophy in contemporary clinical practice, Aliment Pharmacol Ther 35 (2012) 380–390.

[7] M. DeGaetani, C.A. Tennyson, B. Lebwohl, S.K. Lewis, H. Abu Daya, C. Arguelles-Grande, et al., Villous atrophy and negative celiac serology: a diagnostic and therapeutic dilemma, Am J Gastroenterol 108 (2013) 647–653.

[8] I. Aziz, M.F. Peerally, J.H. Barnes, V. Kandasamy, J.C. Whiteley, D. Partridge, et al., The clinical and phenotypical assessment of seronegative villous atrophy; a prospective UK centre experience evaluating 200 adult cases over a 15-year period (2000-2015), Gut 66 (2017) 1563–1572.

[9] A. Schiepatti, F. Biagi, G. Fraternale, C. Vattiato, D. Balduzzi, S. Agazzi, et al., Mortality and differential diagnoses of villous atrophy without celiac antibodies, Eur J Gastroenterol Hepatol 29 (2017) 572–576.

[10] U. Volta, G. Caio, E. Boschetti, F. Giancola, K.J. Rhoden, E. Ruggeri, et al., Seronegative celiac disease: Shedding light on an obscure clinical entity, Dig Liver Dis 48 (2016) 1018–1022.

[11] F. Biagi, G.R. Corazza, Defining gluten refractory enteropathy, Eur J Gastroenterol Hepatol 13 (2001) 561–565.

[12] J.K. Greenson, The biopsy pathology of non-coeliac enteropathy, Histopathology 66 (2015) 29–36.

[13] A.K. Kamboj, A.S. Oxentenko, Clinical and Histologic Mimickers of Celiac Disease, Clin Transl Gastroenterol 8 (2017) e114.

[14] C.L. Jansson-Knodell, I.A. Hujoel, A. Rubio-Tapia, J.A. Murray, Not All That Flattens Villi Is Celiac Disease: A Review of Enteropathies, Mayo Clin Proc 93 (2018) 509–517.

[15] C.L. Jansson-Knodell, J.A. Murray, A. Rubio-Tapia, Management of Small Bowel Villous Atrophy in Patients Seronegative for Celiac Disease, Am J Gastroenterol (2020) Mar 4[Epub ahead of print], doi:10.14309/ajg.0000000000000575.

[16] F. Fernández-Bañares, L. Crespo, C. Núñez, N. López-Palacios, E. Tristán, S. Vivas, S. Farrais, B. Arau, J. Vidal, G. Roy, M. Esteve, Gamma delta(+) intraepithelial lymphocytes and coeliac lymphogram in a diagnostic approach to coeliac disease in patients with seronegative villous atrophy, Aliment Pharmacol Ther 51 (2020) 699–705.

[17] N.R. Lewis, B.B. Scott, Systematic review: the use of serology to exclude or diagnose coeliac disease (a comparison of the endomysial and tissue transglutaminase antibody tests), Aliment Pharmacol Ther 24 (2006) 47–54.

[18] C. Dahle, A. Hagman, S. Ignatova, M. Ström, Antibodies against deamidated gliadin peptides identify adult coeliac disease patients negative for antibodies against endomysium and tissue transglutaminase, Aliment Pharmacol Ther 32 (2010) 254–260.

[19] A. Schiepatti, D.S. Sanders, F. Biagi, Seronegative coeliac disease: clearing the diagnostic dilemma, Curr Opin Gastroenterol 34 (2018) 154–158.

[20] A. Schiepatti, D.S. Sanders, I. Aziz, A. De Silvestri, J. Goodwin, T. Key, L. Quaye, P. Giuffrida, A. Vanoli, M. Paulli, S.S. Cross, P. Vergani, E. Betti, G. Maiorano, R. Ellis, J.A. Snowden, A. Di Sabatino, G.R. Corazza, F. Biagi, Clinical phenotype and mortality in patients with idiopathic small bowel villous atrophy: a dual centre international study, Eur J Gastroenterol Hepatol (2020) accepted for publication.

[21] M.V. Pensieri, F. Pulvirenti, A. Schiepatti, S. Maimaris, S. Lattanzio, I. Quinti, C. Klersy, G.R. Corazza, F. Biagi, The high mortality of patients with common variable immunodeficiency and small bowel villous atrophy, Scand J Gastroenterol 54 (2019 Feb) 164–168.

[22] S. Chetcuti Zammit, A. Schiepatti, I. Aziz, M. Kurien, D.S. Sanders, R. Sidhu, Use of small-bowel capsule endoscopy in cases of equivocal celiac disease, Gastrointest Endosc (2020) Jan 7. pii: S0016-5107(20)30005-5[Epub ahead of print], doi:10.1016/j.gie.2019.12.044.

[23] J.A. Abrams, B. Diamond, H. Rotterdam, P.H. Green, Seronegative celiac disease: increased prevalence with lesser degrees of villous atrophy, Dig Dis Sci 49 (2004) 546–550.

[24] K. Rostami, J. Kerckhaert, R. Tiemessen, B.M. von Blomberg, J.W. Meijer, C.J. Mulder, Sensitivity of antiendomysium and antigliadin antibodies in untreated celiac disease: disappointing in clinical practice, Am J Gastroenterol 94 (1999) 888–894.

[25] K. Rostami, J. Kerckhaert, B.M. von Blomberg, J.W. Meijer, P. Wahab, C.J. Mulder, SAT and serology in adult coeliacs, seronegative coeliac disease seems a reality, Neth J Med 53 (1998) 15–19.

[26] T.T. Salmi, P. Collin, I.R. Korponay-Szabó, K. Laurila, J. Partanen, H. Huhtala, et al., Endomysial antibody-negative coeliac disease: clinical characteristics and intestinal autoantibody deposits, Gut 55 (2006) 1746–1753.

[27] L. Fry, Dermatitis herpetiformis, Baillieres Clin Gastroenterol 9 (1995) 371–393.

[28] L. Fry, P.P. Seah, P.G. Harper, et al., The small intestine in dermatitis herpetiformis, J Clin Pathol 27 (1974) 817–824.

[29] W. Dickey, D.F. Hughes, S.A. McMillan, Disappearance of endomysial antibodies in treated celiac disease does not indicate histological recovery, Am J Gastroenterol 95 (2000) 712–714.

[30] M. Esteve, M. Rosinach, F. Fernández-Bañares, C. Farré, A. Salas, M. Alsina, et al., Spectrum of gluten-sensitive enteropathy in first-degree relatives of patients with coeliac disease: clinical relevance of lymphocytic enteritis, Gut 55 (2006) 1739–1745.

[31] D.X. Henry, D.S. Sanders, K. Basu, A. Schiepatti, J.A. Campbell, Seronegative Coeliac Disease masquerading as Irritable Bowel Syndrome type Symptoms, J Gastrointestin Liver Dis 29 (2020) 111–113 Mar 13.

[32] W. Dickey, D.F. Hughes, S.A. McMillan, Reliance on serum endomysial antibody testing underestimates the true prevalence of coeliac disease by one fifth, Scand J Gastroenterol 35 (2000) 181–183.

[33] P. Collin, K. Kaukinen, H. Vogelsang, I. Korponay-Szabó, R. Sommer, E. Schreier, et al., Antiendomysial and antihuman recombinant tissue transglutaminase antibodies in the diagnosis of coeliac disease: a biopsy-proven European multicentre study, Eur J Gastroenterol Hepatol 17 (2005) 85–91.

[34] M. Maglio, F. Ziberna, R. Aitoro, V. Discepolo, G. Lania, V. Bassi, et al., Intestinal Production of Anti-Tissue Transglutaminase 2 Antibodies in Patients with Diagnosis Other Than Celiac Disease, Nutrients 9 (10) (2017).

[35] D.J. Unsworth, J.A. Walker-Smith, Autoimmunity in diarrhoeal disease, J Pediatr Gastroenterol Nutr 4 (1985) 375–380.

[36] G.R. Corazza, F. Biagi, U. Volta, M.L. Andreani, L. De Franceschi, G. Gasbarrini, Autoimmune enteropathy and villous atrophy in adults, Lancet 350 (1997) 106–109.

[37] C. Catassi, E. Fabiani, M.I. Spagnuolo, G. Barera, A. Guarino, Severe and protracted diarrhea: results of the 3-year SIGEP multicenter survey. Working Group of the Italian Society of Pediatric Gastroenterology and Hepatology (SIGEP), J Pediatr Gastroenterol Nutr 29 (1999) 63–68.

[38] L.M. Marthinsen, H. Scott, J. Ejderhamn, Autoimmune enteropathy in Swedish children, 1985-2002: a call for strict diagnostic criteria, Scand J Gastroenterol 43 (2008) 1102–1107.

[39] O. Baud, O. Goulet, D. Canioni, F. Le Deist, I. Radford, D. Rieu, S. Dupuis-Girod, N. Cerf-Bensussan, M. Cavazzana-Calvo, N. Brousse, A. Fischer, J.L. Casanova, Treatment of the immune dysregulation, polyendocrinopathy, enteropathy, X-linked syndrome (IPEX) by allogeneic bone marrow transplantation, N Engl J Med 344 (2001) 1758–1762.

[40] S.E. Flanagan, E. Haapaniemi, M.A. Russell, R. Caswell, H.L. Allen, E. De Franco, T.J. McDonald, H. Rajala, A. Ramelius, J. Barton, K. Heiskanen, T. Heiskanen-Kosma, M. Kajosaari, N.P. Murphy, T. Milenkovic, M. Seppänen, Å. Lernmark, S. Mustjoki, T. Otonkoski, J. Kere, N.G. Morgan, S. Ellard, A.T. Hattersley, Activating germline mutations in STAT3 cause early-onset multi-organ autoimmune disease, Nat Genet 46 (2014) 812–814.

[41] U. Volta, M.G. Mumolo, G. Caio, E. Boschetti, R. Latorre, F. Giancola, et al., Autoimmune enteropathy: not all flat mucosa mean coeliac disease, Gastroenterol Hepatol Bed Bench 9 (2016) 140–145.

[42] S. Akram, J.A. Murray, D.S. Pardi, G.L. Alexander, J.A. Schaffner, P.A. Russo, et al., Adult autoimmune enteropathy: Mayo Clinic Rochester experience, Clin Gastroenterol Hepatol 5 (2007) 1282–1290.

[43] R. Masia, S. Peyton, G.Y. Lauwers, I. Brown, Gastrointestinal biopsy findings of autoimmune enteropathy: a review of 25 cases, Am J Surg Pathol 38 (2014) 1319–1329.

[44] J.M. Martín-Villa, S. Camblor, R. Costa, A. Arnaiz-Villena, Gut epithelial cell autoantibodies in AIDS pathogenesis, Lancet 342 (1993) 380.

[45] U. Volta, G.L. De Angelis, A. Granito, N. Petrolini, E. Fiorini, M. Guidi, et al., Autoimmune enteropathy and rheumatoid arthritis: a new association in the field of autoimmunity, Dig Liver Dis 38 (2006) 926–929.

[46] A. Di Sabatino, F. Biagi, M. Lenzi, L. Frulloni, M.V. Lenti, P. Giuffrida, et al., Clinical usefulness of serum antibodies as biomarkers of gastrointestinal and liver diseases, Dig Liver Dis 49 (2017) 947–956.

[47] A. Carroccio, U. Volta, L. Di Prima, N. Petrolini, A.M. Florena, M.R. Averna, Autoimmune enteropathy and colitis in an adult patient, Dig Dis Sci 48 (2003) 1600–1606.

[48] R. Mirakian, A. Richardson, P.J. Milla, J.A. Walker-Smith, J. Unsworth, M.O. Savage, et al., Protracted diarrhoea of infancy: evidence in support of an autoimmune variant, Br Med J 293 (1986) 1132–1136.

[49] T. Hibi, M. Ohara, K. Kobayashi, W.R. Brown, K. Toda, H. Takaishi, et al., Enzyme linked immunosorbent assay (ELISA) and immunoprecipitation studies on antigoblet cell antibody using a mucin producing cell line in patients with inflammatory bowel disease, Gut 35 (1994) 224–230.

[50] C. Folwaczny, N. Noehl, K. Tschop, et al., Goblet cell autoantibodies in patients with inflammatory bowel disease and their first-degree relatives, Gastroenterology 113 (1997) 101–106.

[51] F. Biagi, P.I. Bianchi, L. Trotta, et al., Anti-goblet cell antibodies for the diagnosis of autoimmune enteropathy? Am J Gastroenterol 104 (2009) 3112.

[52] G. Malamut, V. Verkarre, C. Callens, O. Colussi, G. Rahmi, E. MacIntyre, C. Haïoun, B. Meresse, N. Brousse, S. Romana, O. Hermine, N. Cerf-Bensussan, C. Cellier, Enteropathy-associated T-cell lymphoma complicating an autoimmune enteropathy, Gastroenterology 142 (2012) 726–729.

[53] R. Ciccocioppo, G.A. Croci, F. Biagi, A. Vanoli, C. Alvisi, G. Cavenaghi, R. Riboni, M. Arra, P.G. Gobbi, M. Paulli, G.R. Corazza, Intestinal T-cell lymphoma with enteropathy-associated T-cell lymphoma-like features arising in the setting of adult autoimmune enteropathy, Hematol Oncol 36 (2018) 481–488.

[54] R. Ciccocioppo, M.L. Russo, M.E. Bernardo, F. Biagi, L. Catenacci, M.A. Avanzini, C. Alvisi, A. Vanoli, R. Manca, O. Luinetti, F. Locatelli, G.R. Corazza, Mesenchymal stromal cell infusions as rescue therapy for corticosteroid-refractory adult autoimmune enteropathy, Mayo Clin Proc 87 (2012) 909–914.

[55] M.A. Park, J.T. Li, J.B. Hagan, et al., Common variable immunodeficiency: a new look at an old disease, Lancet 372 (2008) 489–502.

[56] H. Chapel, M. Lucas, M. Lee, et al., Common variable immunodeficiency disorders: division into distinct clinical phenotypes, Blood 112 (2008) 277–286.

[57] I. Kalha, J.H. Sellin, Common variable immunodeficiency and the gastrointestinal tract, Curr Gastroenterol Rep 6 (2004) 377–383.

[58] G. Luzi, A. Zullo, F. Iebba, V. Rinaldi, L. Sanchez Mete, M. Muscaritoli, et al., Duodenal pathology and clinical-immunological implications in common variable immunodeficiency patients, Am J Gastroenterol 98 (2003) 118–121.

[59] S. Agarwal, L. Mayer, Pathogenesis and treatment of gastrointestinal disease in antibody deficiency syndromes, J Allergy Clin Immunol 124 (2009) 658–664.

[60] J.A. Daniels, H.M. Lederman, A. Maitra, E.A. Montgomery, Gastrointestinal tract pathology in patients with common variable immunodeficiency (CVID): a clinicopathologic study and review, Am J Surg Pathol 31 (2007) 1800–1812.

[61] P.J. Mannon, I.J. Fuss, S. Dill, J. Friend, C. Groden, R. Hornung, et al., Excess IL-12 but not IL-23 accompanies the inflammatory bowel disease associated with common variable immunodeficiency, Gastroenterology 131 (2006) 748–756.

[62] C. Bästlein, R. Burlefinger, E. Holzberg, C. Voeth, M. Garbrecht, R. Ottenjann, Common variable immunodeficiency syndrome and nodular lymphoid hyperplasia in the small intestine, Endoscopy 20 (1988) 272–275.

[63] K. Washington, T.T. Stenzel, R.H. Buckley, M.R. Gottfried, Gastrointestinal pathology in patients with common variable immunodeficiency and X-linked agammaglobulinemia, Am J Surg Pathol 20 (1996) 1240–1252.

[64] A. Khodadad, A. Aghamohammadi, N. Parvaneh, N. Rezaei, F. Mahjoob, M. Bashashati, et al., Gastrointestinal manifestations in patients with common variable immunodeficiency, Dig Dis Sci 52 (2007) 2977–2983.

[65] K. Teahon, A.D. Webster, A.B. Price, J. Weston, I. Bjarnason, Studies on the enteropathy associated with primary hypogammaglobulinaemia, Gut 35 (1994) 1244–1249.

[66] S.F. Jørgensen, H.M. Reims, D. Frydenlund, K. Holm, V. Paulsen, A.E. Michelsen, et al., A Cross-Sectional Study of the Prevalence of Gastrointestinal Symptoms and Pathology in Patients With Common Variable Immunodeficiency, Am J Gastroenterol 111 (2016) 1467–1475.

[67] C. Cunningham-Rundles, C. Bodian, Common variable immunodeficiency: clinical and immuno-logical features of 248 patients, Clin Immunol 92 (1999) 34–48.

[68] P.E. Hermans, J.A. Diaz-Buxo, J.D. Stobo, Idiopathic late-onset immunoglobulin deficiency: clinical observations in 50 patients, Am J Med 61 (1976) 221–237.

[69] R. Hermaszewski, A. Webster, Primary hypogammaglobulinaemia: a survey of clinical manifestations and complications, QJM 86 (1993) 31–42.

[70] I. Quinti, A. Soresina, G. Spadaro, S. Martino, S. Donnanno, C. Agostini, et al., Long-term follow-up and outcome of a large cohort of patients with common variable immunodeficiency, J Clin Immunol 27 (2007) 308–316.

[71] J.M. Woodward, E. Gkrania-Klotsas, A.Y. Cordero-Ng, A. Aravinthan, B.N. Bandoh, H. Liu, et al., The role of chronic norovirus infection in the enteropathy associated with common variable immunode-ficiency, Am J Gastroenterol 110 (2015) 320–327.

[72] M.A. Heneghan, F.M. Stevens, E.M. Cryan, R.H. Warner, C.F. McCarthy, Celiac sprue and immun-odeficiency states: a 25-year review, J Clin Gastroenterol 25 (1997) 421–425.

[73] A.D. Webster, G. Slavin, M. Shiner, T.A. Platts-Mills, G.L. Asherson, Coeliac disease with severe hypogammaglobulinaemia, Gut 22 (1981) 153–157.

[74] F. Biagi, P.I. Bianchi, A. Zilli, A. Marchese, O. Luinetti, V. Lougaris, et al., The significance of duodenal mucosal atrophy in patients with common variable immunodeficiency: a clinical and histopathologic study, Am J Clin Pathol 138 (2012) 185–189.

[75] G. Malamut, V. Verkarre, F. Suarez, J.F. Viallard, A.S. Lascaux, J. Cosnes, et al., The enteropathy associated with common variable immunodeficiency: the delineated frontiers with celiac disease, Am J Gastroenterol 105 (2010) 2262–2275.

[76] E.S. Resnick, Morbidity and mortality in common variable immune deficiency over 4 decades, Blood 119 (2012) 1650–1657.

[77] L.J. Kobrynski, Noninfectious complications of common variable immune deficiency, Allergy Asthma Proc 40 (2019) 129–132.

[78] A. Rubio-Tapia, M.L. Herman, J.F. Ludvigsson, D.G. Kelly, T.F. Mangan, T.T. Wu, et al., Severe spruelike enteropathy associated with olmesartan, Mayo Clin Proc 87 (2012) 732–738.

[79] T.R. Ziegler, C. Fernández-Estívariz, L.H. Gu, M.W. Fried, L.M. Leader, Severe villus atro-phy and chronic malabsorption induced by azathioprine, Gastroenterology 124 (2003) 1950–1957.

[80] M.M. Boscá, R. Añón, E. Mayordomo, R. Villagrasa, N. Balza, C. Amorós, et al., Methotrexate induced sprue-like syndrome, World J Gastroenterol 14 (2008) 7009–7011.

[81] N. Kamar, P. Faure, E. Dupuis, O. Cointault, K. Joseph-Hein, D. Durand, et al., Villous atrophy induced by mycophenolate mofetil in renal-transplant patients, Transpl Int 17 (2004) 463–467.

[82] A.M. Larson, G.B. McDonald, M. Feldman, L.S. Friedman, L.J. Brandt, Gastrointestinal and hepatic complications of solid organ and hematopoietic cell transplantation, Sleisenger and Fordtran's Gas-trointestinal and Liver Disease. Pathophysiology, diagnosis, management, 10th ed., Elsevier Saunders, Philadelphia, 2016, pp. 556–578.

[83] M. Palta, C.G. Willett, B.G. Czito, M. Feldman, L.S. Friedman, L.J. Brandt, Radiation Injury, Sleisenger and Fordtran's Gastrointestinal and Liver Disease. Pathophysiology, diagnosis, management, 10th ed., Elsevier Saunders, Philadelphia, 2016, pp. 664–676.

[84] U. Kaosombatwattana, J. Limsrivilai, A. Pongpaibul, M. Maneerattanaporn, P. Charatcharoenwitthaya, Severe enteropathy with villous atrophy in prolonged mefenamic acid users - a currently under-recognized in previously well-recognized complication: Case report and review of literature, Medicine (Baltimore) 96 (2017) e8445.

[85] F.W.D. Tai, M.E. McAlindon, NSAIDs and the small bowel, Curr Opin Gastroenterol 34 (2018) 175–182.

[86] L. Marthey, G. Cadiot, P. Seksik, P. Pouderoux, J. Lacroute, F. Skinazi, et al., Olmesartan-associated enteropathy: results of a national survey, Aliment Pharmacol Ther 40 (2014) 1103–1109.

[87] M. Herman, A. Rubio-Tapia, E. Marietta, T.T. Wu, J. Murray, Severe enteropathy in a patient on valsartan, Am J Gastroenterol 108 (2013) S302.

[88] J. Cyrany, T. Vasatko, J. Machac, M. Nova, J. Szanyi, M. Kopacova, Letter: telmisartan-associated enteropathy –is there any class effect? Aliment Pharmacol Ther 40 (2014) 569–570.

[89] M. Basson, M. Mezzarobba, A. Weill, P. Ricordeau, H. Allemand, F. Alla, et al., Severe intestinal malabsorption associated with olmesartan: a French nationwide observational cohort study, Gut 65 (2016) 1664–1669.

[90] Y.H. Dong, Y. Jin, T.N. Tsacogianis, M. He, P.H. Hsieh, J.J. Gagne, Use of olmesartan and enteropathy outcomes: a multi-database study, Aliment Pharmacol Ther 47 (2018) 792–800.

[91] E.V. Marietta, A. Cartee, A. Rishi, J.A. Murray, Drug-induced enteropathy, Dig Dis 33 (2015) 215–220.

[92] A. Schiepatti, F. Biagi, D. Cumetti, O. Luinetti, A. Sonzogni, A. Mugellini, et al., Olmesartan-associated enteropathy: new insights on the natural history? Report of two cases, Scand J Gastroenterol 51 (2016) 152–156.

[93] N. Burbure, B. Lebwohl, C. Arguelles-Grande, P.H. Green, G. Bhagat, S. Lagana, Olmesartan-associated sprue-like enteropathy: a systematic review with emphasis on histopathology, Hum Pathol 50 (2016) 127–134.

[94] S. Scialom, G. Malamut, B. Meresse, N. Guegan, N. Brousse, V. Verkarre, et al., Gastrointestinal Disorder Associated with Olmesartan Mimics Autoimmune Enteropathy, PLoS One 10 (2015) e0125024.

[95] A. Rubio-Tapia, N.J. Talley, S.R. Gurudu, T.T. Wu, J.A. Murray, Gluten-free diet and steroid treatment are effective therapy for most patients with collagenous sprue, Clin Gastroenterol Hepatol 8 (2010) 344 -34.

[96] J.D. Levinson, L.J. Nastro, Giardiasis with total villous atrophy, Gastroenterology 74 (1978) 271–275.

[97] T. Marth, V. Moos, C. Müller, F. Biagi, T. Schneider, Infection with Tropheryma whipplei and Whipple's Disease, Lancet Infect Dis 16 (2016) e13–e22.

[98] W.P. Fung, K.K. Tan, S.F. Yu, K.M. Sho, Malabsorption and subtotal villous atrophy secondary to pulmonary and intestinal tuberculosis, Gut 11 (1970) 212–216.

[99] M.S. Kapembwa, P.A. Batman, S.C. Fleming, Griffin GE H.IV enteropathy, Lancet 2 (1989) 1521–1522.

[100] I.S. Brown, A. Bettington, M. Bettington, C. Rosty, Tropical sprue: revisiting an underrecognized disease, Am J Surg Pathol 38 (2014) 666–672.

[101] M.J. Batheja, J. Leighton, A. Azueta, R. Heigh, The Face of Tropical Sprue in 2010, Case Rep Gastroenterol 4 (2010) 168–172.

[102] M. Nooshadokht, B. Kalantari-Khandani, I. Sharifi, H. Kamyabi, N.P.M. Liyanage, L.A. Lagenaur, et al., Stool antigen immunodetection for diagnosis of Giardia duodenalis infection in human subjects with HIV and cancer, J Microbiol Methods 141 (2017) 35–41.

[103] S.A. Fouad, S. Esmat, M.M. Basyoni, M.S. Farhan, M.H. Kobaisi, Molecular identification of giardia intestinalis in patients with dyspepsia, Digestion 90 (2014) 63–71.

[104] F. Biagi, D. Balduzzi, P. Delvino, A. Schiepatti, C. Klersy, G.R. Corazza, Prevalence of Whipple's disease in north-western Italy, Eur J Clin Microbiol Infect Dis 34 (2015) 1347–1348.

[105] P.J. Lappinga, S.C. Abraham, J.A. Murray, E.A. Vetter, R. Patel, T.T. Wu, Small intestinal bacterial overgrowth: histopathologic features and clinical correlates in an underrecognized entity, Arch Pathol Lab Med 134 (2010) 264–270.

[106] S.M. Riordan, C.J. McIver, D. Wakefield, V.M. Duncombe, M.C. Thomas, T.D. Bolin, Small intestinal mucosal immunity and morphometry in luminal overgrowth of indigenous gut flora, Am J Gastroenterol 96 (2001) 494–500.

[107] A. Strocchi, G. Corazza, J. Furne, C. Fine, A. Di Sario, G. Gasbarrini, et al., Measurements of the jejunal unstirred layer in normal subjects and patients with celiac disease, Am J Physiol 270 (1996) G487–G491.

[108] M. Voutilainen, M. Juhola, M. Färkkilä, P. Sipponen, Gastric metaplasia and chronic inflammation at the duodenal bulb mucosa, Dig Liver Dis 35 (2003) 94–98.

[109] P.G. Foukas, L. de Leval, Recent advances in intestinal lymphomas, Histopathology 66 (2015) 112–136.

[110] F. Biagi, P. Lorenzini, G.R. Corazza, Literature review on the clinical relationship between ulcerative jejunoileitis, coeliac disease and enteropathy associated T cell lymphoma, Scand J Gastroenterol 35 (2000) 785–790.

[111] C. Cellier, E. Delabesse, C. Helmer, N. Patey, C. Matuchansky, B. Jabri, et al., Refractory sprue, coeliac disease, and enteropathy-associated T-cell lymphoma. French Coeliac Disease Study Group, Lancet 356 (2000) 203–208.

[112] F. Biagi, A. Marchese, F. Ferretti, R. Ciccocioppo, A. Schiepatti, U. Volta, et al., A multicentre case control study on complicated coeliac disease: two different patterns of natural history, two different prognoses, BMC Gastroenterology 14 (2014) 139.

[113] A. Di Sabatino, F. Biagi, P.G. Gobbi, G.R. Corazza, How I treat enteropathy-associated T-cell lymphoma, Blood 119 (2012) 2458–2468.

[114] F. Biagi, P. Gobbi, A. Marchese, E. Borsotti, F. Zingone, C. Ciacci, et al., Low incidence but poor prognosis of complicated coeliac disease: a retrospective multicentre study, Dig Liver Dis 46 (2014) 227–230.

[115] S. Ondejka, D. Jagadeesh, Enteropathy-associated T-cell lymphoma, Curr Hematol Malig Rep 11 (2016) 504–513.

[116] J. Delabie, H. Holte, J.M. Vose, F. Ullrich, E.S. Jaffe, K.J. Savage, et al., Enteropathy-associated T-cell lymphoma: clinical and histological findings from the Internation Peripheral T-Cell Lymphoma Project, Blood 118 (2011) 148–155.

[117] M.V. Lenti, F. Biagi, M. Lucioni, A. Di Sabatino, M. Paulli, G.R. Corazza, Two cases of monomorphic epitheliotropic intestinal T-cell lymphoma associated with coeliac disease, Scand J Gastroenterol 54 (2019) 965–968.

[118] A.M. Perry, R.A. Warnke, Q. Hu, P. Gaulard, C. Copie-Bergman, S. Alkan, et al., Indolent T-cell lymphoproliferative disease of the gastrointestinal tract, Blood 122 (2013) 3599–3606.

[119] G. Malamut, B. Meresse, S. Kaltenbach, C. Derrieux, V. Verkarre, E. Macintyre, et al., Small intestinal CD4+ T-cell lymphoma is a heterogenous entity with common pathology features, Clin Gastroenterol Hepatol 12 (2014) 599–608.

[120] T. Al-Saleem, H. Al-Mondhiry, Immunoproliferative small intestinal disease (IPSID): a model for mature B-cell neoplasms, Blood 105 (2005) 2274–2280.

[121] H.J. Freeman, Collagenous sprue, Can J Gastroenterol 25 (2011) 189–192.

[122] F.R. Ponziani, V. Gerardi, A. Gasbarrini, Diagnosis and treatment of small intestinal bacterial over-growth, Expert Rev Gastroenterol Hepatol 10 (2016) 215–227.

[123] M.M. Walker, M. Potter, N.J. Talley, Eosinophilic gastroenteritis and other eosinophilic gut diseases distal to the oesophagus, Lancet Gastroenterol Hepatol 3 (2018) 271–280.

[124] F. Biagi, P.I. Bianchi, J. Campanella, G. Zanellati, G.R. Corazza, The impact of misdiagnosing celiac disease at a referral centre, Can J Gastroenterol 23 (2009) 543–545.

[125] R.G. Shidrawi, R. Przemioslo, D.R. Davies, M.R. Tighe, P.J. Ciclitira, Pitfalls in diagnosing coeliac disease, J Clin Pathol 47 (1994) 693–694.

[126] J. Taavela, O. Koskinen, H. Huhtala, M.L. Lähdeaho, A. Popp, K. Laurila, et al., Validation of morphometric analyses of small-intestinal biopsy readouts in celiac disease, PLoS One 8 (2013) e76163.

[127] J. Campanella, F. Biagi, P.I. Bianchi, G. Zanellati, A. Marchese, G.R. Corazza, Clinical response to gluten withdrawal is not an indicator of coeliac disease, Scand J Gastroenterol 43 (2008) 1311–1314.

CHAPTER 5

Nonresponsive and complicated coeliac disease

Hugo A Penny[a], Annalisa Schiepatti[b], David S Sanders[a]
[a]Academic Unit of Gastroenterology, University of Sheffield, Sheffield, United Kingdom
[b]Gastroenterology Unit, IRCCS Pavia, ICS Maugeri, University of Pavia, Italy

5.1 Introduction

Coeliac disease (CD) is a chronic autoimmune condition that develops in genetically susceptible individuals with a reported prevalence of around 1% in Western countries [1]. CD classically manifests as a small intestinal enteropathy, although the manifestations are broad and can involve both the gastrointestinal intestinal (GI) tract and distinct extra-intestinal sites throughout the body [2]. Inflammation and tissue damage in the small intestine results from an abnormal immune response toward ingested gluten. Persisting inflammation in active CD puts individuals at risk of osteoporosis, nutrient deficiencies, and intestinal malignancies. Thus, a life-long gluten-free diet (GFD) is the central tenet to management of this condition [2].

Whilst most individuals will display clinical improvement shortly after commencing a GFD, between 7-30% of patients will continue to experience symptoms and/ or have persisting intestinal inflammation [3]. These individuals are typically classified as having non-responsive CD (NRCD). NRCD can manifest as a lack of initial response to a GFD (primary NRCD), or as a recurrence of signs/ symptoms in an individual who initially responded to a GFD (secondary NRCD) [4].

NRCD covers a broad range of pathologies including ongoing gluten exposure, super-sensitivity to gluten and slow mucosal healing [5]. Some individuals will present with apparent NRCD, but rather will have developed symptomology as a result of an alternative pathology which may be associated with CD, such as microscopic colitis, pancreatic exocrine insufficiency, inflammatory bowel disease (IBD), small intestinal bacterial overgrowth, or lactose or fructose intolerance [6–8]. In addition, a subset of individuals will also have developed a state of persistent mucosal inflammation despite strict adherence to the GFD and be classified as having refractory coeliac disease (RCD). Individuals with RCD have a high risk of developing complications such as ulcerative jejunitis (UJ) and enteropathy associated T-cell lymphoma (EATL) [9,10]. Moreover, those with CD are at increased risk of developing other malignant complications such as small bowel (SB) adenocarcinoma [10]. As these pre-malignant and malignant conditions

Coeliac Disease and Gluten-Related Disorders.
DOI: https://doi.org/10.1016/B978-0-12-821571-5.00005-2

are considered the major complications of CD, they are typically classified together as 'complicated' coeliac disease (CCD) [11]. Differentiating between NRCD, RCD and CCD is important as it enables timely identification and treatment of the causative pathology. In this chapter, we provide an overview of the pathologies underlying NRCD and CCD and highlight a clinical approach to the patient with NRCD/ CCD.

5.2 Dietary indiscretion and mucosal healing

Deliberate or inadvertent ongoing gluten ingestion is the commonest cause of persisting symptoms in coeliac disease and is reported in around 35-50% of cases of NRCD [6,12]. Complete nonadherence is uncommon, but effective adherence is reported to occur in only around 40-90% of cases [13]. Indeed, maintaining complete gluten abstinence is difficult and adherence may be breached because of the widespread use of gluten as a food additive, high costs of the GFD, palatability and the social restrictions imposed by the GFD [14].

Assessing adherence to a GFD is challenging. A detailed dietetic evaluation by a specialist dietitian, including the use of food diaries, is an effective method and can identify inadvertent gluten exposure particularly if individuals lack an understanding of which foodstuffs or products contain gluten. However, this is time-consuming and access to a specialist dietitian to conduct this assessment is limited [15]. Moreover, inadvertent exposure may even occur in the setting of presumed gluten abstinence, which might not be identified on direct questioning [16]. Questionnaires have been developed that can be applied by non-expert personnel to quickly assess dietary adherence to a GFD [17,18]. However, the sensitivity and specificity of such questionnaires, against duodenal histology remain low [15]. Serological markers of coeliac disease (tissue transglutaminase [tTG] and endomysial antibodies [EMA]) have traditionally been used in clinical practice to monitor for adherence. The normalization of circulating tTG titers after institution of a GFD is often taken to reflect dietary adherence to a GFD and considered to correlate with mucosal healing. However, a recent meta-analysis of the diagnostic accuracy of tTG and EMA IgA antibodies for predicting persistent villous atrophy on a GFD demonstrated a specificity of 0.83 (95% CI 0.79-0.87) and 0.91 (95% CI 0.87-0.94) and sensitivity of 0.5 (95% CI 0.41-0.60) and 0.45 (95% CI 0.34-0.57), respectively [19]. Therefore, these serological tests cannot be relied upon to inform on mucosal healing after the institution of a GFD.

Tests that determine the excretion of gluten immunogenic peptides (GIPs) have recently been developed. These tests quantify the amount of GIP in the stool and urine by enzyme-linked immunosorbent assay, or immunocromatography [20]. In one recent study of 44 coeliac patients on a GFD who had faecal and urine GIPs measured, a quarter of the cohort had a least one positive test [21]. Notably, 32% of asymptomatic patients had at least one positive test and the concordance between GIP testing and dietary reporting was only

70% [21]. This demonstrates the potential usefulness of GIP testing to determine ongoing gluten exposure in coeliac patients presenting with persisting symptoms. However, so far these tests have failed to gain widespread acceptance into clinical practice owing to certain limitations. Namely, urine and faecal GIPs are only able to detect gluten ingestion 1-2 days, and 2-4 days, prior to testing, respectively [22]. Therefore, it is possible that the window of gluten exposure is missed by the time individuals are tested for GIPs. This is not the only non-invasive marker of gluten exposure available and the last decade has seen a rise in the development of other commercially available tests, including point of care tests [23]. However, robustly designed studies are awaited to assess their place in clinical practice.

In view of this, repeat duodenal biopsies are currently the most appropriate way to assess for mucosal healing and thus, indirectly inform on effective adherence to the GFD [24]. However, in adults, the rate of mucosal healing after starting a GFD is uncertain and can be highly variable between individuals. Some studies suggest that histological remission occurs in most individuals (68%) within the first year following diagnosis [25]. Others have reported histological remission in 34-65% of individuals up to two years post diagnosis, while it has also been shown that mucosal recovery may take as long as 5 years in some individuals [26–28]. Aside from non-adherence to the GFD, the reasons underlying why the rate of recovery is so variable between people are unclear [26]. Therefore, it is important to be aware that some patients with persistent villous atrophy at repeat biopsy may be slow to respond to the GFD, rather than have dietary indiscretion or true RCD.

5.3 An alternative pathology

An obvious but crucial step in considering the aetiology of persisting symptoms is to re-examine the primary diagnosis of CD. Indeed, case series suggest that around 8% of patients evaluated with a presumed diagnosis of NRCD, did not have an original diagnosis of CD [6,12]. This is particularly important in the setting of a patient with a historical diagnosis of CD presenting with secondary non-responsiveness, because diagnostic testing in CD has advanced greatly over the last two decades, alongside a shift in our understanding of how CD manifests in adults, with a greater emphasis on extra-intestinal signs and symptoms than previously thought [2]. It is also of note that the recent past has seen the recognition of the broader group of disorders that falls under the spectrum of CD, which include 'potential' and 'seronegative' CD [2]. Whilst villous atrophy in the context of negative serology maybe due to CD, other diagnoses must be considered first, as CD only accounts for 30% of serology negative villous atrophy [29]. Other causes include immune-mediated (e.g. autoimmune enteropathy), inflammatory (e.g. eosinophilic gastroenteritis), infectious (e.g. giardiasis), iatrogenic (e.g. NSAIDs) and idiopathic [30]. Therefore, careful review of original index investigations is needed, and if re-examining the primary diagnosis raises any doubt, consideration should be given to an appropriate gluten challenge with repeat serology and duodenal biopsies. Individuals with CD have an increased risk of developing a range of other GI pathologies, including microscopic colitis, pancreatic exocrine insufficiency, IBD, small intestinal bacterial overgrowth and lactose or fructose intolerance [6–8]. Therefore, consideration should be given to these other conditions in individuals who present with ongoing symptoms in the absence of mucosal changes typical of CD. In addition, functional gastrointestinal disorders such as irritable bowel syndrome (IBS) and gastrointestinal dysmotility are more prevalent in patients with CD [31,32] and should not be overlooked as a cause of persisting symptoms in these individuals.

5.4 Refractory coeliac disease

RCD is defined as persistent or recurrent malabsorptive symptoms and/ or signs with ongoing mucosal inflammation despite strict adherence to a GFD [33]. RCD has a reported prevalence of 0.3-4% of patients with CD [35–37] and is reported to account for 8-23% of NRCD cases [6,12,36]. However, the true prevalence of RCD is likely to be lower, as the diagnosis of RCD is challenging.

Individuals with RCD can be subcategorized as having either RCD Type 1 (RCD1) or Type 2 (RCD2), based on the abnormal expansion of a subset of small intestinal intraepithelial lymphocytes (IELs), which is detected in RCD2 [38]. These IELs are often described as being "aberrant," as they lack the usual expression of cell surface markers such as CD3 and CD8, but do express intracellular CD3 proteins [39]. Notably, this IEL subset is present in normal individuals, uncomplicated CD and RCD1, but at lower frequencies to that found in RCD2 [39,40]. The frequency of the aberrant IEL subset within the small intestines can be detected using techniques such as flow cytometry or immunohistochemistry (IHC), although flow cytometry is considered a more accurate approach at differentiating IEL subsets than IHC [41]. In RCD2, there is evidence that aberrant IELs display clonal rearrangement of their T cell receptor (TCR) [39,40,42]. However, clonal rearrangement of the TCR is not unique to RCD2 and has been described transiently in uncomplicated CD at time of diagnosis, and also in individuals with CD who continue to ingest gluten [43,44]. Clonality alone is therefore not an adequate indicator of RCD2 [41].

Unlike in RCD2, the immune cell changes associated with mucosal inflammation in RCD1 are the same as that in individuals with ongoing dietary gluten ingestion. Therefore, it can be difficult to differentiate whether persistent mucosal inflammation is due to RCD or related to ongoing exposure to dietary gluten [45]. In addition, research suggests that some patients may be sensitive to small traces of gluten (less than 20 parts per million – the recommended threshold below which food products may be considered gluten free in many countries), which compounds the difficulty in differentiating between ongoing exposure to dietary gluten and RCD [45]. The Gluten Contamination Elimination Diet (GCED) has been designed to remove any possible trace sources of dietary gluten exposure and can be used to help differentiate between the two pathologies [46]. Its use may prevent patients from receiving unnecessary corticosteroid and immunosuppressive treatment for an incorrect diagnosis of RCD [45].

RCD2 is predominantly diagnosed in adults aged 50 or above, although younger cases have been observed [47]. Individuals with RCD2 have a high risk of developing UJ and/ or EATL [34,35]. UJ is a rare condition characterized by chronic idiopathic ulcerations affecting the small bowel, unrelated to drugs, ischaemia, infections or other known causes [48]. Its clinical presentation alongside RCD2 can vary and can include GI perforation, obstruction and/ or bleeding [49]. By contrast, EATL is a T-cell non-Hodgkin lymphoma and can be considered as the lymphomatous transformation of the IEL phenotype in RCD2. Patients typically present with constitutional symptoms such as generalized malaise, anorexia and weight loss, as well as severe malabsorption [50]. Inflammation in RCD2 can be patchy, and the lesions associated with UJ and EATL are commonly located in the jejunum and distal SB [51,52]. Therefore, endoscopic inspection of the entire length of the SB is essential in suspected RCD. Small bowel capsule endoscopy (SBCE) is a useful minimally invasive screening tool used in this manner. The features that define sinister lesions from benign ones on SBCE are not clear, so ulcerative lesions should be biopsied using more invasive enteroscopic techniques and undergo detailed histological evaluation [53]. In addition, EATL can present solely at extraintestinal sites, which would not be detected by endoscopy [9]. Therefore, individuals should have cross-sectional imaging performed. Consideration should also be given to positron emission tomography (PET) scanning, to detect spread and stage of disease [54].

5.4.1 Treatment in refractory coeliac disease

Treatment depends on the type of RCD, with the broad aims of correcting the effects of malabsorption and promoting mucosal healing. The mainstay of treatment in RCD1 is nutritional support, corticosteroids +/- azathioprine (or derivatives), and most patients with RCD1 achieve clinical remission and mucosal healing with this approach [55,56]. As such, the 5-year survival is around 80-100% [34,35]. In contrast, treatment options are limited in RCD2 and as a result morbidity and mortality is higher in this subgroup. Progression to EATL is reported to occur in 33-67% of RCD2 cases; EATL has a 5-year survival of only around 10% [34,35].

While there is no evidence that EATL can be prevented, early identification and treatment of RCD2 may delay its onset. Unlike in RCD1, immunosuppressants such as azathioprine are not recommended in the treatment of RCD2 due to the risk of progression to lymphoma in this group [4]. Rather, treatment is initially commenced with oral steroids, such as budesonide [59,60]. The use of the adenosine nucleoside analogue cladribine has been shown to induce clinical and histological improvement, although it does not prevent the development of EATL [61,62]. Excessive production of interlukin(IL)-15 by small intestinal enterocytes is considered to be a key mechanism driving the proliferation of aberrant IELs in RCD2/ EATL [40]. However, anti-IL-15 monoclonal antibody therapy was recently shown to have no significant effect at reducing the percentage of aberrant IELs in individuals with RCD2 [63]. IL-15 elicits some of its actions via the janus kinase signaling pathway and the therapeutic targeting of this pathway in RCD2/ EATL is the subject of current interest [64]. Immunoablative chemotherapy followed by autologous haemopoietic stem cell transplantation (aHSCT), is now indicated in a range of autoimmune and inflammatory diseases [23,65–67]. Current recommendations are that aHSCT can be considered as an option for individual patients after careful discussions of risks and benefits and performed in accredited specialist centers with major experience and appropriate infrastructure [23,67]. However, further observational and prospective studies with greater numbers are warranted to define the risk:benefit ratio of aHSCT and identify where it should feature in the treatment algorithm for RCD2 [23].

5.5 Other complications of coeliac disease

It is generally considered that individuals with CD have an increased risk of SB adenocarcinoma [68]. This is a rare malignancy (0.6-0.7 per 100,000 of the general population per year) which mostly affects individuals in their seventh decade of life [50,69]. Early studies suggest a 60-80-fold relative risk of developing SB adenocarcinoma in CD [69–71]. A more recent study identified 5 cases of SB adenocarcinoma in 770 patients with CD (a frequency of 0.65%), although the study comprised a patient cohort from a specialist tertiary center and thus may be subject to referral bias [72]. The factors underlying carcinogenesis in CD are unknown. Various mechanisms have been proposed, including that SB adenocarcinoma may follow an "adenoma-carcinoma" sequence [50]. However, to date this has not been substantiated. Like EATL, SB adenocarcinoma is most commonly diagnosed in $\geq$50 year olds, often around the time of index CD diagnosis [9,11], and the risk of both conditions has been found to fall $\geq$1 year after CD diagnosis (albeit a non-significant trend for SB adenocarcinoma) [9]. This suggests that the GFD and subsequent mucosal healing may have a protective role for developing these malignancies [9], although this has not yet been proven. While the threshold of suspicion for malignancy in older individuals diagnosed with CD should be low, those diagnosed with SB adenocarcinoma should be screened for CD, to prevent delaying the institution of a GFD, which may help recovery from surgical intervention. If complete surgical resection can be accomplished, the prognosis in SB adenocarcinoma is better than if EATL is diagnosed [73].

Besides EATL and SB adenocarcinoma, there is a documented risk of other malignancies in CD, but the quality of the studies evaluating these associations is poor. This can be attributed in part to small patient numbers and the inclusion of cases often from periods when the diagnosis of CD was not as accurate as it is presently [50]. Indeed, there are studies suggesting associations between CD and oropharyngeal, oesophageal, and colorectal cancer, as well as abdominal B cell lymphoma. Other studies assessing the increased risk of liver and pancreatic cancer have produced mixed results [74]. The results of a meta-analysis of 17 studies investigating the risk of malignancies in CD supported a higher risk for oesophageal cancer (pooled OR 3.72, 95% CI 1.90–7.28) and small intestinal carcinoma (pooled OR 14.41, 95% CI 5.53–37.60), but not other GI cancers, pancreatic or liver cancer [74]. A more recent study has corroborated the increased risk of SB adenocarcinoma and EATL in CD but that the risk of other forms of lymphoma is not increased in CD [9].

A small number of reports have also detailed an association between CS and CD [75,76]. However, the precise relationship is unclear, as most have been case reports and there is a lack of large observational studies in this area. Little is known about the pathophysiology of CS, but it is characterized histologically by detection of a sup-epithelial thickened collagen band [77,78]. Other histological features include villous atrophy, intraepithelial lymphocytosis and patchy epithelial detachment [77–79]. CS mainly affects middle-aged to elderly women, presenting with symptoms of malabsorption [79]. As with RCD2, small bowel ulceration and GI malignancies can co-occur with, or complicate CS [79]. Early case studies reported progressive, irreversible malabsorption leading to a dismal outlook for individuals with CS, while more recent studies have reported more positive outcomes [79]. Treatment often follows an empirical course, including the use of steroids, immunosuppressants and/ or biologics [79]. However, there is no unified treatment strategy for individuals with CS, mostly due to a lack of robust studies with suitable patient numbers owing to the rarity of this condition.

5.6 Clinical approach

Collectively, NRCD and CCD cover a broad range of pathologies. Differentiating between the causes of NRCD and CCD is important as it enables timely identification and treatment of the causative pathology. A stepwise approach is recommended to do achieve this, an outline of which is illustrated in Fig. 5.1.

5.7 Concluding remarks

Up to a third of individuals with CD present with NRCD. Dietary indiscretion is the commonest cause of NRCD, yet currently there is no reliable objective assessment of ongoing gluten exposure in these patients. The diagnosis and management of RCD is

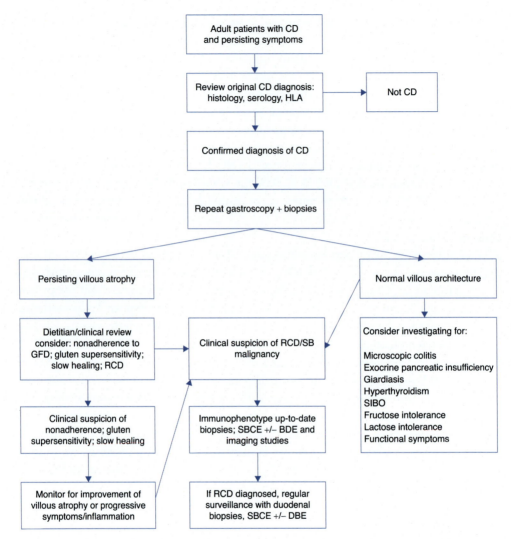

Figure 5.1 *Flow chart outlining the assessment of adults with NRCD.* Currently, there is no consensus on the interval for monitoring individuals considered to have persisting villous atrophy secondary to ongoing gluten exposure or slow healing. UK experts recommend a 12-monthly monitoring interval for patients with RCD1 once treatment response is established and a 6-monthly interval for patients with RCD2 [5]. **Permission process not started.** Select figure and click icon in top toolbar to start permission process or indicate no permission is required.

challenging, and patients should be referred to a specialist center with multidisciplinary experience for support with diagnosis and treatment. Novel therapeutic strategies are required to provide realistic treatment options in RCD2 and EATL to impact the poor outcomes in this group.

Supplementary materials

None

Acknowledgments

No funding was obtained for this review. DSS receives an educational grant from Schaer (a gluten-free food manufacturer). Dr Schaer did not have any input in drafting of this manuscript.

Author contributions

HAP, AS and DSS wrote and revised the manuscript.

Conflicts of interest

None

References

[1] P. Singh, A. Arora, T.A. Strand, et al., Global Prevalence of Celiac Disease: Systematic Review and Meta-analysis, Clin. Gastroenterol. Hepatol. 16 (6) (2018) 823–836 e822.

[2] B. Lebwohl, D.S. Sanders, P.H.R. Green, Coeliac Disease, Lancet 6:391 (10115) (2018) 70–81.

[3] S. O'Mahony, P.D. Howdle, M.S. Losowsky, Review article: management of patients with non-responsive coeliac disease, Aliment. Pharmacol. Ther. 10 (5) (1996) 671–680.

[4] E.M.R. Baggus, M. Hadjivassiliou, S. Cross, et al., How to manage adult coeliac disease: perspective from the NHS England Rare Diseases Collaborative Network for Non-Responsive and Refractory Coeliac Disease Baggus EMR, et al, Frontline Gastroenterol. 0 (2019) 1–8.

[5] P.D. Mooney, K.E. Evans, S. Singh, D.S. Sanders, Treatment failure in coeliac disease: a practical guide to investigation and treatment of non-responsive and refractory coeliac disease, J. Gastrointestin Liver Dis. 21 (2) (2012) 197–203.

[6] A.S. Abdulkarim, L.J. Burgart, J. See, J.A. Murray, Aetiology of nonresponsive celiac disease: results of a systematic approach, Am. J. Gastroenterol. 97 (2002) 2016–2021.

[7] M.I. Pinto-Sanchez, C.L. Seiler, N. Santesso, et al., Association Between Inflammatory Bowel Diseases and Celiac Disease: A Systematic Review and Meta-Analysis, Gastroenterol. S0016-5085 (20) (2020) 30609   -0, doi:10.1053/j.gastro.2020.05.016.

[8] K.D. Fine, R.L. Meyer, E.L. Lee, The prevalence and causes of chronic diarrhea in patients with celiac sprue treated with a gluten-free diet, Gastroenterol. 112 (6) (1997) 1830–1838.

[9] T. van Gils, P. Nijeboer, L.I.H. Overbeek, M. Hauptmann, et al., Risks for lymphoma and gastrointestinal carcinoma in patients with newly diagnosed adult-onset celiac disease: Consequences for follow-up, United European Gastroenterol. J. 6 (10) (2018) 1485–1495.

[10] S.D. Rampertab, K.A. Forde, P.H. Green, Small bowel neoplasia in coeliac disease, Gut. 52 (8) (2003) 1211–1214.

[11] F. Biagi, A. Marchese, F. Ferretti, et al., A multicentre case control study on complicated coeliac disease: two different patterns of natural history, two different prognoses, BMC Gastroenterol. 14 (2014) 139.

[12] D.H. Dewar, S.C. Donnelly, S.D. McLaughlin, M.W. Johnson, H.J. Ellis, P.J. Ciclitira, Celiac disease: management of persistent symptoms in patients on a gluten-free diet, World J. Gastroenterol. 18 (12) (2012) 1348–1356.

[13] N.J. Hall, G. Rubin, A. Charnock, Systematic review: adherence to a gluten-free diet in adult patients with coeliac disease, Aliment. Pharmacol. Ther. 30 (4) (2009) 315–330.

[14] M.L. Moreno, A. Cebolla, A. Munoz-Suano, et al., Detection of gluten immunogenic peptides in the urine of patients with coeliac disease reveals transgressions in the gluten-free diet and incomplete mucosal healing, Gut. 66 (2) (2017) 250–257.

[15] M.S. Lau, P.D. Mooney, M.A. Rees, et al., OWE-022 Gluten free diet adherence assessment using CDAT and BIAGI questionnaires in patients with coeliac disease, Gut. 67 (2018) A160–A161.

[16] B.A. Lerner, L.T. Phan Vo, S. Yates, et al., Detection of Gluten in Gluten-Free Labeled Restaurant Food: Analysis of Crowd-Sourced Data, Am. J. Gastroenterol. 114 (5) (2019) 792–797.

[17] F. Biagi, P.I. Bianchi, A. Marchese, et al., A score that verifies adherence to a gluten-free diet: a cross-sectional, multicentre validation in real clinical life, Br. J. Nutr. 108 (10) (2012) 1884–1888 Nov 28.

[18] D.A. Leffler, M. Dennis, J.B. Edwards George, et al., A simple validated gluten-free diet adherence survey for adults with celiac disease, Clin. Gastroenterol. Hepatol. 7 (5) (2009) 530 May–6,536. e1–2.

[19] J.A. Silvester, S. Kurada, A. Szwajcer, C.P. Kelly, D.A. Leffler, D.R. Duerksen, Tests for Serum Trans-glutaminase and Endomysial Antibodies Do Not Detect Most Patients With Celiac Disease and Persistent Villous Atrophy on Gluten-free Diets: a Meta-analysis, Gastroenterol. 153 (3) (2017) 689–701 e681.

[20] Á. Cebolla, M.L. Moreno, L. Coto, C. Sousa, Gluten Immunogenic Peptides as Standard for the Evaluation of Potential Harmful Prolamin Content in Food and Human Specimen, Nutrients 10 (2018) 1927.

[21] A.F. Costa, E. Sugai, M.P. Temprano, et al., Gluten immunogenic peptide excretion detects dietary transgressions in treated celiac disease patients, World J. Gastroenterol. 25 (11) (2019) 1409–1420.

[22] M.L. Moreno, A. Cebolla, A. Munoz-Suano, et al., Detection of gluten immunogenic peptides in the urine of patients with coeliac disease reveals transgressions in the gluten-free diet and incomplete mucosal healing, Gut. 66 (2) (2017) 250–257.

[23] M.S. Lau, P.D. Mooney, W.L. White, et al., The Role of an IgA/IgG-Deamidated Gliadin Peptide Point-of-Care Test in Predicting Persistent Villous Atrophy in Patients With Celiac Disease on a Gluten-Free Diet, Am. J. Gastroenterol. 112 (12) (2017) 1859–1867.

[24] H.A. Penny, E.M.R. Baggus, A. Rej, et al., Non-Responsive Coeliac Disease: A Comprehensive Review from the NHS England National Centre for Refractory Coeliac Disease, Nutrients 12 (1) (2020) 216.

[25] H. Pekki, K. Kurppa, M. Maki, et al., Predictors and Significance of Incomplete Mucosal Recovery in Celiac Disease After 1 Year on a Gluten-Free Diet, Am. J. Gastroenterol. 110 (7) (2015) 1078–1085.

[26] L.M. Sharkey, G. Corbett, E. Currie, J. Lee, N. Sweeney, J.M. Woodward, Optimising delivery of care in coeliac disease - comparison of the benefits of repeat biopsy and serological follow-up, Aliment. Pharmacol. Ther. 38 (10) (2013) 1278–1291.

[27] A. Rubio-Tapia, M.W. Rahim, J.A. See, B.D. Lahr, T.T. Wu, J.A. Murray, Mucosal recovery and mortality in adults with celiac disease after treatment with a gluten-free diet, Am. J. Gastroenterol. 105 (6) (2010) 1412–1420.

[28] P.J. Wahab, J.W. Meijer, C.J. Mulder, Histologic follow-up of people with celiac disease on a gluten-free diet: slow and incomplete recovery, Am. J. Clin. Pathol. 118 (3) (2002) 459–463.

[29] I. Aziz, M.F. Peerally, J.H. Barnes, et al., The clinical and phenotypical assessment of seronegative villous atrophy; a prospective UK centre experience evaluating 200 adult cases over a 15-year period (2000-2015), Gut. 66 (9) (2017) 1563–1572.

[30] A. Schiepatti, D.S. Sanders, F. Biagi, Seronegative coeliac disease: clearing the diagnostic dilemma, Curr. Opin. Gastroenterol. 34 (3) (2018) 154–158.

[31] S.M. Barratt, J.S. Leeds, K. Robinson, et al., Reflux and irritable bowel syndrome are negative predictors of quality of life in coeliac disease and inflammatory bowel disease, Eur. J. Gastroenterol. Hepatol. 23 (2) (2011) 159–165.

[32] V.P. Bentivoglio, L. Raguseo, M. Tripaldi, et al. Increased prevalence of abdominal pain-functional gastrointestinal disorders in pediatric celiac patients. DOI:https://doi.org/10.1016/j.dld.2017.09.068.

[33] J.F. Ludvigsson, D.A. Leffler, J.C. Bai, et al., The Oslo definitions for coeliac disease and related terms, Gut. 62 (1) (2013) 43–52.

[34] A. Rubio-Tapia, D.G. Kelly, B.D. Lahr, A. Dogan, T.T. Wu, J.A. Murray, Clinical staging and survival in refractory celiac disease: a single center experience, Gastroenterol. 136 (1) (2009) 99–107 quiz 352-103.

[35] G. Malamut, P. Afchain, V. Verkarre, et al., Presentation and long-term follow-up of refractory celiac disease: comparison of type I with type II, Gastroenterol. 136 (1) (2009) 81–90.

[36] B. Roshan, D.A. Leffler, S. Jamma, et al., The incidence and clinical spectrum of refractory celiac disease in a north american referral center, Am. J. Gastroenterol. 106 (5) (2011) 923–928.

[37] S.A. Rowinski, E. Christensen, Epidemiologic and therapeutic aspects of refractory coeliac disease - a systematic review, Dan. Med. J. 63 (12) (2016).

[38] C. Cellier, E. Delabesse, C. Helmer, et al., Refractory sprue, coeliac disease, and enteropathy-associated T-cell lymphoma, Lancet 356 (2000) 203–208.

[39] C. Cellier, N. Patey, L. Mauvieux, et al., Abnormal intestinal intraepithelial lymphocytes in refractory sprue, Gastroenterol. 114 (3) (1998) 471–481.

[40] J. Ettersperger, N. Montcuquet, G. Malamut, et al., Interleukin-15-Dependent T-Cell-like Innate Intraepithelial Lymphocytes Develop in the Intestine and Transform into Lymphomas in Celiac Disease, Immunity 45 (3) (2016) 610–625.

[41] W.H. Verbeek, M.S. Goerres, B.M. von Blomberg, et al., Flow cytometric determination of aberrant intra-epithelial lymphocytes predicts T-cell lymphoma development more accurately than T-cell clonality analysis in Refractory Celiac Disease, Clin. Immunol. 126 (1) (2008) 48–56.

[42] E. Bagdi, T.C. Diss, P. Munson, P.G. Isaacson, Mucosal intra-epithelial lymphocytes in enteropathy-associated T-cell lymphoma, ulcerative jejunitis, and refractory celiac disease constitute a neoplastic population, Blood 94 (1) (1999) 260–264.

[43] H. Liu, R. Brais, A. Lavergne-Slove, et al., Continual monitoring of intraepithelial lymphocyte immunophenotype and clonality is more important than snapshot analysis in the surveillance of refractory coeliac disease, Gut. 59 (4) (2010) 452–460.

[44] R. Celli, P. Hui, H. Triscott, et al., Clinical Insignficance of Monoclonal T-Cell Populations and Duodenal Intraepithelial T-Cell Phenotypes in Celiac and Nonceliac Patients, Am. J. Surg. Pathol. 43 (2) (2019) 151–160.

[45] M.M. Leonard, P. Cureton and A. Fasano, Indications and Use of the Gluten Contamination Elimination Diet for Patients with Non-Responsive Celiac Disease Nutrients, 2017, 9, 1129; doi:10.3390/nu9101129.

[46] J.R. Hollon, P.A. Cureton, M.L. Martin, E.L. Puppa, A. Fasano, Trace gluten contamination may play a role in mucosal and clinical recovery in a subgroup of diet-adherent non-responsive celiac disease patients, BMC Gastroenterol. 13 (2013) 40.

[47] S.A. Rowinski, E. Christensen, Epidemiologic and therapeutic aspects of refractory coeliac disease - a systematic review, Dan. Med. J. 63 (12) (2016).

[48] F. Biagi, P. Lorenzini, G.R. Corazza, Literature Review on the Clinical Relationship between Ulcerative Jejunoileitis, Coeliac Disease, and Enteropathy-associated T-Cell Lymphoma, Scand. J. Gastroenterol. 35 (8) (2000) 785–790.

[49] R. Rai, T.M. Bayless, Isolated and diffuse ulcers of the small intestine, in: M. Feldman, B.F. Scharschmidt, M.H. Sleisenger (Eds.), Sleisenger & Fordtran's gastrointestinal and liver disease, 6th ed., WB Saunders, Philadelphia, 1997, pp. 1771–1778.

[50] C. Catassi, I. Bearzi, G.K.T. Holmes, Association of celiac disease and intestinal lymphomas and other cancers, Gastroenterol. 128 (2005) S79–S86.

[51] L.J. Egan, S.V. Walsh, F.M. Stevens, et al., Celiac-associated lymphoma. A single institution experience of 30 cases in the combination chemotherapy era, J. Clin. Gastroenterol. 21 (2) (1995) 123–129.

[52] C. Tomba, R. Sidhu, D.S. Sanders, et al., Celiac Disease and Double-Balloon Enteroscopy: What Can We Achieve? The Experience of 2 European Tertiary Referral Centers, J. Clin. Gastroenterol. 50 (2016) 313–317.

[53] M. Pennazio, C. Spada, R. Eliakim, et al., Small-bowel capsule endoscopy and device-assisted enteroscopy for diagnosis and treatment of small lbowel disorders: European Society of Gastrointestinal Endoscopy (ESGE) Clinical Guideline, Endoscopy 47 (2015) 352–376.

[54] M. Hadithi, M. Mallant, J. Oudejans, J.H. van Waesberghe, C.J. Mulder, E.F. Comans, 18F-FDG PET versus CT for the detection of enteropathy-associated T-cell lymphoma in refractory celiac disease, J. Nucl. Med. 47 (10) (2006) 1622–1627.

[55] A. Al-Toma, W.H. Verbeek, C.J. Mulder, Update on the management of refractory coeliac disease, J. Gastrointestin Liver Dis. 16 (1) (2007) 57–63.

[56] A. Rubio-Tapia, J.A. Murray, Classification and management of refractory coeliac disease, Gut. 59 (4) (2010) 547–557.

[57] A. Al-Toma, W.H. Verbeek, M. Hadithi, B.M. von Blomberg, C.J. Mulder, Survival in refractory coeliac disease and enteropathy-associated T-cell lymphoma: retrospective evaluation of single-centre experience, Gut. 56 (10) (2007) 1373–1378.

[58] S. Daum, R. Ipczynski, M. Schumann, U. Wahnschaffe, M. Zeitz, R. Ullrich, High rates of complications and substantial mortality in both types of refractory sprue, Eur. J. Gastroenterol. Hepatol. 21 (1) (2009) 66–70.

[59] P. Brar, S. Lee, S. Lewis, I. Egbuna, G. Bhagat, P.H. Green, Budesonide in the treatment of refractory celiac disease, Am. J. Gastroenterol. 102 (10) (2007) 2265–2269.

[60] S.S. Mukewar, A. Sharma, A. Rubio-Tapia, T.T. Wu, B. Jabri, J.A. Murray, Open-Capsule Budesonide for Refractory Celiac Disease, Am. J. Gastroenterol. 112 (6) (2017) 959–967.

[61] G.J. Tack, W.H.M. Verbeek, A. Al-Toma, et al., Evaluation of cladribine treatment in refractory celiac disease type II, World J. Gastroenterol. 17 (2011) 506–513.

[62] A. Al-Toma, M.S. Goerres, J.W.R. Meijer, et al., Cladribine therapy in refractory celiac disease with aberrant T cells, Clin. Gastroenterol. Hepatol. 4 (2006) 1322–1327.

[63] C. Cellier, G. Bouma, T. van Gils, et al., Safety and efficacy of AMG 714 in patients with type 2 refractory coeliac disease: a phase 2a, randomised, double-blind, placebo-controlled, parallel-group study, Lancet Gastroenterol. Hepatol. (2019).

[64] S. Yokoyama, P.-Y. Perera, T.A. Waldmann, T. Hiroi, L.P. Perera, Tofacitinib, a janus kinase inhibitor demonstrates efficacy in an IL-15 transgenic mouse model that recapitulates pathologic manifestations of celiac disease, J. Clin. Immunol. 33 (2013) 586–594.

[65] J.A. Snowden, M. Badoglio, M. Labopin, et al., Evolution, trends, outcomes, and economics of hematopoietic stem cell transplantation in severe autoimmune disease, Blood Advances 1 (27) (2017) 2742–2755.

[66] J.A. Snowden, M. Badoglio, T. Tobias Alexander, The rise of autologous HCT for autoimmune diseases: what is behind it and what does it mean for the future of treatment? An update on behalf of the EBMT Autoimmune Diseases Working Party, Expert Review of Clinical Immunology (2019), doi:10.1080/1744666X.2019.1656526.

[67] R.F. Duarte, M. Labopin, P. Bader, et al., Indications for haematopoietic stem cell transplantation for haematological diseases, solid tumours and immune disorders: current practice in Europe, 2019, Bone Marrow Transplant. 54 (2019) 1525–1552.

[68] A.J. Daveson, R.P. Anderson, Small bowel endoscopy and coeliac disease, Best Pract. Res. Clin. Gastroenterol. 26 (3) (2012) 315–323.

[69] C.M. Swinson, G. Slavin, E.C. Coles, C.C. Booth, Coeliac disease and malignancy, Lancet 1 (1983) 111–115.

[70] M. Silano, U. Volta, A.M. Mecchia, M. Dessì, R. Di Benedetto, M. De Vincenzi, Delayed diagnosis of coeliac disease increases cancer risk, BMC Gastroenterol. 7 (2007) 8.

[71] P.H.R. Green, S.N. Stavropoulos, S.G. Panagi, S.L. Goldstein, D.J. Mcmahon, H. Absan, A.I. Neugut, Characteristics of adult celiac disease in the USA: results of a national survey, Am. J. Gastroenterol. 96 (2001) 126–131.

[72] G. Caio, U. Volta, F. Ursini, et al., Small bowel adenocarcinoma as a complication of celiac disease: clinical and diagnostic features, BMC Gastroenterol. 19 (2019) 45.

[73] H.J. Freeman, Malignancy in adult celiac disease, World J. Gastroenterol. 15 (13) (2009 Apr 7) 1581–1583.

[74] Y. Han, W. Chen, P. Li, J. Ye, Association Between Coeliac Disease and Risk of Any Malignancy and Gastrointestinal Malignancy. Medicine (Baltimore). 94 (38) (2015) e1612.

[75] W.M. Weinstein, D.R. Saunders, G.N. Tytgat, et al., Collagenous sprue— an unrecognized type of malabsorption, N. Engl. J. Med. 283 (1970) 1297–1301.

[76] M.E. Robert, M.E. Ament, W.M. Weinstein, The histologic spectrum and clinical outcome of refractory and unclassified sprue, Am. J. Surg. Pathol. 24 (2000) 676–687.

[77] A.A. Maguire, J.K. Greenson, G.Y. Lauwers, et al., Collagenous sprue: a clinicopathologic studyof 12 cases, Am. J. Surg. Pathol. 33 (2009) 1440–1449 7.

[78] A. Rubio-Tapia, N.J. Talley, S.R. Gurudu, et al., Gluten-free diet and steroid treatment are effective therapy for most patients with collagenous sprue, Clin. Gastroenterol. Hepatol. 8 (2010) 344–349 8.

[79] H.J. Freeman, Collagenous sprue, Can. J. Gastroenterol. 25 (2011) 189–192.

CHAPTER 6

New perspectives on the diagnosis of adulthood coeliac disease

Carolina Ciacci[a], Fabiana Zingone[b]

[a]Coeliac Center, Gastroenterology Unit, Department of Medicine, Surgery, Dentistry, Scuola Medica Salernitana, University of Salerno
[b]Coeliac Center, Gastroenterology Unit, Department of Surgery, Oncology and Gastroenterology, University of Padua, Padua, Italy

6.1 Introduction

The present chapter deals with an up-to-date in diagnosing coeliac disease (CeD), considering the availability of novel tests and approaches.

Currently, most national and international guidelines agree that in case of suspicion of CeD, for both adults and children, the search for IgA anti-transglutaminase (a-tTg) antibodies is the first diagnostic step [1-3]. The test has high sensitivity and specificity [4]. In addition to these antibodies, the search for total IgA immunoglobulins should be associated, as 2% of celiac patients have an IgA deficiency, compared to 0.2% of the general population, which would search IgA a-tTG falsely negative [5,6]. Anti-endomysial IgA antibodies (EMA) testing is based on immunofluorescence; therefore, it is more expensive than the IgA a-tTG test and operator-dependent. Because of that, EMA is mostly a confirmation test for the positive IgA a-tTG antibodies. EMA testing is indicated in low levels of a-tTG IgA, being more sensitive and specific [7].

The IgA and IgG class anti-deamidated antibodies (DGP) are antibodies recently identified and valuable for children under the age of two years or those of class IgG in IgA deficiency subjects. Combination tests provide better results capable of detecting CeD close to 100% [5,7].

In recent years, the debate among CeD experts on the need to perform a biopsy in adults at the time of diagnosis has been very heated, particularly after the advent of children's guidelines that recognize a subgroup of patients in which it is avoidable [8].

However, the increasing incidence of new cases of CeD, the need for precision testing, and the need to lower the diagnosis costs prompted the research for further tests, and diagnostic algorithms are now under evaluation.

6.2 Neoepitopes

A recent study tested a complex of gliadin deaminated and transglutaminase (tTG-DGP complex) synthesized peptides as a diagnostic marker of CeD and of the healing mucosa

Coeliac Disease and Gluten-Related Disorders.
DOI: https://doi.org/10.1016/B978-0-12-821571-5.00013-1

during a gluten free diet (GFD) [9]. To this end, the authors used the serum of a retrospective cohort of 90 coeliac subjects at diagnosis and 79 controls associated with a prospective cohort of 82 coeliac subjects at diagnosis and 217 controls. By evaluating the immune reactivity induced by this complex and the serum of both cohorts, the authors distinguish CeD patients from controls with a sensitivity of 99% and a specificity of 100%. Furthermore, 17 subjects with enteropathy without CeD demonstrated an immune reaction comparable to that of controls. Moreover, there was an absence of an immune response against the tTG-DGP complex of the sera of 4 selected IgA deficient patients.

However, the use in clinical practice requires further studies that evaluate the gain in diagnostic terms compared to the use of only antibodies also in terms of costs. It will be necessary to define its usefulness for diagnostic purposes in those subjects who begin a GFD before completing or performing the correct diagnostic procedure, and in the rare cases of seronegative patients with duodenal atrophy without IgA deficiency. (Table 6.1)

6.3 Gluten tetramers

CeD is characterized by HLA-DQ2/8-restricted responses of CD4+ T cells to gluten peptides. DQ2-restricted gluten epitopes are recognized explicitly by intestinal T cells isolated from CeD patients. Major histocompatibility complex tetramers consist of four MHC molecules loaded with a single peptide bond to a streptavidin molecule coupled with a fluorogenic marker. Multivalent engagement of MHC molecules leads to a stable binding of the tetramer to the T-cell surface, allowing direct visualization of the T-cells. T-cells bound HLA-DQ-gluten tetramers and are quantified by flow cytometry [10-12].

Therefore, HLA-DQ-gluten tetramers can detect gluten-specific T-cells in blood samples of suspected CeD but even in subjects on a GFD without a confirmed CeD after a gluten challenge. Recently Sarna et al. [13] reported this test's ability to identify CeD patients on a GFD even without gluten challenge. The authors tested this diagnostic test in 62 subjects with CeD patients on a GFD, in 19 subjects without CeD on a GFD due to self-reported gluten sensitivity, in 10 untreated subjects with CeD on a gluten-containing diet and 52 presumed healthy individuals on a gluten-containing diet. The authors found 97% sensitivity (95% CI 0.92-1.00) and 95% specificity (95% CI 0.84-1.00) in identified CeD subjects on a GFD vs subjects without CeD on a GFD. The values identified subjects with CeD on a gluten-containing diet show 100% sensitivity (95% CI 1.00-1.00]) and 90% specificity (95% CI 0.83-0.98) vs controls.

6.4 Point-of-care tests

The CeD specific autoantibody detection is usually performed in well-equipped laboratories by trained operators. These test sites, however, are not available everywhere. There is, therefore, a high demand to test patients for CeD qualitatively or semi-quantitatively

Table 6.1 Novel tests and approaches for celiac disease diagnosis.

New approach	Brief description
Neoepitopes	A complex of gliadin deaminated and transglutaminase (tTG-DGP complex) synthesized peptides
Gluten Tetramers	HLA-DQ-gluten tetramers detect gluten-specific T cells in blood samples
Point-of-care tests (antibodies)	Based on the lateral flow method, test the transglutaminase of red blood cell's ability to bind to anti-transglutaminase (a-tTG) antibodies in whole blood
Point-of-care tests (HLA)	Test the presence of HLA-DQ2.5, HLADQ8, HLA-DQ2.2, and HLA-DQA1*05 on minimally processed blood and saliva
Autoantibodies in saliva and feces	An enzyme-linked immunomagnetic electrochemical assay measures IgA a-tTG anti-transglutaminase (a-tTG) antibodies in saliva and feces
VOC biomarkers in urine and feces	Gas chromatography coupled with Mass spectrometry assesses Volatile Organic Compounds (VOC) derived from altered gut microbiota fermentation, urine, breath, and feces
Deposits of IgA tTG2	Search for deposits of IgA a-tTG anti-transglutaminase (a-tTG) antibodies in the small bowel mucosa by immunohistochemistry
EMA biopsy	Search for anti-endomysial IgA antibodies (EMA) in the medium of cultured intestinal biopsy before and after gliadin-challenge
Intestinal Epithelial lymphocytes count	Flow cytometry of intestinal epithelial lymphocytes which shows increased numbers of intraepithelial lymphocytes (IELs) inactive coeliac disease

in the doctor's office or even at home. The tests categorize as Point of Care (POC) tests. They are all based on the lateral flow method, constructed on the ability of the transglutaminase of red blood cells to bind to a-tTG antibodies in whole blood. They have good sensitivity and specificity and little invasivity as the blood derives from a finger prick. Lau et al. concluded that although POC tests had comparable sensitivities to serology, its low specificity may increase unnecessary investigations [14]. Also, in a pre-endoscopy setting, POC tests did not show more accurate results compared to serology. [15].

A recent study has evaluated the diagnostic accuracy of a POC test (Simtomax) among 622 patients referred for fertility treatment in two Danish fertility clinics, finding a sensitivity of 42.9% (95% CI 9.9-81.6) and a specificity of 86.8% [16]. Therefore, these tests have not yet gained widespread acceptance.

Recently, researchers developed a coeliac disease-loop-mediated isothermal amplification (CD-LAMP), a LAMP assay based on the same system. The assay enables rapid identification of the signature CD risk genotypes, HLA-DQ2.5, HLADQ8, HLA-DQ2.2, and HLA-DQA1*05 has been developed and validated that can be performed on minimally processed blood and saliva samples [17]. However, future studies evaluating the accuracy of this test are needed. A novel rapid assay for IgA a–tTG antibodies detection (LFRET) has been recently analyzed, finding a good accuracy [18].

The results on the usefulness of these POC tests in adults thus far are conflicting, and therefore these tests have not yet gained widespread acceptance.

6.5 Detection of CeD autoantibodies in saliva and feces

Saliva is an excellent specimen for screening analysis because it is easily obtained by non-invasive techniques [19].

The first test available to measure IgA a–tTG antibodies in saliva consists of a fluid-phase radioimmunoassay method [19]. Despite a high sensitivity and specificity [20] the test suffers from problems associated with the use of radioisotopes, cost of scintillation fluids, radioactive waste disposal, and restriction of its use to institutes having the permission for handling radioisotopes.

Adornetto et al. [21] described an enzyme-linked immunomagnetic electrochemical assay for measuring a–tTg IgA in saliva, based on magnetic beads to support the immuno-logical chain reaction and differential pulse voltammetry as the detection technique. This method showed high clinical specificity and sensitivity, bypassing the problems intrinsic to the radioimmunoassay method and the unpleasant blood sample collection required for routine serum sample analysis. Although these results are encouraging, there is still not enough evidence to recommend their use.

CeD specific antibodies have also been searched in feces. Di Tola et al. [22] found that the ROC curve showed a diagnostic significance in IgA a–tTG antibodies (AUC=0.862, p<0.0001), IgA DGP (AUC=0.822, p<0.0001) and IgA/IgG a–tTG/DGP (AUC=0.783, p=0.0003) faecal tests. However, the sensitivity of faecal IgA antibodies against TG2 of 76% makes it not suitable for accurate screening for CeD.

6.6 Detection of CeD-specific VOC biomarkers in urine and feces

Various metabolic pathways produce Volatile Organic Compounds (VOCs) [23]. In particular, the gut microbiome altered in several gastrointestinal disorders causes altered gut fermentation patterns recognizable by VOC analysis in urine, breath, and feces. Gas chromatography, coupled with mass spectrometry (GC–MS), a "gold standard" in VOC analysis, was applied to distinguish CeD and irritable bowel syndrome. The authors identified a compound, 1,3,5,7-cyclooctatetraene, in urine that could relate to CeD [24]. This compound was absent in samples of diarrhea-predominant irritable bowel syndrome.

Moreover, the authors evaluated the effect of a GFD on the exhaled breath [24,25]. The analysis of VOCs in urine might be several benefits over the other biological fluids. Urine collection is non-invasive and does not cause discomfort even with multiple sampling. Moreover, VOCs' concentration in urine is higher than in blood, as urine is pre-concentrated in the kidney, facilitating metabolites' detection [26].

6.7 No biopsy strategy

Since the early '60s, CeD diagnosis relies on the combination of serology and duodenal biopsy. However, in the most recent years, serology's high sensitivity and specificity allowed the pediatricians to implement a diagnostic strategy in children without performing duodenal histology. The ESPGHAN 2012 study demonstrated that symptomatic children showing high IgA a–tTG titers had villous atrophy and crypt hyperplasia invariably [27]. Therefore, the consensus concluded that symptoms, high IgA a–tTG, confirmed by EMA and HLA testing, allowed a precise diagnosis avoiding the endoscopy and biopsy. Recently, ESPGHAN guidelines have been updated and the authors established that in case of IgA a–tTG ≥10 times the upper limit of normal and EMA positivity, the no-biopsy diagnosis may be applied, whilst HLA DQ2-/DQ8 determination and symptoms are instead not obligatory criteria [3].

Based on the established safety and accuracy of CeD diagnosis based on serology, several studies indicated that the same strategy might help achieve adult diagnoses [28-30].

Among immunologic markers, interferon-γ-secreting T cells reactive to gluten were detected in the peripheral blood of CeD patients after short-term consumption of gluten-containing food, by using the enzyme-linked immunospot (ELISPOT) assays or by flow cytometry tetramer technology. The main limitations of this technique's extensive use for clinical practice are limited sensitivity and specificity compared to available serology tests and the high cost of ELISPOT and tetramers immune assays [31,32].

Others reported that in CeD patients, a single gluten challenge is followed by an increased level of serum IL-2, and to a lesser extent, IL-8 and IL-10 after 4 hours [33].

Recently Sanders et al. conducted a multicenter study supporting a no–biopsy approach to diagnose adult CeD [34]. The authors showed that IgA a–tTG levels of ≥10 × ULN had a high predictive value at identifying adults with intestinal changes diagnostic of CeD.

6.8 Empowering the intestinal biopsy

A small percentage of CeD patients show seronegative coeliac disease (SCD), that is villous atrophy in the presence of negative serology at diagnosis. It is frequently associated with misclassification, consequently with inappropriate treatments and long-term morbidity [35]. Two recent studies primarily assessed the prevalence of SCD in a cohort of CeD patients. Both these articles reported a prevalence of SCD, accounting for 2% of their total CeD population [36,37]. In these cases, the positivity of genetic CeD markers, HLA-DQ2 and -DQ8, is a mandatory requirement to suspect the diagnosis of SCD. Furthermore, both clinical and histological improvement should be proved after an adequate period of GFD. Thus, before posing a diagnosis of SCD, it is mandatory to rule out other causes of villous atrophy since this condition can be found in other enteropathies unrelated to gluten ingestions to avoid an unnecessary lifelong GFD. Other papers reported that SCD is rare, mainly diagnosed in older adults and associated with a severe histological and clinical involvement [38] or in patients with late-stage disease (with possible refractory CeD or lymphoma) [39]. In terms of the epidemiology of SCD, some studies show that SCD prevalence is higher in men and older subjects and patients with complicated CeD [36,37,40].

Deposits of IgA a-tTG2 antibodies have been found in the small bowel mucosa of SCD patients, but not in other forms of enteropathy, such as autoimmune [39].

A-tTG2 antibodies are produced primarily at an intestinal level by specific B lymphocytes and, it has been proven that these antibodies are synthesized in CeD patients [41] in which IgA a-tTG2 deposits are mainly stored below the epithelial layer and around blood vessels [42]. Once produced, a-tTG2 antibodies deposit in the small intestinal mucosa, even before they can be detected in the bloodstream [43]. Furthermore, serum a-tTG antibodies are thought to result from "spill-over" from the intestine [42]. Therefore, these autoantibodies at the gut level reinforce the diagnosis in borderline cases, mainly in SCD [42]. In this setting, also the EMA assay in cultured intestinal mucosa biopsies before and after an in-vitro gliadin-challenge may be an additional tool to either confirm or exclude the presence of a gluten-related enteropathy [44-46]. Another possible approach to better-identified complex cases could be the flow cytometry of intestinal epithelial lymphocytes IELs, which showed increased IELs in active CeD and a 97% specificity for CeD diagnosis [47].

Table 6.1 summarizes the novel tests and approaches for CeD diagnosis. Finally, there are endoscopic techniques to enhance the diagnosis of CeD, but these are still limited by availability, expertise, tolerability, and cost [48,49].

6.9 Conclusions

The scientific community points to a simplification of the CeD diagnosis. The golden standard of confirming the diagnosis by finding mucosal damage, set in the pre-serology era, is now old and avoidable in most cases. About ten years ago, pediatricians took the first step in children avoiding the biopsy. Nowadays, the increased sensitivity and specificity of the available plasma biomarkers and those in the pipeline suggest a paradigm change in adult CeD diagnosis. There is, however, the need for a few other steps, such as a consensus of the prominent experts of the CeD to change the guidelines at least for a subgroup of young adults without alarm signs of any overlapping illness.

References

[1] J.C. Bai, C. Ciacci, World Gastroenterology Organisation Global Guidelines: Celiac Disease February 2017, J. Clin. Gastroenterol. 51 (9) (2017) 755–768.

[2] A. Al-Toma, U. Volta, R. Auricchio, G. Castillejo, D.S. Sanders, C. Cellier, et al., European Society for the Study of Coeliac Disease (ESsCD) guideline for coeliac disease and other gluten-related disorders, United. Eur. Gastroenterol. J. 7 (5) (2019) 583–613.

[3] S. Husby, S. Koletzko, I. Korponay-Szabó, K. Kurppa, M.L. Mearin, C. Ribes-Koninckx, et al., European Society Paediatric Gastroenterology, Hepatology and Nutrition Guidelines for Diagnosing Coeliac Disease 2020, J. Pediatric. Gastroenterol. Nutr. 70 (1) (2020) 141–156.

[4] U. Volta, A. Fabbri, C. Parisi, M. Piscaglia, G. Caio, F. Tovoli, et al., Old and new serological tests for celiac disease screening, Expert. Rev. Gastroenterol. Hepatol. 4 (1) (2010) 31–35.

[5] K. Grossmann, N. Röber, R. Hiemann, S. Rödiger, P. Schierack, D. Reinhold, et al. Simultaneous detection of celiac disease-specific IgA antibodies and total IgA. Auto Immun Highlights 7 (1) (2016) 2.

[6] N. Wang, L. Truedsson, K. Elvin, B.A. Andersson, J. Rönnelid, L. Mincheva-Nilsson, et al., Serological assessment for celiac disease in IgA deficient adults, PloS. One. 9 (4) (2014) e93180.

[7] H.A. Penny, S.A. Raju, DS. Sanders, Progress in the serology-based diagnosis and management of adult celiac disease, Expert. Rev. Gastroenterol. Hepatol. 14 (3) (2020) 147–154.

[8] K.J. Werkstetter, I.R. Korponay-Szabó, A. Popp, V. Villanacci, M. Salemme, G. Heilig, et al., Accuracy in Diagnosis of Celiac Disease Without Biopsies in Clinical Practice, Gastroenterology. 153 (4) (2017) 924–935.

[9] R.S. Choung, S. Khaleghi Rostamkolaei, J.M. Ju, E.V. Marietta, C.T. Van Dyke, J.J. Rajasekaran, et al., Synthetic Neoepitopes of the Transglutaminase-Deamidated Gliadin Complex as Biomarkers for Diagnosing and Monitoring Celiac Disease, Gastroenterology. 156 (3) (2019) 582–591 e1.

[10] Y. van de Wal, Y.M. Kooy, P.A. van Veelen, S.A. Peña, L.M. Mearin, O. Molberg, et al., Small intestinal T cells of celiac disease patients recognize a natural pepsin fragment of gliadin, Proc. Nat. Acad. Sci. USA. 95 (17) (1998) 10050–10054.

[11] J.D. Altman, P.A. Moss, P.J. Goulder, D.H. Barouch, M.G. McHeyzer-Williams, J.I. Bell, et al., Phenotypic analysis of antigen-specific T lymphocytes. Science. 1996. 274: 94-96, J. Immunol. (Baltimore, Md: 1950). 187 (1) (2011) 7–9.

[12] H. Quarsten, S.N. McAdam, T. Jensen, H. Arentz-Hansen, Ø. Molberg, K.E. Lundin, et al., Staining of celiac disease-relevant T cells by peptide-DQ2 multimers, J. Immunol. (Baltimore, Md: 1950). 167 (9) (2001) 4861–4868.

[13] V.K. Sarna, K.E.A. Lundin, L. Mørkrid, S.W. Qiao, L.M. Sollid, A. Christophersen, HLA-DQ-Gluten Tetramer Blood Test Accurately Identifies Patients With and Without Celiac Disease in Absence of Gluten Consumption, Gastroenterology. 154 (4) (2018) 886–896 e6.

[14] M.S. Lau, P.D. Mooney, W.L. White, M.A. Rees, S.H. Wong, M. Hadjivassiliou, et al., Office-Based Point of Care Testing (IgA/IgG-Deamidated Gliadin Peptide) for Celiac Disease, Am. J. Gastroenterol. 113 (8) (2018) 1238–1246.

[15] P.D. Mooney, M. Kurien, K.E. Evans, I. Chalkiadakis, M.F. Hale, M.Z. Kannan, et al., Point-of-care testing for celiac disease has a low sensitivity in endoscopy, Gastrointestinal. Endoscopy. 80 (3) (2014) 456–462.

[16] L. Grode, T. Møller Jensen, T. Parkner, I.E. Agerholm, P. Humaidan, B. Hammer Bech, et al., Diagnostic Accuracy of a Point-of-Care Test for Celiac Disease Antibody Screening among Infertile Patients, Inflam. Intestinal. Dis. 4 (3) (2019) 123–130.

[17] M. Erlichster, J.A. Tye-Din, M.D. Varney, E. Skafidas, KP. Rapid, Loop-Mediated Isothermal Amplification Detection of Celiac Disease Risk Alleles, J. Mol. Diag. JMD. 20 (3) (2018) 307–315.

[18] J. Rusanen, A. Toivonen, J. Hepojoki, S. Hepojoki, P. Arikoski, M. Heikkinen, et al., LFRET, a novel rapid assay for anti-tissue transglutaminase antibody detection, PloS. One. 14 (11) (2019) e0225851.

[19] G. Lac, Saliva assays in clinical and research biology, Pathologie-biologie. 49 (8) (2001) 660–667.

[20] M. Bonamico, R. Nenna, M. Montuori, R.P. Luparia, A. Turchetti, M. Mennini, et al., First salivary screening of celiac disease by detection of anti-transglutaminase autoantibody radioimmunoassay in 5000 Italian primary schoolchildren, J. Pediatric. Gastroenterol. Nutr. 52 (1) (2011) 17–20.

[21] G. Adornetto, L. Fabiani, G. Volpe, A. De Stefano, S. Martini, R. Nenna, et al., An electrochemical immunoassay for the screening of celiac disease in saliva samples, Anal. Bioanal. Chem. 407 (23) (2015) 7189–7196.

[22] M. Di Tola, M. Marino, R. Casale, V. Di Battista, R. Borghini, A. Picarelli, Extension of the celiac intestinal antibody (CIA) pattern through eight antibody assessments in fecal supernatants from patients with celiac disease, Immunobiology. 221 (1) (2016) 63–69.

[23] A.H. Jalal, F. Alam, S. Roychoudhury, Y. Umasankar Prospects and challenges of volatile organic compound sensors in human healthcare. ACS Sens. 3 (7) (Jul 27, 2018) 1246–1263.

[24] R.P. Arasaradnam, E. Westenbrink, M.J. McFarlane, R. Harbord, S. Chambers, N. O'Connell, et al., Differentiating coeliac disease from irritable bowel syndrome by urinary volatile organic compound analysis–a pilot study, PloS. One. 9 (10) (2014) e107312.

[25] A. Baranska, E. Tigchelaar, A. Smolinska, J.W. Dallinga, E.J. Moonen, J.A. Dekens, et al., Profile of volatile organic compounds in exhaled breath changes as a result of gluten-free diet, J. Breath. Res. 7 (3) (2013) 037104.

[26] G.A. Mills, V. Walker, Headspace solid-phase microextraction procedures for gas chromatographic analysis of biological fluids and materials, J. Chromatogr. A. 902 (1) (2000) 267–287.

[27] S. Husby, S. Koletzko, I.R. Korponay-Szabo, M.L. Mearin, A. Phillips, R. Shamir, et al., European society for pediatric gastroenterology, hepatology, and nutrition guidelines for the diagnosis of coeliac disease, J. Pediatric. Gastroenterol. Nutr. 54 (1) (2012) 136–160.

[28] V. Fuchs, K. Kurppa Serology-based criteria for adult coeliac disease have excellent accuracy across the range of pre-test probabilities. Aliment. Pharmacol. Ther. 49 (3) (2019) 277–84.

[29] R. Tortora, N. Imperatore, P. Capone, G.D. De Palma, G. De Stefano, N. Gerbino, et al., The presence of anti-endomysial antibodies and the level of anti-tissue transglutaminases can be used to diagnose adult coeliac disease without duodenal biopsy, Aliment. Pharmacol. Ther. 40 (10) (2014) 1223–1229.

[30] L.J. Virta, MM. Saarinen Declining trend in the incidence of biopsy-verified coeliac disease in the adult population of Finland, Aliment. Pharmacol. Ther. 2005-2014. 46 (11-12) (2017) 1085–93.

[31] S. Picascia, R. Mandile, R. Auricchio, R. Troncone, C. Gianfrani, Gliadin-specific t-cells mobilized in the peripheral blood of coeliac patients by short oral gluten challenge: clinical applications, Nutrients. 7 (12) (2015) 10020–10031.

[32] A. Camarca, G. Radano, R. Di Mase, G. Terrone, F. Maurano, S. Auricchio, et al., Short wheat challenge is a reproducible in-vivo assay to detect immune response to gluten, Clin. Experiment. Immunol. 169 (2) (2012) 129–136.

[33] M. Brottveit, A.C. Beitnes, S. Tollefsen, J.E. Bratlie, F.L. Jahnsen, F.E. Johansen, et al., Mucosal cytokine response after short-term gluten challenge in celiac disease and non-celiac gluten sensitivity, Am. J. Gastroenterol. 108 (5) (2013) 842–850.

[34] H.A. Penny, S.A. Raju, M.S. Lau, L.J. Marks, E.M. Baggus, J.C. Bai, et al., Accuracy of a no-biopsy approach for the diagnosis of coeliac disease across different adult cohorts, Gut. (2020).

[35] A. Schiepatti, D.S. Sanders, F. Biagi, Seronegative coeliac disease: clearing the diagnostic dilemma, Curr. Opin. Gastroenterol. 34 (3) (2018) 154–158.

[36] U. Volta, G. Caio, E. Boschetti, F. Giancola, K.J. Rhoden, E. Ruggeri, et al., Seronegative celiac disease: Shedding light on an obscure clinical entity, Dig. Liver. Dis. Official J. Italian Soc. Gastroenterol. Italian Assoc. Study Liver. 48 (9) (2016) 1018–1022.

[37] A. Schiepatti, F. Biagi, G. Fraternale, C. Vattiato, D. Balduzzi, S. Agazzi, et al., Short article: Mortality and differential diagnoses of villous atrophy without coeliac antibodies, Eur. J. Gastroenterol. Hepatol. 29 (5) (2017) 572–576.

[38] M. DeGaetani, C.A. Tennyson, B. Lebwohl, S.K. Lewis, H. Abu Daya, C. Arguelles-Grande, et al., Villous atrophy and negative celiac serology: a diagnostic and therapeutic dilemma, Am. J. Gastroenterol. 108 (5) (2013) 647–653.

[39] T.T. Salmi, P. Collin, I.R. Korponay-Szabó, K. Laurila, J. Partanen, H. Huhtala, et al., Endomysial antibody-negative coeliac disease: clinical characteristics and intestinal autoantibody deposits, Gut. 55 (12) (2006) 1746–1753.

[40] I. Aziz, M.F. Peerally, J.H. Barnes, V. Kandasamy, J.C. Whiteley, D. Partridge, et al., The clinical and phenotypical assessment of seronegative villous atrophy; a prospective UK centre experience evaluating 200 adult cases over a 15-year period (2000-2015), Gut. 66 (9) (2017) 1563–1572.

[41] R. Marzari, D. Sblattero, F. Florian, E. Tongiorgi, T. Not, A. Tommasini, et al., Molecular dissection of the tissue transglutaminase autoantibody response in celiac disease, J. Immunol. (Baltimore, Md: 1950). 166 (6) (2001) 4170–4176.

[42] M. Borrelli, M. Maglio Intestinal anti-transglutaminase 2 immunoglobulin A deposits in children at risk for coeliac disease (CD): data from the PreventCD study. Clin. Exp. Immunol. 191(3) (2018) 311–7.

[43] I.R. Korponay-Szabó, T. Halttunen, Z. Szalai, K. Laurila, R. Király, J.B. Kovács, et al., In vivo targeting of intestinal and extraintestinal transglutaminase 2 by coeliac autoantibodies, Gut. 53 (5) (2004) 641–648.

[44] L. De Leo, M. Bramuzzo, F. Ziberna, V. Villanacci, S. Martelossi, G.D. Leo, et al., Diagnostic accuracy and applicability of intestinal auto-antibodies in the wide clinical spectrum of coeliac disease, EBioMedicine. 51 (2020) 102567.

[45] A. Carroccio, G. Iacono, D. D'Amico, F. Cavataio, S. Teresi, C. Caruso, et al., Production of anti-endomysial antibodies in cultured duodenal mucosa: usefulness in coeliac disease diagnosis, Scandinavian. J. Gastroenterol. 37 (1) (2002) 32–38.

[46] A. Picarelli, M. Di Tola, L. Sabbatella, M.C. Anania, T. Di Cello, R. Greco, et al., 31-43 amino acid sequence of the alpha–gliadin induces anti-endomysial antibody production during in vitro challenge, Scandinavian. J. Gastroenterol. 34 (11) (1999) 1099–1102.

[47] J. Valle, J.M.T. Morgado, J. Ruiz-Martín, A. Guardiola, M. Lopes-Nogueras, A. García-Vela, et al., Flow cytometry of duodenal intraepithelial lymphocytes improves diagnosis of celiac disease in difficult cases, United Eur. Gastroenterol J. 5 (6) (2017) 819–826.

[48] G. Cammarota, P. Fedeli, A. Gasbarrini, Emerging technologies in upper gastrointestinal endoscopy and celiac disease, Nat. Clin. Pract. Gastroenterol. Hepatol. 6 (1) (2009) 47–56.

[49] P. Iovino, A. Pascariello, I. Russo, G. Galloro, L. Pellegrini, C. Ciacci, Difficult diagnosis of celiac disease: diagnostic accuracy and utility of chromo-zoom endoscopy, Gastrointestinal endoscopy. 77 (2) (2013) 233–240.

CHAPTER 7

Non-dietary therapies for celiac disease

Shakira Yoosuf, Amelie Therrien, Daniel A. Leffler
Division of Gastroenterology, Department of Medicine, Beth Israel Deaconess Medical Center, Boston, MA, United States

7.1 Introduction

The gluten-free diet (GFD) remains the mainstay of the management of celiac disease (CeD) and the need for alternative therapies remains unmet. Research has shown that almost two-thirds of CeD patients are interested in novel medical therapies, and that this interest does not vary with the duration of their time on a GFD [1]. Some of the limitations of the gluten-free foods include their poor availability in some markets and social environments, higher costs, as well as their altered palatability and nutritive profiles [2]. Further, even in patients that adhere to the GFD, inadvertent gluten exposures are frequent, translating into persistent symptoms and decreased healing of intestinal villi. In a few others, non-response may be seen due to refractory CeD [3].

For a variety of reasons, we currently find ourselves in interesting times with regard to research in CeD There has been a substantial growth in our knowledge of CeD pathogenesis over the past two decades, with molecular medicine and immunological techniques expediting the discovery of several potential disease markers as well as therapeutic targets. Some of the molecular pathways that drive CeD have been targeted in drug trials, however success in clinical trials has been mixed. Similarly, there is much scope for research on improving outcome measures which would help in disease activity monitoring and quantifying drug efficacy in clinical trials. This chapter aims to provide a state-of-the-art summary of the latest non-dietary therapies of CeD, followed by various endpoints that may be employed to study CeD.

7.2 Aims of research and therapy

In a study that surveyed CeD patients, 87% of participants showed interest in potential new adjunct drugs that would allow them to have a GFD without fear of cross-contamination, whereas 65% reported wanting drugs that would replace the GFD altogether [1]. Most drugs developed so far have been tested for application as an adjunct to the GFD; however it is also encouraging to see that some of the new immune-altering drugs in the pipeline show potential to replace or substantially reduce the need for a GFD.

Coeliac Disease and Gluten-Related Disorders.
DOI: https://doi.org/10.1016/B978-0-12-821571-5.00011-8

Indications where these drugs may be used either as an adjunct or replacement for GFD are discussed in the forthcoming section.

7.2.1 Reduction of symptoms among Celiac Disease patients on a GFD

A recent FDA-requested listening session involving CeD patients and their caregivers reflected on how impactful, symptoms related to CeD were on daily life. These symptoms included both gastrointestinal (GI) and non-GI symptoms after gluten ingestion. Patients generally felt that maintaining a strict GFD was the best method for preventing CeD symptoms. Nonetheless, they were open to treatments for accidental gluten exposures, although there was hesitancy. They were generally not open to treatments that were required to be taken regularly that did not promote intestinal healing [4].

7.2.1.1 Acute symptoms following gluten exposure

A significant proportion of CeD patients respond to a GFD, but have occasional gluten exposures associated with acute reactions. Such patients may develop upon gluten exposure, symptoms that are different in quality and intensity than their initial symptoms at the time of the diagnosis [5]. In a study of 105 CeD patients adherent to a GFD, a suspected symptomatic reaction to gluten was reported by 66%, with the median time to symptom onset being around 1h and median symptom duration being around 24h. Common symptoms included abdominal pain (80%), diarrhea (52%), fatigue (33%), headache (30%) and irritability (29%) [6]. For some individuals, symptoms are severe, debilitating, and cause significant anxiety and hyper-vigilance regarding risk of cross-contamination, with potential effects on their quality of life. Factors associated with severe acute symptoms include strict adherence to the GFD, coexisting IBS, anxiety and depression [5].

While the pathogenesis of acute reactions is incompletely understood at present, recent oral gluten challenge and gliadin-based vaccine studies have demonstrated a definite temporal association between rise in serum interleukin (IL)-2 levels and onset of symptoms, especially nausea and vomiting, suggesting that these reactions result from a cytokine release syndrome. In an oral gluten challenge study, Goel at al. found elevations in IL-2, chemokine (C-C motif) ligand-20 (CCL20), IL-6, chemokine (C-X-C motif) ligand (CXCL)-9, CXCL-8, interferon (IFN)-γ, IL-10, IL-22, IL-17A, tumor necrosis factor (TNF)-α, CCL-2 and amphiregulin. IL-2 and IL-17A were the earliest to rise (2h), with peak levels of most cytokines observed at 4h. IL-2 increased most (median: 57-fold), followed by CCL20 (median: 10-fold) [7]. In another oral gluten challenge study by Cartee et al., IL-2 and IL-8 were found to precede the appearance of symptoms and correlated with symptom severity as well, with IL-2 being the earliest and most prominent marker among all cytokines. This may be consistent with gliadin-specific CD4+ cell activation that is known to occur in response to the gluten challenge, but additional involvement of the innate immune system cannot be ruled out. With increasing clarity on the role of these cytokines and other mediators, specific therapies intercepting them would prevent acute symptoms during accidental gluten exposures [8]. An interesting experiment yet to be performed, could be to pre-treat these individuals with an anti-IL-2 agent such as daclizumab or basiliximab to evaluate if IL-2 is really central in the development of acute symptoms.

7.2.1.2 Non-Responsive Celiac Disease

Persistence of symptoms of CeD despite 6-12 months of a strict GFD, also known as Non-Responsive Celiac Disease (NRCD) occurs in around 12-19% of patients with CeD, with the most common cause being inadvertent gluten exposure [3,9,10]. Notably, the sensitivity to gluten exposure is highly variable from one individual to another, with some experiencing symptoms even at amounts as low as 10 mg of gluten [11]. Refractory Celiac Disease (RCD) is yet another etiology of NRCD; most RCD patients are type 1, wherein neither the T-cell immunophenotype nor TCR clonality is affected, in contrast to RCD type 2. Interestingly, a study showed that some patients initially considered to have RCD, could be reclassified as having NRCD resulting from trace gluten exposure, based on their improvement after undergoing a gluten contamination elimination diet [12]. It is thus possible that most individuals considered as RCD type 1, are in fact, ultrasensitive to low-grade gluten contamination.

This ubiquity of gluten exposure is further proven by recent studies that have used monoclonal antibodies to identify gluten immunogenic peptides (GIP) either in the food, urine or stools. One recent study showed that CeD patients with incomplete mucosal healing despite a GFD had detectable GIP in their urine, with their levels correlating well with the level of gluten detected in the diet [13]. Another study showed detectable stool GIP in 30% of CeD subjects on a GFD, without any correlation with either serology levels or dietary transgression per the food questionnaires. Among patients with symptoms, the rate of positive stool GIP went up to 67% [14]. A subsequent study in a South American population did however show correlation of the presence of stool and urine GIP with dietary transgressions [15]. Furthermore, the rate of GIP positivity in either the urine or stools among treated CeD individuals rises significantly with repetitive testing, being 67% over 10 days [16] and 88.7% over 4 weeks [17]. An add-on therapy that eliminates immune reactions related to ingested traces (<1g) of gluten would therefore be highly useful to mitigate persistent symptoms in NRCD patients as well as risk of long term complications, including possibly RCD [18].

7.2.2 Replacing the GFD

7.2.2.1 Burdens of a GFD

Following a strict GFD is strenuous psychologically, socially and economically, for patients and caregivers [19]. Although the costs of the gluten-free foods are deescalating, they still remain 20-300% more expensive than their wheat-based counterparts [20], which further adds to the complexity of decision-making in planning meals. Some of the other practical issues include difficulty with label-reading, limited availability of gluten-free options in social situations like traveling/ restaurants/ gatherings etc., and stigmas attached to following a special diet such as the GFD [21–23]. Also, although the treatment of CeD improves overall well-being, an association has been shown between reaching complete mucosal recovery, and coexistence of anxiety disorders, highlighting hypervigilance among these individuals [24]. Such hypervigilant individuals, while doing well in terms of knowledge and disease state, may still feel that their CeD is inadequately treated [25]. Preoccupation with the GFD and social restrictions contribute to a decreased quality of life [26].

In addition to economic and psychosocial costs, nutritional burdens can stem from the higher caloric indices and fat content, as well as low fiber content in the GFD, since currently many of the gluten-free options available in the market are highly processed foods [27, 28]. This has resulted in higher rates of obesity and fatty liver diseases [29, 30], as well as constipation being more frequent following CeD diagnosis [31]. Thus, a study that evaluated the treatment burden of patients suffering from various chronic diseases, found CeD patients to perceive treatment burdens as high as those of chronic kidney disease patients on hemodialysis [32].

7.2.2.2 Evidences of outcomes based on mucosal healing

A sizable proportion of CeD patients on a GFD have ongoing exposures to gluten. Inadequate knowledge about the GFD has been noted in a significant proportion of subjects [6], suggesting that gluten exposure may be more frequent than what is reported, especially in the absence of symptoms. This is further substantiated by recent gluten detection studies that used the highly sensitive and specific G12 monoclonal antibody to show that gluten exposure is detectable in up to 20% of asymptomatic patients adherent to a GFD, despite normal tTG [33]. Furthermore, a double blind placebo-controlled gluten challenge study by Cartee et al. suggests that a significant proportion of CeD patients may not be able to accurately report gluten exposures [8].

While many non-dietary therapies being developed currently are likely to find application as adjuncts to a GFD to treat symptomatic patients, the occurrence of asymptomatic gluten exposures, with their potential attendant risk of delayed mucosal healing, necessitates therapies that can replace the GFD altogether. A recent study performed in Canada among treated CeD patients adherent to a GFD for almost two years, revealed gluten either in the food, urine or stools of 66% of the participants over a period of 10 days. The experiment showed not only that gluten contamination is pervasive, but it is also consequential since the presence of detected gluten was associated with residual villous atrophy two years after CeD diagnosis in these patients [34]. While we currently have incomplete knowledge of the natural history of CeD, we may wonder whether long term trace gluten ingestion may cumulatively add a risk of refractory CeD or CeD-associated malignancy. In light of the above findings, novel therapies are potentially useful in asymptomatic patients as well, especially among those that have detectable levels of gluten in their food or urine/stool.

Finally, complete mucosal recovery is an established goal for CeD therapies, but considering the patchiness and extensiveness of the disease [35], mucosal healing cannot be reliably assessed with current standard testing. Although, emerging video capsule endoscopy techniques may provide important data in addition to or in place of duodenal biopsy [3]. As such, normal duodenal biopsies may be interpreted as a *lack of detectable histologic disease*. Moreover, a review of the biopsies from the CeliAction study showed that normal Vh/Cd ratio of ≥3.0 may be only rarely attained despite at least one year on a GFD (only 8% had normal Vh/Cd ratio), but subtle villous atrophy may be missed by conventional pathology review [36]. The time to reach mucosal healing is highly variable from one individual to another and may depend on the extensiveness of the initial disease, age of the subject and amount of residual gluten in the diet. The Reset CeD Study Group recently reported that among the individuals with well controlled CeD included in one of the Nexvax2 trials, only 6% were either Marsh 0 or 1 and all the individuals with serial biopsies from the bulb, first, second and third part of the duodenum had at least one biopsy with Marsh 3a histology [37, 38]. Nevertheless, mucosal healing has been associated with improvement of bone density, as well as a lower risk of lymphoma [39]. Irrespective of the symptoms or attitudes of patients toward the GFD, therapy that would accelerate or allow complete mucosal recovery is a worthy goal.

7.3 Therapeutic targets and drugs

This section details different drugs being developed based on their mechanism of targeting specific parts of the pathogenetic pathway of CeD, which involves an initial phase where gluten comes in contact with innate and adaptive immune systems, followed by an immunological response that mediates the enteropathy and many other manifestations of CeD.

7.3.1 Preventing contact between gluten and the immune system
7.3.1.1 Therapeutic agents
7.3.1.1.1 Glutenases
Glutenases are bacterial-/fungal- derived enzymes that can digest the immunogenic peptides of gluten and render them non-toxic. Most of these glutenases are enzymes targeting glutamine and proline, which are also the principal amino acids found in immunogenic motifs of gluten. These glutenases may find application as add-on therapy for CeD patients already on a GFD.

7.3.1.1.1.1 Latiglutenase The glutamine-targeting glutenase EP-B2 (endoprotease B, isoform-2) is derived from barley seed endosperm, and has a Cys-His-Asn catalytic

triad in its active site. Like other cysteine proteases, it is optimally active at low pH, resistant to pepsin but lyzed at physiological concentrations of trypsin, and has good specificity for the sequence QXP, which is abundant in the 33-mer as well as other immunotoxic gluten sequences. These factors make it a good fit for therapy in CeD as a gastric active enzyme [40]. One of the concerns with any gastric-based glutenase is the possibility of immunotoxic residue reaching the duodenum if gastric emptying occurs before complete digestion. It is therefore reasonable to combine this enzyme with other glutenases which are gastric- stable but commence action when the food chyme reaches the more alkaline duodenum, preempting any immunotoxicity in the intestinal mucosa. Proline-specific endoproteases (PEP) from the microbe *Sphingomonas capsulata* (SC-PEP) is one such enzyme that is active in acidic pH, and is largely unaltered in the presence of pepsin [41]. It is a serine protease with a larger β-propeller domain and a smaller, N-terminal catalytic domain that acts at the carboxy end of the gluten protein [42, 43]. Combination of EP-B2 with SC-PEP is complementary, with EP-B2 efficiently digesting the 33-mer peptides into smaller, not necessarily non-toxic proline containing fragments, and PEP then digesting the proline-glutamine links in these smaller oligopeptides.

Several clinical trials have been conducted to assess the effectiveness of this enzyme mixture in making dietary gluten safe for patients with CeD. The proprietary enzyme cocktail ALV003, (aka IMGX003 and latiglutenase) uses a 1:1 combination of EP-B2 (aka ALV001) plus PEP (aka ALV002). A randomized control trial (RCT) studied CeD patients on GFD, who received a 3-day challenge with 16g of oral gluten daily, alongwith either ALV003 or placebo (NCT00859391). The ALV003 group showed significantly lower immunological activation, as seen on peripheral T cell IFN-γ responses to gliadin [44]. A phase 2a RCT (NCT00959114) that included 41 CeD patients on GFD, found that after a 6-week gluten challenge with 2g bread crumbs daily, Vh:Cd ratio deteriorated significantly less in patients treated with ALV003 than with placebo. However, intraepithelial CD3+ lymphocytes remained unchanged in the ALV003-treated patients compared to placebo-treated patients. This study was significant for having used bread crumbs to simulate real life gluten intake as opposed to pre-digested gluten [45]. Although the aforementioned clinical trials were promising, the results of the most recent phase 2b trial called CeliAction (NCT01917630) were disappointing. This was the largest trial to dateinCeD and included 490 CeD patients who were symptomatic despite GFD for a year (NRCD), in a dose-ranging, placebo-controlled double-blinded study. In a modified intention-to-treat analysis, no significant differences were observed on histological, serological or symptomatic endpoints [46]. In a post-hoc analysis of data, it was found that there was significant reduction in the abdominal symptoms in those patients who were seropositive for celiac antibody. The authors concluded that seronegative patients did not experience any symptomatic improvement, possibly because their symptoms may be attributable to non-CeD causes [47].

A phase 2 trial, named CeliacShield is underway to further test the effect of latiglutenase on histopathological parameters of 80 well-treated CeD patients, who have been adherent to a GFD for at least a year (NCT03585478). Solutions for Celiac is a multicenter phase 2b study (NCT04243551) that will test efficacy of latiglutenase in NRCD patients [48].

7.3.1.1.1.2 Kuma Kuma030 is a novel, engineered glutenase whose structure is based on a naturally occurring enzyme from the acidophilic microbe *Alicyclobacillus sendaiensis*, called kumamolisin-As (KumaWT). It is a serine endoprotease, with optimal activity over the pH range of 2–4/37°C and therefore adaptable for use in the gastric environment [49]. The initially designed enzyme called KumaMax or Kuma010, had 116 times higher proteolytic activity, and 877 times higher specificity for target gliadin oligopeptides. Kuma030 is a more efficacious form of the enzyme, and in comparison to SC PEP-EP B2, Kuma030 seemed to be more efficient at much lesser concentrations, achieving >99.9% gliadin degradation thereby reducing the gluten content to 3ppm as detected by ELISA (Enzyme Linked Immunosorbent Assay), well-below the 20ppm threshold for "gluten-free" labeling [50]. When gluten-sensitive T cells isolated from patients with CeD were incubated with gliadin pre-treated with Kuma030, a dose dependent reduction in IFN-γ production and T cell proliferation was observed [51].

A phase 1 trial involving Kuma 062 (TAK-062), an improved version of this enzyme was recently completed on healthy adult volunteers (NCT03701555). Doses of 300, 600 and 900 mg of TAK-062 were taken 10 minutes before a gluten challenge containing 1 to 6 g of gluten. Interestingly, the gluten challenge consisted of a smoothie that included lactose-free ice-cream, egg-whites, orange juice, whole wheat, lime juice, vanilla extract. TAK-062 consistently achieved degradation of $\geq$95.5% (95% confidence interval: 92.6–98.3%) of immunogenic gliadin fractions in the stomach, independent of the use of PPI therapy or the formulation of the drug (capsule vs. liquid) [52]. Phase 2 studies evaluating the ability of TAK-062 to improve symptoms and duodenal histology in CeD patients are planned to begin shortly.

7.3.1.1.2 Binding agents

Agents that bind with high specificity to dietary gluten in the gastrointestinal lumen prior to absorption could serve as potential add-on therapies to the GFD. IgY is a novel molecule that can bind and eliminate toxic gluten immunogenic peptides; it is an antibody harvested from the yolk of eggs laid by hens that have been super immunized against gliadin. Gujral et al. showed that in gut-like in vitro conditions, these antibodies can neutralize gliadin fraction. Using 50% mannitol, they also successfully created an acid-resistant capsule form of these IgY antibodies, called AGY. In vitro testing showed that gliadin absorption decreased substantially across the intestinal epithelial cells with the addition of AGY; further, it was found to be more effective in the presence of food [53].

AGY was subsequently tested in a phase 1 trial; capsules were administered along with meals for 6 weeks, and 10 patients completed the study, with no safety concerns identified. Most patients had lower scores on the Celiac Symptom Index, improved quality of life, lowered antibodies, and lowered lactulose:mannitol ratio when taking AGY as compared to the baseline run-in period [54]. It has potentially minimal toxicity, when administered orally to humans as it does not get absorbed into the systemic circulation to cause systemic immune activation. However, it would have to be avoided in individuals with egg allergy.

7.3.1.1.3 Probiotics

Gut microbiota may modulate the digestion of dietary gluten, influence tolerance to dietary antigens and intestinal permeability. They also promote maturation of the intestinal mucosa and express pro- or anti- inflammatory peptides that regulate the activity of innate and adaptive immune systems [55, 56]. In an elegant study from Canada, duodenal biopsies from patients with active CeD were found to show increased proteolytic activity against gluten substrates, that correlates with increased Proteobacteria abundance, including Pseudomonas. In mice expressing CeD risk genes, *P.aeruginosa* elastase was found to synergize with gluten to induce more severe inflammation, with moderate villus blunting, showing that proteases expressed by opportunistic pathogens modulate development of sensitivities to dietary antigens [57]. There is also evidence to suggest the association of specific gut microbiota with certain genotypes involved in the development of CeD, e.g., a study found that among infants with a first-degree relative with CeD, those carrying a high risk haplotype HLA-DQ2, had higher proportions of Firmicutes and Proteobacteria and lower proportions of *Actinobacteria*, when compared to those with a low genetic risk (non-HLA-DQ2/8). However, maternal HLA status was not reported and modulation in the infant's microbiota may also be the reflection of the mother's [58]. Similarly, another study called Celiac Disease Genomic, Environmental, Microbiome, and Metabolomic Study (CDGEMM) is underway in the USA, Italy and Spain, to identify the role of gut microbiome and the metabolome in the risk of developing CeD in a birth cohort of children [59].

In this context, probiotics have been studied as complementary therapies for CeD in animal models, although there is paucity of literature from human subjects. Probiotics exert an influence on survival of intestinal pathogens and commensals, as well as on immunity. In a study, administration of the strain *Lactobacillus casei* ATCC 9595 to HLA-DQ8-expressing transgenic mouse models that were immunized and challenged with intra-gastric gliadin, was associated with decreased enteropathy-like response mediated by CD4+T cells, weight loss and was associated with complete recovery of villous blunting [60]. Lindfors et al. also demonstrated the inhibitory influence of *Bifidobacterium lactis* on the toxic effects of gluten-derived peptides in intestinal cell culture (Caco-2), including decreased gliadin-induced intestinal permeability [61]. *Saccharomyces boulardii KK1* strain has also been shown to hydrolyze gliadin toxic peptides, and administration of the strain was associated with decreased mucosal damage and proinflammatory cytokine profile in mice models [62].

A human trial that involved administration of *Lactobacillus casei* along with *Lactobacillus plantarum, Bifidobacterium animalis* subsp. Lacti, *B. breve* Bbr8 LMG P-17501 and *B. breve* B110 LMG P-17500 for 6 weeks caused a significant decrease in IBS-SSS (Irritable Bowel Syndrome- Severity Scoring System) and GSRS (Gastrointestinal Symptom Rating Scale) scores compared to placebo among CeD patients with persistent IBS-type symptoms [63]. In a human trial conducted with *Bifidobacterium infantis* in the formulation Bifidobacterium Natren Life Start (NLS); there was an improvement of symptoms, as well as circulating MIP-1β, a potent chemoattractant for monocytes, granulocytes and lymphocytes, although the number of participants in the study was small [64]. In a later, larger trial with an improved strain of the same bacterium, called Bifidobacterium Natren Life Start Super Strain (NLS-SS), 6 weeks of administration of the probiotic resulted in decreased mucosal expression of innate immune markers, such as α-defensin-5, macrophage count and Paneth cell count in response to an oral gluten challenge [65]. Another formulation VSL#3, now known as De Simone formulation, is a proprietary blend of 450 billion viable lyophilized bacteria from 9 strains; it was tested in a RCT with 42 NRCD participants; a 12-week course did not demonstrate any change in the gastrointestinal microbial counts, and hence more studies are required to understand the value of the probiotics in CeD [66].

Some yet to be completed trials on probiotics in CeD subjects include a proof of concept clinical trial of the safety and activity of *Bifidobacterium longum* NCC 2705 (NCT03775499), a double-blind RCT on the probiotic "Pentabiocel" which is a mixtureof five different bacterial strains (NCT03857360), and a trial on the effect of a 4-month administration of VIVOMIXX®, a probiotic containing 8 different strains of bacteria on nutritional parameters and gut metabolomes (NCT04160767).

7.3.1.1.4 Tight junction modulators

Tight junctions are apical, intercellular junctions that regulate the passage of molecules via the paracellular transport pathway, and prevent passage of pathogens/ antigens in normal conditions. Increased paracellular permeability is an early event in the pathogenesis of CeD [67], and allows gluten immunogenic peptides to pass through. Human protein zonulin, which is similar to the zonula occludens toxin (ZOT) expressed by Vibrio cholera, has been postulated to be a regulator of epithelial permeability and is highly expressed in the mucosa and blood of patients with celiac disease. Gliadin binds to the chemokine receptor CXCR3 on intestinal epithelium releasing zonulin and activates the MyD88 dependent pathway to increase permeability [68]. Larazotide acetate (aka

AT-1001, INN-202), an octapeptide drug reported to antagonize zonulin has been found to promote assembly of actin and E- cadherin around tight junctions of Madin-Darby canine kidney (MDCK) type II cells, thus promoting cell junction integrity [69]. It was also studied in a non-transformed porcine jejunal cell line modelof leaky intestinal barrier, as well as among Yorkshire-cross pigs [70, 71]. In this model, Larazotide appeared to protect the tight junction barrier during anoxia/reoxygenation injury by inhibiting the activation of myosin light chain kinase, which is a regulator of tight junction opening and closing [72]. Larazotide's effect was studied in transgenic mice models with pre-sensitization to gluten. Larazotide countered the intestinal barrier disintegration, decreased the macrophage count in the lamina propria, and kept the trans-membrane conductance intact, compared to mice subjected to gluten challenge without larazotide [73].

Phase 2 trials (NCT00362856) of larazotide acetate using oral challenges with 2.5g of gluten showed that 12mg doses of larazotide decreased the cytokine response and intestinal permeability, as assessed by urinary LAMA (lactulose to mannitol) extraction fraction, although this difference was not statistically significant [74]. A phase 2a double blind RCT, tested 4 doses of the drug (0.25, 1, 4, and 8mg) on 86 patients undergoing a gluten challenge, with outcomes being urinary LAMA ratio (primary outcome measure), the Gastrointestinal Symptom Rating Scale (GSRS), Psychological General Well-Being Index as well as adverse event profile. There was no significant effect of larazotide on the primary efficacy endpoint, however, there was a lack of increase in the GSRS at two doses of 0.25 and 4mg, but not in the other dose groups of 1 and 8mg. Larazotide was generally well tolerated by patients, with significant adverse events being headache and urinary tract infections in more than 5% of the patients; however there were no dropouts on account of these events [75]. In Phase 2b trials, 1, 4, and 8mg doses of larazotide were tested, with no statistically significant difference found in LAMA levels. The 1mg but not 4mg or 8mg doses brought a reduction in gastrointestinal symptoms induced by the gluten challenge [76]. In a subsequent, larger phase 2 multicenter gluten challenge study that tested 0.5, 1, and 2mg doses of larazotide (NCT01396213) in 342 patients over 12 weeks, primary end point of reduction of symptoms was met in the 0.5mg dose but not in the higher doses. There was specifically a significant improvement in the abdominal symptoms domain of the CeD PRO and the weekly number of symptomatic days. There was also an improvement in the non-GI domain symptoms. The pattern that has therefore emerged in all the above studies is that only lower doses had an effect, implying an inverse-dose relationship. Peptide aggregation at higher doses of the drug, reducing activity in vivo may be one possible explanation for this observation [67]. Therefore, a multicenter randomized double-blind, placebo-controlled phase 3 trial is on-going and is expected to include 630 NRCD patients; enteric coated, lower doses of 0.25mg TID and 0.5mg TID for 16 weeks will be used (NCT03569007).

7.3.1.1.5 Pancreatic enzyme supplementation

CeD–associated mucosal damage and resultant loss of enterokinase activity may decrease pancreatic enzyme activation and function, leading to steatorrhea. This could occur even in the absence of a demonstrable decrease in pancreatic enzyme levels, assessed using standard tests like fecal elastase-1; this is considered to be one of the potential factors leading to NRCD symptoms [77]. Observational studies that support the use of Pancreatic Enzyme Supplements (PES) also include a previous uncontrolled open-label trial that demonstrated significant reduction of diarrhea in 18 out of 20 NRCD subjects, all having low fecal elastase-1 levels suggestive of exocrine pancreatic insufficiency (EPI) [78]. However, there is paucity of literature to support empirical use of PES in NRCD patients, especially among those who do not have low fecal elastase-1 levels. In fact, a small, cross-over RCT that assessed the efficacy of PES in NRCD patients failed to demonstrate any improvement in gastrointestinal symptoms (NCT02475369, unpublished).

7.3.1.1.6 Agents that have not reached human clinical trials

7.3.1.1.6.1 Polymeric binders May be used to eliminate ingested gluten traces, similar to those used in hyperkalemia and other diseases by sequestering toxic compounds in the gastrointestinal tract [79]. Poly (hydroxyethyl methacrylate-co-styrene sulfonate) or P(HEMA-co-SS), is a non-absorbable, co-polymer of hydroxyethylmethacrylate(HEMA), and sodium 4-styrene sulfonate (SS). Liang et al. demonstrated the mechanism of the sequestration of alpha gliadin by the experimental form of this polymer viz. polymer BL-7010, using spectroscopic and light scattering methods [79]. Further in vitro studies proved the prevention of toxic effect of gliadin on cell permeability in the presence of the polymer [80]. Subsequent mice studies also revealed selectivity in sequestration of gliadin and hordein as compared to other nutrients [81]. In order to test practical applicability of industrial grade preparation of the polymer, McCarville and colleagues studied the effectiveness of two batches of BL-7010 together, original polymerA and the polymer B which is an industrial preparation form of polymer; mice models showed that polymers bind avidly to the gliadin, with no interaction with vitamins, pepsin, and pancreatin and minimal interaction with albumin. Moreover, systemic absorption was negligible and repeated toxicity studies showed safety in the mice models. While polymer A was completely effective in abrogating villous damage, polymer B was slightly less effective [82]. These transgenic mice had depleted CD4+CD25+Foxp3 cells, which mediate immune tolerance; then at age 6 weeks onwards they were sensitized with pepsin trypsin digested gliadin and cholera toxin (CT); post-sensitization, mice were administered gluten plus either polymer A or polymer B. Results of a dose-ranging Phase 1 cross-over RCT (NCT01990885) on well-treated CeD patients, to determine human safety and pharmacokinetic profile are yet to be published.

7.3.1.1.6.2 Single chain fragment Variable Is a fusion protein that contains the Fab (antigen binding site of antibody fragment) sites of the variable and light chains of the antibody. This Single chain fragment variable (scFv) has been previously explored for cancer immunotherapy and is also being tested *in vitro* for its effectiveness in neutralizing gliadin, for use in CeD [83]. Hens immunized with gliadin yield mRNA with the sequence for scFv [84], which are then used to produce recombinant scFV in E. coli bacteria in larger amounts. To improve binding affinity, two scFv may be joined with a peptide linker to create a tandem scFv (tscFv). The tscFv was tested by Eggenreich et al.; scFv had the highest affinity for digested gliadin, followed by wheat, and spelt flour. There was no binding with rice and millet flours, indicating the specificity of binding [83]. Further testing in human subjects would require gastric resistant preparations of this product.

7.3.1.2 Limitations of this approach (will probably not replace a GFD)
7.3.1.2.1 Glutenases
Glutenases function only under specific thermo-chemical conditions, and may fail to completely eliminate all immunogenic epitopes; therefore they are at best employed as an adjunct to GFD to prevent any possible immune-activation by trace amounts of dietary gluten, especially in NRCD. Some of the factors that may compromise their function include pH, sensitivity to/ lysis by digestive enzymes (trypsin/pepsin), rate of gastric emptying according to the type of food intake, amount of gluten intake, specificity toward peptides/length of oligopeptide. Finally, a recent report described the occurrence of tTG-2 as well as anti-tTG2 antibodies in the stomach [85]. Whether there is a clinically significant gluten-mediated immune reaction in the stomach remains to be established.

7.3.1.2.2 Tight junction modulators
The zonulin inhibitor larazotide has yielded conflicting results so far. Its exact mechanisms of action remain unclear and prior studies lack histology follow-up. Moreover, there are other pathways than the paracellular pathway that may mediate gluten absorption, such as transcytosis, where gliadin peptides form immune complexes with sIgA that bind the CD71 receptor on the epithelial surface [86]. Interestingly, CD71 receptor is the transferrin receptor and thus, may be upregulated in patients with iron deficiency anemia complicating their CeD. This process therefore is more likely to occur in active CeD. In addition to antigen capture by M cells, other possible transduction mechanisms include binding of gliadin peptides with EGF receptor, GM1 ganglioside, and CXCR3 [86]. Moreover, stimulation of CXCR3 by gluten triggering the secretion of zonulin may not be the only mechanism by which tight junctions are dissembled [86]. Indeed, IFN-γ produced by activated lymphocytes may also alter the expression of tight junctions [87]. IFN-γ also increased transcytosis of gliadin peptides [86].

7.3.2 Prevention of immune response

Some drugs that are capable of modulating the immune response to gluten in the diet would be ideal as add on agents to a GFD for preventing the downstream inflammatory damage that may be triggered by gluten traces. However there are now also therapies with the potential to subvert the need for a GFD altogether. They are discussed in the following sections.

7.3.2.1 Therapeutic agents
7.3.2.1.1 tTG inhibitors

Inhibitors of human tissue transglutaminase-2 (tTG-2) would prevent conversion of gliadin to the more immunogenic, deamidated gliadin peptides, and could be potentially useful therapeutic agents. tTG-2 is a multi-functional enzyme that catalyzes the linkage of glutamine and lysine side-chains to modify proteins, in the presence of ionic calcium and thiol. The enzyme is known to be associated with pathogenesis of not only CeD, but also some cancers [88] and, Parkinson's [89], Alzheimer's [90], and Huntington's diseases [91]; hence tTG-2 inhibitors are being tested in many of the aforementioned conditions. Irreversible inhibitors that covalently bind to the cysteine in the active site of tTG-2 have been developed, using compounds that bear the moieties -diazo-5-oxonorleucine (DON) [92] and 3-halo-4,5-dihydroisoxazole [93]. A prototype dihydroisoxazole called 1b, showed good oral bioavailability, efficient tTG-2 inhibition in small intestinal tissue, and low toxicity in animal studies [93]. Recently, two other dihydroisoxazoles, ERW1041E and CK805 were shown to prevent the development of villous atrophy in a novel CeD mouse model, suggesting that tTG-2 may have a crucial role in the development of the enteropathy [94].

Similarly, tTG-2 inhibitors have been designed with gliadin peptide-like sequences, wherein glutamine residues are substituted with the DON moiety [95]. Gluten peptide analog ZDON has been shown to have a very high specificity for tTG-2; three potential tTG-2 inhibitors based on this structure, ZED1098, ZED1219, and ZED1227 (Zedira pharmaceuticals) have potential therapeutic application in CeD. They covalently bind to active site cysteine of tTG-2 using the Michael Acceptor warhead [96]. Among these, the ZED1227 has been studied in human trials. A phase 1a study on healthy male volunteers using a single dose of ZED1227 were followed by a phase 1b multidose study on 96 healthy volunteers using four ascending doses over a seven-day period, as per reports from Zedira pharmaceuticals [97]. This was followed by a multi-dose multicenter phase 2a trial (EUDRA CT No. 2017-002241-30), wherein participants are randomized to either ZED1227 or placebo in addition to a 6-week gluten challenge. The primary outcome was the Vh:Cd ratio measured morphometrically and the secondary endpoints included the CDSD and ICDSQ, attenuation of CD3+ IELs, Marsh stages, nutritional biomarkers as well as additional biomarkers of the disease. Results of the studies are yet to be published.

The ubiquity of tissue transglutaminases in the body where they participate in diverse pathways like coagulation factor synthesis, epithelial integrity, etc. renders it important for tTG-2 inhibitors to be highly specific for the gluten-binding site of the enzyme. Some of the concerns with the use of tTG-2 inhibitors include the potential cross-inhibition of other transglutaminase enzymes; however, mice studies suggest that it is safe in doses up to 1000mg/kg, and also decreases the activity of tTG-2 and inflammation of bowel mucosa [98].

Further, assome gluten peptides are potentially immunogenic even without deamidation by tTG-2, combining tTG-2 inhibitors with other pharmaceutical agents that eliminate immunogenic peptides before they enter lamina propria would be logical. Moreover, for the enzyme to be amenable to blockade, the open configuration requires permissive effects of calcium ions and thus, epithelial stress. Also, optimal timing of administration of the medication with gluten exposure remains to be determined.

7.3.2.1.2 IL-15 inhibitors

IL-15 is an essential growth factor for the proliferation of IELs. It mediates the inflammatory response that leads to intestinal epithelial damage. IL-15 is overexpressed in both the lamina propria and epithelium in active CeD. Interestingly, it is not upregulated in potential CeD. It contributes to reprogram CD8+ T cells into NK-like CD8+ effectors (NKG2D) that resist activation-induced apoptosis. It also promotes the up-regulation of MIC-A, MIC-B, ULBP and HLA-E on enterocytes [99]. IL-15 has a tripartite receptor complex with one subunit (IL-15Ra) that is specific to IL-15 and also binds CD122 (that signals for IL-2 and IL-15RB) and CD132 (receptor for the common γc chain) with a high affinity.

AMG 714, now known as PRV-015 is a human monoclonal antibody against IL-15; it was originally developed for rheumatoid arthritis, in which a clinical response rate of 65% was shown during phase 1b and 2b studies, with no NK cell induction [100]. It was recently evaluated in a Phase 2a double blind RCT involving a gluten challenge in individuals with CeD on a GFD for at least 12 months. The drug was administered subcutaneously every two weeks for 10 weeks (total 6 doses of either 150 or 300 mg) in individuals undergoing a 10-week gluten challenge (2g or 4g). The primary endpoint was the percentage change from baseline to week 12 in Vh:Cd ratio. Secondary endpoints included CD3-positive IEL density and symptom scores. The primary endpoint was not met, but there was a lesser increase in the CD3+ IELS among the group who received 300mg, as well as decrease in theoccurrence of diarrhea with both dosages. In addition, the 300mg dose but not the 150mg dose showed a treatment difference in the change of the CeD-PRO compared to placebo. The most common adverse event was injection site reactions [101]. The drug subsequently was renamed PRV-015 and is expected to undergo phase 2b testing shortly [102].

Other anti–IL-15 therapies are also in the therapeutic pipeline of several companies, with few having published preclinical data in CeD. One of them is an anti–human IL-15 antibody 04H04. It was evaluated in a rhesus macaque model of CeD (Macaca mulatta). Indeed, this species may share with humans CeD characteristics, such as positive serology and enteropathy. The monkeys underwent a gluten challenge for at least 2 months and were subsequently treated with human anti–IL-15 antibody 04H04 intravenously weekly for 28-90 days. The drug-induced improvement of the intraepithelial lymphocytosis and villous height/crypt depth ratios. Furthermore, there was a reduction of intestinal CD8+ and CD4+ IFN γ producing T cells [103]. Additional preclinical studies have been announced.

Considering recent evidence suggesting the requirement ofa synergistic action between IL-15 and IL-21 to induce activation and proliferation of intraepithelial CTL, it is possible that only acting on IL-15 may not be enough to prevent enteropathy [104].

7.3.2.1.3 Cathepsins

Cathepsins are intracellular proteases that mediate apoptosis. A type of cathepsin called cathepsin S is expressed specifically in antigen presenting cells (APCs) where it mediates the proteolysis of the invariant chain. The latter is an intracellular molecule that prevents intracytoplasmic, self- antigen loading during the early stages of development of the MHC-II molecules of APCs. Once the MHC-II complex matures intracellularly, cathepsin S cleaves the invariant chain to permit antigen loading. Hence cathepsin S is important for the MHC II to function normally to present processed antigens to CD4+ cells, a process that is central to autoimmunity [105]. The cathepsin inhibitor, RO5459072 (also called RG7625), has been developed to target this pathway and is being tested in CeD, Sjogren's syndrome and other autoimmune diseases. A phase 1, proof of concept study (NCT02295332) showed that this drug resulted in a decrease in maturation of MHC-II bearing B cells and dendritic cells [106]. The effects of RO5459072 on the immune response to gluten challenge in CeD patients has been investigated in a phase 1, placebo-controlled RCT (NCT02679014). Volunteers with previously diagnosed CeD were randomized to receive either 100mg of RO5459072 or placebo twice daily for 28 days. The results of the study are yet to be published.

7.3.2.1.4 Adhesion molecules and chemokines

7.3.2.1.4.1 CCR9 antagonists Immunoreactive T cells are attracted to the intestinal mucosa by various mechanisms, including CCL25 a chemokine secreted by the intestinal epithelium that binds to the CCR9 receptor expressed on T cells. Antagonists to CCR9 include CCX8037 and GSK-1605786. The latter is also known as CCX-282, Traficet-EN or Vercinon and is being developed for potential use in Crohn's disease and CeD. This molecule was characterized and tested in vitro by Walters et al. on Molt-4-T cell lines which endogenously express CCR9. Stimulation of Molt-4 cells with CCL25 resulted in release of intracellular ionic calcium which in turn promoted chemotaxis. The antagonist of CCL25 prevented its interaction with CCR9, and prevented release of calcium and recruitment of inflammatory T cells. The same study also found the results to be replicable in mouse models where intestinal inflammation was attenuated in response to the drug. Furthermore, the drug CCX282 was found to be highly selective for CCL25–CCR9 interaction [107]. Tubo et al. also found similar results in mouse models; interestingly they also inoculated this drug into inflamed skin in mice, and found no attenuation of inflammatory response in the draining lymph nodes. This suggests that this drug could selectively act onthe intestinal immune system and spare other immune organs thereby avoiding generalized immunosuppression [108]. A RCT (NCT00540657) to study the efficacy of administration of 250mg of oral CCX282 twice daily in comparison to placebo was completed in 2007 in patients with CeD on 24 months of GFD. The outcome measures included effects on Vh:Cd ratio and markers of intestinal inflammation, serology and symptoms upon gluten exposure. The results of this study are yet to be published.

7.3.2.1.4.2 Vedolizumab Vedolizumab is a humanized monoclonal antibody that prevents the interaction of anti-$\alpha4\beta7$ integrin with mucosal addressin cell adhesion molecule-1 (MAdCAM-1) on endothelial cells. The latter is found exclusively on the intestinal and colonic mucosa, enabling vedolizumab to prevent chemotaxis of memory T cells from the circulation into the mucosa. A phase 2 clinical trial (NCT02929316) of vedolizumab in CeD patients was however terminated prematurely.

7.3.2.1.5 Hookworm therapy

Chronic low grade helminthiasis, while inducing specific immune response to itself, also diminishes Th1 cell response to other antigens, which can in turn suppress autoimmunity [109]. In another experimental approach, the effect of hookworm (*Necator americanus*) therapy on CeD was investigated; the basis of the therapy is related to the hygiene hypothesis and the observation that there has been a simultaneous decrease in infectious diseases and an increase in autoimmune diseases globally [110]. Daveson et al. found that inoculation of hookworms into the skin increased microbial richness in the fecal samples in the same patients. It is possible therefore that helminthic infection modulates CeD pathogenesis through an unknown mechanism similar to probiotics [109, 111]. In a small study involving 12 subjects, participants infected with 20 hookworm larvae underwent gluten micro-challenge over 52 weeks, with non-worsening of Vh:Cd ratio [112]. However, larger studies with standardized gluten challenge protocols would be needed to draw definitive conclusions on the efficacy of hookworm therapy.

7.3.2.1.6 Budesonide for acute reactions

As mentioned in the preceding sections, symptomatic reactions to accidental gluten exposure are frequent among CeD patients [6]. In some individuals, the reaction may be severe and debilitating [113]. Budesonide is a micronized topical steroid with a high first pass metabolism. It has direct anti-inflammatory properties on the duodenal mucosa [114], triggers translation of anti-inflammatory genes and modulates Th1 and Th2 immunity [115]. It has been studied in the treatment of celiac crisis [116] and refractory celiac disease [117]. A small case-series also reported efficacy of short courses of budesonide to decrease the severity and duration of symptomatic reactions in some individuals [113].

7.3.2.1.7 Agents that have not reached human clinical trials yet

7.3.2.1.7.1 γc Receptor Antagonist Increased levels of IL-15RA and IL-21 were found in the mucosa of individuals with untreated CeD [118]. These cytokines stimulate transcriptional activity, proliferation and cytolytic properties of intraepithelial cytotoxic T cells and their conversion to NKG-2D cytotoxic lymphocytes, as well as inhibit apoptosis of cytotoxic lymphocytes and block T-reg activity. In addition, IL-21 promotes B cell function and immunoglobulin production. Hence IL-15 and IL-21 are involved in both B and T cell mediated damage. IL-15 and IL-21 also act synergistically for the expression of Granzyme-B that is an effector molecule involved in villous damage. A recent study showed that IL-15 and IL-21 are strongly co-expressed in active but not in potential CeD [119]. This suggests that blockade of both the cytokines may be essential in preventing active CeD.

BNZ-2 is a peptide that blocks the cytokine binding site on the γc receptor. It was designed to selectively prevent binding of IL-15 and IL-21, but not IL-2, 4, 7 and 9 [104]. This molecule's structure was in fact modeled after IL-15 and 21 structures, and has equipotency for both [120]. Ex-vivo experiments with epithelial cells and intra-epithelial cytotoxic lymphocytes (CTL) from untreated CeD patients were promising, showing that BNZ-2 prevented 1) IL-15/IL-21–mediated transcriptional alterations, 2) IE-CTL proliferation, 3) increase in IFN-γ transcript levels. Interestingly, the latter was also assessed in duodenal organ culture generated from biopsies from patients with active CeD that were stimulated with PT-gliadin [104]. Studies in humanized NOG-IL-15 transgenic mice that are prone to GI inflammation have suggested that BNZ-2 preserved epithelial integrity in them and prevents CD8 cell homing to the lamina propria by downregulation of integrin β-7. Pharmacokinetic studies in mice and monkeys show that a single injection of BNZ-2 may induce a rise in bioavailability over days thereby making a weekly once subcutaneous injection a suitable dosing regimen. Pharmacodynamic studies in monkeys show that there were dose dependent decreases in the IL-15- dependent CD8 cells and IL21- partly dependent B cells with no effect on T Regs. A PEGylated form of BNZ-2 will be investigated in a Phase 1 multicenter placebo controlled double blind clinical trial in a near future.

7.3.2.1.7.2 Cystic fibrosis transmembrane conductance regulator modulator

Cystic fibrosis is caused by mutations involving the cystic fibrosis transmembrane conductance regulator (CFTR). Interestingly, malfunction of the CFTR triggers the activation of transglutaminase-2 (tTG-2). Mice bearing CFTR defects have also been found to have increased sensitivity to enteropathogenic effects of oral gliadin. Similarly, P31-43 gliadin peptide inhibits CFTR in mouse intestinal epithelial cells, causing a local stress response that could contribute to the immunopathology of CeD [121]. Ivacaftor, an FDA approved pharmacological potentiator of CFTR has been found to attenuate pro-inflammatory effects of gliadin in preclinical CeD models including cultured human epithelial cells as well as gliadin-sensitive BALB/C mice and non-obese diabetic (NOD) mice that are particularly susceptible to oral gliadin challenge. Interestingly, this drug has also been found to modulate adaptive immune response, and induces tolerance to gliadin in peripheral blood mononuclear cells (PBMC) from CeD patients that are co-cultured with gliadin-challenged intestinal epithelial cells [122]. Another CFTR potentiator, genistein has also been found to similarly mitigate the proinflammatory response to gliadin in preclinical models [123]. These findings suggest that other CFTR potentiators like Vrx-532, as well as TG-2 inhibitors may be explored in similar settings.

7.3.2.1.7.3 JAK inhibitor The protein tyrosine kinases JAK1 and JAK3 are linked intracellularly to the tripartite IL-15 receptor complex. Considering the pivotal effect of IL-15 in the pathogenesis of CeD(see section on IL-15 inhibitors), acting on the signal transduction following the binding of this cytokine with the receptor appears attractive. Tofacitinib is a pan-JAK inhibitor that is approved by the FDA for treatment of rheumatoid arthritis and ulcerative colitis. It was investigated in a CeD mice model involving IL-15 producing transgenic mice. Administration of tofacitinib resulted in improvement of the intestinal histology, decrease in the amount of CD8+ NKG2D T cells, as well as increase in visceral adiposity [124]. Studies are still in the preclinical stage.

7.3.2.1.7.5 Elafin Elafin is a protease inhibitor that acts on elastases [125] that has been shown to improve colonic inflammation [126]. In a study, expression of elafin in the small intestinal epithelium was found to be lower in patients with active CeD compared to controls. Further, in vitro, it was found to significantly slow the deamidation of the 33-mer peptide to its more immunogenic form in intestinal epithelial tissues. Also, treatment of gluten-sensitive mice with elafin delivered using *L. lactis* vector normalized inflammation and maintained ZO-1 expression [127]. Similarly elafin has also been found to be underexpressed in keratinocytes of patients with active DH, thereby potentially reducing its ability to decrease deamidation of tTG-3 [128]. A subcutaneous formulation of elafin is currently being tested for pulmonary arterial hypertension therapy (NCT03522935), similar formulations could find application in human trials for CeD as well.

7.3.2.1.7.6 HLA blockers HLA blockers prevent the interaction of antigen presenting cells (APCs) that present gluten immunogenic epitopes mounted on their surface MHC-II ligands to the T Cell Receptor (TCR) of CD4+T helper cells. This would prevent the immunotoxic cascade mediated by adaptive immunity in CeD. Cyclic peptides have been designed for competitively blocking the gliadin-binding groove of HLA-DQ2. These have the gliadin like epitopes-LQPFPQPELPY, KQPFPEKELPY, or LQLQPF-PQPEKPYPQPEKPY and are cyclized to enhance blockade of the HLA grooves, using sulfide or polyethylene glycol bridges. Similarly, dimeric peptides with gliadin scaffolds have also demonstrated effective blockade. However, it has not translated into reduced T cell activation [129]. Ubiquitous requirement of HLA in various immune responses and a potential interference by the peptides in that function is one of the main concerns regarding use of HLA blockers. PV-267 and PV-623 are HLA-DR2b- blocking drugs from pre-clinical testing stages in Multiple Sclerosis, and it remains to be determined if similar drugs would be safe and efficacious in CeD models [130].

7.3.2.2 Limitations of this approach
7.3.2.2.1 Complexity of the immune response; blocking one pathway may not completely stop the process

The digestive tract is one of the most complex systems in the human body. The whole spectrum of the interactions between our microbiota and the food we eat, as well as our immune system remain mostly unknown. There are many different pathways to the digestion of gluten, its absorption, as well as the immune response and it is likely that acting on only one pathway may not completely prevent the immune activation and the enteropathy. This may be useful to complement the effects of a GFD in individuals with NRCD. However, to replace the GFD, perhaps utilizing a combination of therapies with complementary mechanisms of action may be more successful.

7.3.2.2.2 Extra-intestinal manifestations

Further, most of our understanding about the pathogenesis of CeD is derived from the perspective of intestinal damage as the end point. Since most extra-intestinal manifestations of CeD like liver disease and DH respond in parallel to the resolving enteropathy, it is likely that novel therapeutics will be useful in treating these extra-intestinal manifestations as well. Nonetheless, there are a few extra-intestinal features like neuropathy and arthropathy that show more unpredictable responses to the GFD, it is therefore possible that they are caused by as yet undiscovered immune/ non-immune mechanisms that operate in CeD. Better understanding of mechanisms would enhance drug discovery for all possible manifestations of CeD.

7.3.2.2.3 Role of B cells

T cells have occupied much of the recent focus on development of immune based therapies. This is because gluten-specific T cells are key actors in priming the B cells to become plasma cells and produce anti-tTG antibodies and anti-deamidated gliadin peptides antibodies [131]. Recent evidence suggests that the plasma cells may also harbor

MHCII and gluten-specific B Cell Receptor (BCR) and/or tTG-2 specific BCR. In fact, one study found plasma cells to be the most abundant cells that express gluten specific MHC-II in CeD patients on gluten-containing diets, but not among those on the GFD [132]. The precise cross-talk mechanisms between B-cell as an APC and gluten-specific T cells have not been established yet and further understanding will determine whether B cells should be considered as a therapeutic target as well.

7.3.3 Drugs to induce immune tolerance

Mucosal tolerance is believed to be induced by two potential mechanisms; clonal anergy of reactive lymphocytes and induction of regulatory T cells. Clonal anergy is triggered by high doses of antigen that would result in defective crosstalk with the APC. T-regs are key actors in building immune tolerance to external antigens. They can either suppress autoreactive T cells in a one-on-one manner or have a bystander suppression effect by secreting anti-inflammatory cytokines. Previous studies have suggested intrinsic T-reg dysregulation in celiac disease [133]. Other proinflammatory cytokines may also inhibit the T-regs. Improving immune tolerance to gluten has been a pathway exploited initially by a vaccine (Nexvax2), and more recently with nanoparticle technology.

7.3.3.1 Therapeutic agents
7.3.3.1.1 Nexvax2

Nexvax2 is an epitope-specific immunotherapy, meaning that it only uses specific peptides and not the whole antigen [37]. It includes a mix of three peptides that include the immunodominant epitopes for gluten-specific T cells. The goal of this therapy is to eventually induce CD4-positive T cell anergy and stimulate suppressive regulatory T cells, which could lead to long term durable immune tolerance. This happens by the way that the T cells are stimulated by the exact epitope that would bind on the MHC class II on the APC, but since no APCs are involved, there is no co-stimulation by molecules such as CD80 or CD86 that would stimulate pro-inflammatory T-helper cells survival and proliferation [134].

Nexvax2 had promising results in a Phase 1 trial, as all the participants who received Nexvax2 150 mg and had a subsequent gluten challenge had a negative IFN-γ release assay, suggesting inhibition of the CD4 T cells. Moreover, there were no substantive safety issues [37]. However, a recent Phase 2 double blind RCT failed to demonstrate a significant reduction of symptoms in the group receiving Nexvax2, assessed with the CeD PRO. There was also no difference between the group in the rise of IL2 during the gluten challenge. There was however a trend toward an improved villus height/crypt depth ratio in the group receiving Nexvax2. This project nevertheless contributed to collect insightful data on the immediate immune response to gluten in celiac disease, notably an association between IL-2 secretion and GI symptoms, such as nausea and vomiting [135].

7.3.3.1.2 Immune Tolerance Inducing Nanoparticles

Nanoparticle technology is currently being investigated as a method of inducing immune tolerance in various autoimmune disorders, especially type 1 diabetes and multiple sclerosis [136]. Like Nexvax2, this is a peptide-specific immunotherapy, referred to as Tolerogenic Immune Modifying Particles (TIMPs). For use in CeD the particles are made of a negatively charged biocompatible and biodegradable polymer matrix of poly(DL-lactide-co-glycolide (PLGA). TIMPs bind to monocytes and are brought to the spleen and liver, where they are processed by tissue-resident APC, mainly splenic marginal zone macrophages expressing macrophage receptor with collagenous structure (MARCO). MARCO receptors are scavenger receptors, involved in the uptake of self-antigenic debris and apoptotic cells, and participate in tolerogenic and anti-inflammatory responses. This property is taken advantage of to induce tolerogenesis to gliadin by encapsulating them in TIMPs. TIMPs alter APC transcriptional activity and induce an upregulation of surface MHC class II presentation of specific peptide and a downregulation of CD80 and CD86 [137]. Other immune effects include upregulation of PD-L1 and production of regulatory cytokines (IL-10 and TGF-B) [138]. It also triggers CD4+ T cells anergy as well as the differentiation of Th1 cells into TR1 cells [136, 139].

The drug TIMP-Glia, also known as CNP-101 and TAK-101, contains approximately 2.9mcg encapsulated gliadin protein per mg of polymer. Further, the gliadin particles are optimized by including peptides from alpha, beta and γ gliadin fractions, and by fully encapsulating the peptides for safety and prevention of premature disintegration prior to uptake by MARCO receptors. Intravenous administration of gliadin protein encapsulated in TIMP was initially investigated in mice models. On day 0, wild type mice (donor mice) were immunized with gliadin to induce CD4+CD62LCD44^{high} memory T cells. This was followed by adoptive transfer of these memory cells into 16 Rag1-/- mice on Day 1. Mice were randomized to receive 2.5mg injections of either TIMP-Glia or the more inert TIMP- Lys (TIMP with encapsulated lysosomes) on days 10 and 24, while on gluten containing feeds. The remaining mice did not receive either injection, and were randomized to either receive gluten containing or gluten free feeds. Results showed efficacy of TIMP-Glia in preventing gluten induced inflammatory damage on the gut epithelium, as well as weight loss compared to TIMP-Lys and the other mice that did not receive any intravenous injections while on a gluten-containing diet [140].

In humans, the agent was also administered intravenously in a Phase 1 trial testing dose range from 0.1 to 8 mg/kg and no serious adverse events were reported. A Phase 2a proof of concept RCT was recently completed, where 29 well-treated CeD patients were given either the medication (8mg/kg) or placebo once weekly for two weeks and then started on a 14-day gluten challenge. A significant reduction of T cell response was observed during the gluten challenge, assessed by IFN-γ ELISPOT. Moreover, the subjects receiving the medication did not have any change in the Vh/Cd ratio before and after the gluten challenge, compared to those receiving the placebo. There were significantly less circulating gut-homing effector memory CD4+ and CD8+ T cells in the group receiving the medication. There was however no difference in cytokine levels and T cell proliferation in between the 2 groups. Although, there was a transient increase in complement levels in both intervention groups, this was not associated with any adverse events. Finally, there was no difference in adverse events between the drug and placebo groups, and no serious adverse events, overall [141].

KAN-101 is another immune-tolerance inducing intravenous medication originally developed by KanyosBio, and is currently being tested on patients for phase 1 clinical trials (NCT04248855), although data on the nature of this technology is limited at this point of time.

7.3.3.1.3 Agents that have not reach human clinical trials yet

In addition to the peptide-couple nanoparticles technology mentioned earlier, pMHC-coated nanoparticles have also been developed and studied in murine models of other auto-immune diseases. This technology could involve either MHC class I or II, although pMHC class II nanoparticles would be preferable in CeD to help differentiating pathogenic effector T cells into antigen-specific T reg [142]. Other protein and peptide-based immunotherapies include whole antigen therapy, altered peptide ligands and soluble peptide-MHC complexes [136]. In addition to vaccine and nanoparticles, other vectors are being studied, including bacteria.

7.3.3.1.3.1 AG017 Lactobacillus and lactococcus lactis are non-pathogenic bacteria, that can survive through the entire GI tract and resist bile and gastric acids. They can be genetically modified to secrete bioactive molecules, either constitutively or as induced under certain conditions. The molecule of interest may be produced intracellularly and secreted or anchored on the wall of the bacteria [143]. This technology was applied to CeD by engineering *Lactococcus lactis* to secrete a deamidated DQ8 gliadin epitope (LL-eDQ8d) [144]. The induction of tolerance was investigated in a NOD AB DQ8 transgenic mice model. The administration of LL-eDQ8d decreased the proliferative capacity in the lymph nodes and the lamina propria. This engineered bacteria stimulated the production of IL10, TGF-β and induced Foxp3+ T regs [144]. A human Phase 1b/IIa trial involving *L. lactis* that would be engineered to express a DQ2.5 specific gliadin peptide in combination with an immune- modulating cytokines has been announced.

7.3.3.1.3.2 Other immunotherapy platforms Other experimental modalities for immune tolerance induction have been tested in diseases like type 1 diabetes and multiple sclerosis. These include DNA- based vaccines, tolerogenic dendritic cell therapy, T-cell transfers, including chimeric antigen receptors (CAR) T cells, and other approaches to delete clonal populations of CD4+ or CD8+ cells, induce T-reg cells and anergic T-cells [136]. However, not all are suitable for CeD and some may have safety issues (instability of

T reg in vivo, lack of specificity of CAR T cells). Nevertheless, CeD remains an attractive model to perfect antigen-specific therapies considering its limited amount of epitopes, that are already characterized.

7.3.3.2 Limitations of this approach
7.3.3.2.1 Persistence of gluten-reactive memory T cells for years

Although these approaches dampen the immune response following a gluten challenge, we do not have data on the long-term efficacy of these agents, especially considering that gluten reactive memory T cell may be present for years in the blood of individuals with treated celiac disease [145] To date, there is no example of induced immunotolerance in any human disease; there is much to learn about short-term and long-term efficacy and safety.

7.3.3.2.2 Reprogramming of IELs

Furthermore, evidence shows that CeD patients experience a longstanding reshaping of the intraepithelial lymphocytes with cytolytic properties, even on a GFD. IL-15 is persistently secreted by the epithelium of CeD patients, as well as other stress markers, such as HSP70 and HSP27, even on a GFD [118]. Moreover, there is evidence of a permanent reshaping of the IEL TCR repertoire in CeD, independently of the GFD, promoting the expansion of a gluten-sensitive, IFNγ producing Vδ1+, to the detriment of the physiological Vγ4+/Vδ1+ [146]. Thus, persistent epithelial stress and rapid reactivity to gluten exposures may be possible, even though gluten-specific CD4 T cells are suppressed.

7.3.3.2.3 Need for a combination of therapies?

As for inflammatory bowel diseases, it is possible that a combination of therapies acting on different targets may be the key for CeD. However, we must keep in mind the lessons learned from the IBD world in that the immune system may modulate and adapt to therapies, as observed from secondary loss of response to anti-TNF therapies despite adequate drug levels and no anti-drug antibodies.

7.3.4 Prevention of development of celiac disease among predisposed individuals

The ultimate goal of research in CeD would be to be able to predict who among the individuals carrying a predisposing HLA will actually develop CeD (or RCD). As for inflammatory bowel disease, to fully understand the steps toward the development of CeD (or complicated CeD) could allow the consideration of early interventions to prevent the disease. To this end, many pediatric cohorts are currently being followed worldwide to better understand the environmental, genetic, microbial and immunological factors associated with the progression toward the development of an enteropathy.

Among the implicated infectious causes, Reovirus has been studied in animal models to demonstrate that viral infections can shift the mucosal immunity toward proinflammatory responses to antigens such as gluten [147]. In a similar study, certain strains of reoviruses that succeeded in causing a prolonged infection by subverting apoptosis of intestinal epithelial cells, were more likely to cause this loss of tolerance [148]. Human studies also yielded an association between CeD and reoviral infections [149].

In a Danish cohort of 1931 children with celiac risk HLA alleles, the rate ratio of CeD was found to be 1.94 times per episode of acute rotavirus infection [150]. Another multicenter study that prospectively examined 6327 genetically predisposed children aged 1-4 years, found that GI infections were associated with an increase in the occurrence of CeD over the following three months by 33%, and that rotavirus vaccination was found to decrease the incidence of CeD those that had GI infections, implicating wild type rotavirus infections [151]. The TEDDY study group identified a cumulative effect of enteroviral infections and gluten intake between 1-2 years of age to be associated with CeD incidence [152]. Interestingly, rotaviral infections were also associated with celiac crisis in a small case series [153]. As such, vaccines against certain types of virus may become a potential preventive measure.

Recent elegant studies have brought new insights regarding the roles of B cells in CeD, suggesting for instance that they could serve as APC [154]. As such, there could a reciprocal interaction between gliadin-targeting CD4+ cells and B cells, leading to their activation and capacity to recognize, process, and present TG-2- gliadin complexes back to CD4+ T cells [155, 156]. Interestingly, some of these pathways may be triggered and maintained by stimulation from foreign antigens like viruses (EBV) and bacteria [157]. Moreover, an interesting experiment also showed that bacterial-derived gliadin peptide mimics could induce activation ofHLADQ2.5 restricted, gliadin-reactive T-cells obtained from CeD patients in vivo, suggesting that pseudomonas fluorescens could play a role in CeD development [158]. These provide interesting insights about how the microbiota may induce and potentiate the immune response to gluten and could be the target of early interventions.

7.4 Clinical trial endpoints

As there are different goals of therapy (clinical, histological or immunological response), the choice of the primary endpoint of a clinical trial may be especially difficult. The FDA requires 'clinically meaningful endpoints' which are endpoints that are direct measures of how patients feel, function, and survive, which at this time precludes improvement of the enteropathyas the primary basis for approval [159]. Moreover, there are no validated surrogate markers for either the resolution of the enteropathy, or the dampening of the response.

Table 7.1 PROs used in various Celiac Disease trials.

Drug	PRO	Endpoint	References
Bifidobacteriumnatren life start	GSRS	Secondary	Smecuol 2013 [167]
Larazotide Phase 2a	GSRSCeD-GSRS	Secondary	Leffler 2012 [75]
Larazotide Phase 2a	GSRS	Secondary	Kelly 2013 [168]
Larazotide Phase 2b	CeD-GSRSCeD-PRO	PrimarySecondary	Leffler 2015 [67]
Nexvax 2 Phase 2	CeD-PRO	Primary	NCT03644069
AMG 714 Phase 2	GSRS	Secondary	Lahdeaho 2019 [169]
Latiglutenase Phase 2	CDSD and ICDSQ	Exploratory	Murray 2017 [46]

Most current trials are still early phase, with only one on-going Phase 3 trial involving larazotide.

7.4.1 Clinical symptom scores

To assess symptomatic response to gluten challenge in patients with CeD/ suspected CeD and already on a GFD, especially under research settings, questionnaire-based tools have been developed. These are called Patient Reported Outcomes(PRO), which are assessments based on a report that comes directly from the patient without interpretation [160]. Below are some of the common PROs used in CeD (Table 7.1).

GSRS contains 15-questions to assess common gastrointestinal symptoms, organized into 5 sub-domains viz. reflux, abdominal pain, constipation, diarrhea, and indigestion) [161]. However, it was originally developed for IBS and peptic ulcer disease [162], and not optimized for CeD. Hence, a modified version called the CeD-GSRS has been validated for use in newer clinical trials in CeD; it incorporates abdominal pain, diarrhea and indigestion domains of the GSRS [67]. An obvious limitation of both these questionnaires is that they do not assess extra-intestinal symptoms.

A questionnaire that is specific for CeD but also incorporates extra-intestinal symptoms is the Celiac Symptom Index(CSI); it contains 16-questions grouped into 2 sub-domains (specific symptoms and general health). However a drawback remains that this questionnaire was developed and validated before the FDA issued Study Endpoints and Label Development (SEALD) guidelines on developing [163]. Hence the more accepted PROs in CeD trials are CeD-GSRS and the Celiac Disease Patient-Related Outcomes (CeD-PRO); the latter is described in the following section.

The CeD-PROdeveloped by Alba Pharmaceuticals according to FDA's guidelines. Validation work was performed within a Phase 2b clinical trial on the efficacy of Larazotide in NRCD [67]. It includes three CeD symptom domains (diarrhea, abdominal pain and non GI symptoms), with grading 10 symptoms on a 0 to 10 intensity scale. It was also chosen as the primary endpoint in the Phase 2 trial investigating the safety, efficacy and tolerability of Nexvax2 [37].

The Celiac Disease Symptom Diary (CDSD) is another PRO [46] developed initially by Alvine Therapeutics in accordance with FDA guidance. It was tested in a cohort of 202 patients [164] with further validation work performed during a Phase 2b study of Latiglutenase. It includes questions on abdominal pain, bloating, constipation, diarrhea, fatigue, and nausea and is scored from 0-70 [164].

Celiac Disease Assessment Questionnaire (CDAQ) is a cross-over between CeD symptom questionnaire and a Health Related Quality of Life tool (HRQol), the latter of which is assessed using domains like social isolation, dietary burden, worry, etc.; this questionnaire is novel in that it examines the quality of life better than the previous PROs discussed, and therefore gives a broad picture of overall health over the past 4 weeks [165].

Since there is no wide variety of symptoms in GI disorders to begin with, most CeD related PROs do have a significant overlap in capturing different GI complaints like diarrhea, abdominal pain, bloating and nausea, with resultant similarities in scores. However the recall periods of these questionnaires vary, with CeD-PRO and the CDSD having a 24-hour recall compared to 1 week for the CeD-GSRS and GSRS and 4 weeks for the CSI and CDAQ.

A systematic review examined the effect of different doses and durations of gluten challenge on symptomatic endpoints in pediatric and adult populations. It found that 43-80% of the adults with diagnosed or suspected CeD developed symptoms and the percentage of symptoms and their severity increased throughout the gluten challenge. Overall, symptomatic endpoints were found to have a low positive predictive value [166]. Due to the considerable variability in the time to onset of symptoms and their low positive predictive value as an independent outcome measure in gluten challenge, they are usually utilized in combination with other objective biomarkers.

7.4.2 Quality of Life measures

Gastrointestinal symptoms in CeD can contribute to a decreased quality of life (QOL). In fact, even in patients that are asymptomatic at CeD diagnosis, a GFD may bring about significant improvement in QOL [170]. In a study that explored the impacts of CeD and its treatment on the quality of life, central themes identified were physical functioning, sleep, fears and anxiety, day to day management of CeD, daily activities, social functioning and relationships [171].

Some of the questionnaires commonly used in CeD include the Psychological General Well-Being Index (PGWBI) [172,173], SF-36 [174,175] as well as tools that have been validated specifically for QOL in CeD, such as Celiac Disease-related quality of life (CD-QOL) survey [176,177] and the Celiac Disease Questionnaire.

PGWBI is a score derived from six domains: anxiety, depressed mood, self-control, positive well-being, general health and vitality. SF-36 as the name implies is a 36- item self-report, and measures eight domains: physical functioning, role limitations due to physical health, role limitations due to emotional problems, energy/fatigue, emotional well-being, social functioning, pain and general health. Meanwhile, CD-QOL is another research tool that uses a 5-point Likert scale system, with 20 items grouped into 4 subdomains (limitations, dysphoria, health concerns, and inadequate treatment) [178]. It has been validated as a quality of life measure for CD patients [179]. The Celiac Disease Questionnaire is a 28-Item self-report, with 7-point Likert scale responses in four domains: emotional and social problems, disease-related worries and GI symptoms [180]. Based on prospective qualitative research by Leffler et al., two other QOL tools were developed in accordance with FDA guidelines, to complement the daily CDSD: the Impact of Celiac Disease Symptoms Questionnaire (ICDSQ) and the Impact of Adhering to a Gluten-Free Diet Questionnaire (IGFDQ). The ICDSQ is a 14-item scale organized into four domains: daily activities, social activities, emotional well-being and physical functioning, answered using a 5-point Likert scale. While the ICDSQ assesses the impact of CeD on quality of life, the IGFDQ assesses the burden on the GFD on a patient's quality of life [171,181]. CDDUX is a CeD-related quality of life questionnaire validated in the pediatric population, but does not elucidate gastrointestinal symptoms [182].

7.4.3 Serology

Although celiac serologies (anti-tissue transglutaminase IgA and anti-deamidated gliadin peptides) have a utility in clinic to grossly assess the adherence to a GFD, they are neither a surrogate marker of villous atrophy, nor a marker of immune response to gluten exposure [183]. However, for CeD trials incorporating gluten challenges, CeD patients that are seronegative at baseline are generally sought for inclusion. Similarly, increased antibody titers toward the end of the trial may signify increased disease activity [178].

7.4.4 Histology scoring system

While CeD patients are treated with the hope of complete intestinal healing, it is noteworthy that we lack an evidence-based definition of complete mucosal recovery. Further, the patchy distribution and suboptimal precision of the regular fiberoptic endoscopic techniques can lead to sampling errors, when the most diseased segment is not biopsied. Good sampling techniques (one bite per pass), proper orientation of the specimen on the microscopic slides and evaluation by an expert pathologist are essential steps as well.

Over time, various histologic scoring systems have been used. The initial Marsh score was published in 1992 and included Type 1 "infiltrative," type 2 "hyperplastic" and type 3 "destructive lesions" [184]. It was subsequently modified by Oberhuber et al to include three degrees of villous atrophy [185]. Corazza and Villanaci attempted to

Table 7.2 Clinical trials that used villous height/crypt depth ratio (Vh:Cd) as an outcome.

Drug	Endpoint	Reference
Nexvax 2 Phase 2	Secondary	NCT03644069
AMG 714 Phase 2	Primary	Lahdeaho 2019 [169]
Latiglutenase Phase 2	Primary	Murray 2019 [164]
CNP-101 Phase 2	Secondary	NCT03738475

simplify the grading system to minimize inter–observer disagreement (either normal villi but increased IELs (Grade A), atrophic villi (villus to crypt ratio <3:1) but still detectable (Grade B1) or atrophic and flat villi that are not detectable (Grade B2) [186]. Ensari had a similar approach [187]. However, these scoring systems are not sufficiently responsive to change for use in clinical trials.

To this end, quantitative histology has been developed for clinical trials. Compared to the other scores that are categorical, it provides continuous data that have superior responsiveness and reliabilty. It includes the villus–height to crypt depth ratio and density of IELs [188]. This morphologic assessment can be used to generate a result comparable to earlier qualitative assessments as withthe Quantitative-Mucosal Algorithmic Rule for Scoring Histology (Q-MARSH), with ranges of the Vh:Cd ratio corresponding to grades in the modified marsh classification [36]. The recent Tampere consensus established that a change of 0.4 in the ratio could represent a measurable and likely clinically significant difference, as well as a change of 30% or more in the T cells IELs densities [178]. Since then, it has been used in multiple clinical trials (Table 7.2).

7.4.5 Video Capsule Endoscopy- based classification of severity

While traditional endoscopic biopsies are the standard of care to diagnose CeD at present, they are invasive, costly, potentially limited by suboptimal sampling due to incorrect techniques or by patchy disease [189]. Video capsule endoscopy (VCE) overcomes these limitations by allowing more distal access to the small intestine and better delineation of full extent of the disease, including complicated CeD. Therefore, it may complement traditional endoscopies in patients who have a clinical picture strongly suggestive of CeD but have negative biopsies and can be used in patient unwilling or unable to undergo EGD. Also, CeD enteropathy usually follows a distal to proximal small intestinal healing gradient in response to a GFD [190]; this implies that VCE may be more sensitive tools than conventional endoscopies to monitor patients on a GFD since the latter only access the duodenum. Tissue sampling is not possible with this tool, however, it has an eight-fold magnification which is higher than that of conventional endoscopy and allows visualization of individual villi. It is therefore possible to discern mild and severe villous atrophy on the capsule images manually, but the time- consuming and subjective nature of the interpretation are potential drawbacks. Hence, there are now attempts to utilize computer aided techniques; this would enhance their applicability to monitor mucosal status in diagnostic and research settings. Preliminary studies exploited the differences in the brightness and textures of grayscale images to distinguish celiac from control participants' mucosa [191]. Novel sophisticated approaches have been used in recent studies to explore the use of computer-based deep learning methods with fairly good predictive values of>80%, however, sample sizes have been small [192].

7.4.6 IFN-γ ELISPOT

The identification of IFNγ releasing T cells antedates the characterization of gluten specific T cells [193] and the degree of release modulates during a gluten challenge [194]. IFNγ is readily released by activated T cells in CeD and was recently used as the primary endpoint of a Phase 2 trial examining the efficacy of nanoparticles to inhibit the immune activation in CeD [141].

7.4.7 IL-2

IL-2 is a cytokine predominantly secreted by activated T cells. Data obtained from the Nexvax2 phase 1 and 2 trials have shown that the circulating blood levels of this cytokine increase as early as 2 hour after gluten consumption, making it a potential early marker of gluten-induced T cell activation [195,196]. It has also been found to correlate with onset of acute symptoms after gluten exposure in CeD patients on a GFD. In a study of asymptomatic CeD patients on a GFD, Non Celiac Gluten Sensitivity (NCGS) patients and healthy controls, a gluten challenge led to acute proportionate rise in IL-2 that was concordant with the time of onset of symptoms in CeD group, and this phenomenon was specific to CeD [8].

7.4.8 Gluten-reactive T cells (CD4, CD8)

Over the last 15 years, identification and characterization of gluten specific T cells have been refined, now using a baculovirus expression vector that can provide both the DQ2.5 MHC fragment and many corresponding gliadin peptides [197]. Although this testis limited by the amount of blood needed as well as time and laboratory expertise needed, the frequency and activation of gluten specific gut-homing effector memory T cells have started to be used in clinical trials (NCT03738475).

7.4.9 Markers of intestinal permeability and damage

It is well-established that intestinal permeability is increased in CeD. Lactulose/mannitol ratio (LAMA) is a non-invasive way to assess and compare intestinal permeability. This endpoint has been used in trial investigating the efficacy of larazotide [76,161,198] and probiotics but has challenges related to interference from food, intestinal motility and high inter- and intra- subject variability [167]. Serum Zonulin, intestinal fatty- acid binding protein, citrulline have also been used as experimental biomarkers, in addition to products of simvastatin metabolism that has been proposed to monitor CeD-related mucosal healing [199–202]. Table 7.3 provides a list of clinical trials in CeD that have used various endpoints discussed in the above sections.

7.5 Clinical trial pitfalls

7.5.1 Lack of standardized gluten challenge protocols

One of the unique features of CeD is that the trigger of the enteropathy is known and we can safely induce disease activity among study subjects by means of a gluten challenge. However, there is considerable heterogeneity among the methods of gluten challenge used in drug trials and the amount, duration and type of gluten used may have an effect on the immune response during the trial (Table 7.4). Thus, gluten challenge studies have shown a wide range in the serological and histological response to gluten, potentially also compounded by inter-individual variability in gluten sensitivity. Moreover, adherence of the participants to per protocolgluten may be highly variable., Thankfully, we may now objectively assess gluten intake through use of GIP testing in the urine and stools. Preliminary data showed that despite being allegedly taking a standard daily dose of gluten, some participants did not have detectable gluten in their urine and stool, suggesting at least some lack of adherence with the study intervention.

Finally, studies involving a gluten challenge may be subject to selection bias, since subjects that are asymptomatic or have mild symptoms are more likely to be willing to participate compared to CeD individuals with severe, significant symptoms.

7.5.2 Factors related to Patient- Reported Outcomes

Gastrointestinal symptoms are in general not specific to any gastrointestinal disorder, and CeDis no exception. Thus, an individual with coexisting IBS may present an elevated symptom score, despitehavingadequate CeD control. Moreover, wheat flour and other gluten containing products may be high in FODMAP, causing GI symptoms independently of immune activation. An elegant study has demonstrated that nausea and vomiting may be more specific to the immune activation due to gluten exposure in treated CeD patients [205]. GI symptoms are a major component of CeD PROs, with various extraintestinal manifestations not represented consistently. For instance, some include brain fog and rash and others not. Additionally, joint pain is not part of any PRO. Thus, a patient with significant extraintestinal manifestations upon gluten exposures may appear as very mildly symptomatic or asymptomatic according to the PROs. For this reason, additional PROs assessing non-GI symptoms should be considered, depending on the study design and population assessed.

Table 7.3 Published trials and primary/secondary endpoints.

Drug	Primary endpoint	Secondary endpoint	Exploratory endpoints	Reference
Bifidobacteriumnatren life start	LAMA	GSRS, Proinflammatory cytokines		Smecuol 2013 [167]
ALV003 (latiglutenase) Phase 2a	Change in Vh/Cd ratio	Change in intestinal intraepithelial lymphocyte numbers/phenotype Change in serological markers of celiac disease		Lahdeaho 2014 [45]
Latiglutenase Phase 2b	Change in Vh/Cd ratio	Change in IELs frequency Serology titers CDSD		Murray 2017 [164]
Larazotide Phase 2a Phase 2b	LAMA CeD GSRS	GSRS Serology CeD-PRO (GI and abdominal domain scores)	GSRS Bristol Stool Form Scale Changes in celiac serologies	Leffler 2012 [161] Kelly 2013 [76] Leffler 2015 [198]
Nexvax2	CeD-PRO	Serum IL-2 Villusheight		NCT03644069
AMG714	Vh/Cd ratio	CD3+ IELs density GSRS CeD-GSRS Bristol Stool Form Scale Celiac serologies		Lahdeaho 2019 [169]
CNP 101	IFN γ ELISPOT	Vh/Cd ratio IELS density Gliadin specific T cell proliferation Gliadin-specific cytokine secretion Gut-homing CD4, CD8, and $\gamma\delta$ cells		NCT03738475

Table 7.4 Characteristics of Gluten Challenges in published therapeutic trials.

Study (drug)	Type of gluten	Dose	Equivalence in terms of slice of bread / day	Duration	Reference
AN–PEP Phase ½	Toast Bolletje®	7g	2,5 slices	2 weeks	Tack2013 [203]
AN–PEP Phase 0	Dry meals containing gluten powder	4g	1.5 slices	Once	Salden 2015 [204]
ALV003 Phase 2	Breadcrumbs	2g	1.5 slice	6 weeks	Lahdeaho 2014 [45]
Larazotide Phase 2a	Capsules of gluten (45% gliadin, 45% glutenins, 10% globulins)	2.4g (800mg TID)	1 slice	2 weeks	Leffler 2012 [161]
Larazotide Phase 2a	Capsules of gluten taken	2.7g (900mg TID)	1 slice	6 weeks	Kelly 2013 [76]
Nexvax 2 Phase 1	Cookies	9g (3 cookies of 3g each/day)	3 slices	3 days	Goel et al. Lancet Gastroenterol-Hepatol 2017
Nexvax 2 Phase 2	Vital Wheat gluten	6g	2 slices	2 weeks	Daveson et al APT 2020
AMG714 Phase 2	Cookies	2-4g (1 cookie of 1-2g twice daily)	1 slice	10 weeks	Lahdeaho2019 [169]
CNP 101 Phase 2	Vital Wheat Gluten	12g for 3 days 6g for 11 days	4 slices 2 slices	2 weeks	NCT03738475

7.5.3 Trial effect and the role of GIP studies

Negative results from some well-designed gluten challenge trials have shed light on the possibility of trial effect [17,164]. The latter is a source of bias, whereby CeD patients enrolled in a trial may improve their adherence to the GFD, potentially confounding the actual beneficial effect of the drug compared to control.

Prior duration and compliance with the GFD: yet another pitfall of clinical trials is that there is considerable heterogeneity with respect to the duration of GFD that included participants have been on, along with their degree of adherence, which has traditionally been assessed by self-report. Also, despite adherence, trace/inadvertent gluten ingestion cannot be ruled out by self report. Moreover, the baseline GFD of each participants may contain various amount of unsuspected gluten [17,34]. (GIP testing offers an objective tool to resolve some of these uncertainties. As Phase 3 trials must simulate "in real life," quantification of the gluten contamination in the diet at the baseline and throughout the trial would be attractive and may be used along side the study; it is encouraging to note that some of the CeD trials are already incorporating GIP testing to strengthen their design [206].

7.5.4 Is histopathology reliable as an outcome measure?

7.5.4.1 Interobserver variation for histologic assessment

For any diagnostic method, the greater the number of categories (as many as seven in the Marsh-Oberhuber system), the lower the interobserver agreement, and accuracy [207]. These inaccuracies are usually minimized in the clinical setups that employ expert gastrointestinal pathologists that are trained to read celiac biopsies. Similar to the centralized endoscopy reading and histologic evaluations in IBD trials, an unique blinded expert pathologist to perform the histologic assessment would be best. [208,209].

7.5.4.2 Technical aspects of biopsy and processing

Being invasive by nature, biopsies are best done once and done right. A prerequisite for the correct evaluation of duodenal histopathology is proper collection and handling of biopsies by the endoscopist followed by processing and orientation prior to reading the slide. Biopsy forceps can cause crush artifacts, and hence care must be taken to limit the reading of the edges of the specimen [210].Similarly correct orientation of biopsy is crucial to avoid misinterpretation of flat mucosa as normal and vice versa. Tangentially cut sections could give false impressions of flat mucosa [188].

7.5.4.3 Complete normalization may take years

Similar to the wide variation in the time to onset of mucosal damage, which can be as less as 2 weeks in gluten challenge studies, time to heal is equally variable. Overall, only 1/3 of adults have normal villous architecture (a healthy, healed intestine) after 2 years on a GFD and 2/3 after 5 years on a GFD [211–213]. Furthermore, these data were obtained by assessing only the proximal duodenal mucosa. The degree of mucosal involvement of individuals included in trials is thus very heterogenous.

7.5.4.4 Patchiness of disease

Histological abnormalities associated with CeD can be patchy, which is why there have been attempts to standardize a certain minimum target in terms of sites and number of biopsies. Unfortunately, the ideal of taking four or more biopsies [77] to diagnose CeDwas met in only 39% of patients being evaluated for suspected CeD in one study [214]. Similarly, adding biopsies of the duodenal bulb from either the 9- or the 12-o'clock position increases the diagnostic yield because 10% may have villous atrophy exclusively in the bulb [215,216]. Lack of adherence to the recommendation of obtaining only a single biopsy specimen with each pass of the biopsy forceps could also potentially alter the histologic interpretation in trials, however there is no sufficient evidence to support this practice at this time [217].

7.5.4.5 Representation of HLA-DQ8 patients

It is also noteworthy that the tetramer-based studies and immune-tolerizing techniques for CeD studied thus far have been developed for and tested in HLA-DQ2.5 positive individuals; similar approaches in HLA-DQ8 positive CeD patients remain to be explored. However, much work has been accomplished on HLA-DQ8 and HLA-DQ2 mice models, that offer a viable platform for investigating pathogenetic pathways of CeD and pre-clinical drug testing. Some of the research is described in the next section.

Mouse models

HLA-DQ8 mice [73] are among the earliest of animal models for gluten-sensitivity; they were originally obtained by injecting human DQ8 cosmids into endogenous HLA class II knock-out mice embryos (AB^0) and further crossing the progeny (AB^0DQ8) with mice that were similarly transgenic with human CD4 molecule expression ($hCD4^+$) in lieu of murine CD4. The resultant mice showed higher T-cell response on receiving injected gluten challenges, than similar transgenic mice that expressed HCD4/HLA-DQ6, while no response was seen with rice challenge and in AB^0 mice. Verdu et al. later used similar models to show that upon gluten exposure, mice tend to have increased cholinergine-induced gut hypercontractility and ion transport, which may potentially explain gluten sensitivity in humans in the absence of CeD [218].

The above models were improved with NOD/DQ8 mice (Non-Obese Diabetic, human DQ8- expressing), by introducing a predisposition toward autoimmunity. These mice have been used to study dermatitis herpetiformis [219] and CeD [220], as well as to test novel potential therapies like elafin [127], gliadin sequestering therapy [82] and the gliadin based immunotherapy- TIMP-Glia [140]. A study by Caminero et al. on NOD mice also showed the potential incremental role of amylase trypsin inhibitors to gluten- induced damage as well as the tempering role of microbiota in CeD pathogenesis [221].

Abadie et al. later reported their work on DQ8-Dd-villin-IL15tg mice, that elucidated the role of IL15 in pathogenesis of different phenotypes. Transgenic DQ8-Dd-IL-15tg mice (that overexpress IL-15 in the lamina propria and mesenteric lymph nodes, but not in IECs), on being fed with gluten, developed TH1 immunity to gluten and anti-DGP antibodies without altering the cytolytic phenotype of IELs. By contrast, DQ8-villin-IL-15tg mice (that overexpress IL-15 in IECs) were unable to develop anti-gliadin IgG and anti-DGP antibodies; however, they displayed an expansion of IELs with high levels of granzyme B and perforin expression. Notably, neither DQ8-Dd-IL-15tg nor DQ8-villin-IL-15tg mice developed villous atrophy. DQ8-Dd-villin-IL-15tg mice were fed gluten for 30 days, approximately 75% of them developed small intestinal tissue

Table 7.5 Summary of exploratory therapeutic approaches, drugs and their potential applications in celiac disease.

Symptomatic improvement on a GFD		Replacing GFD as a treatment for CeD		
NRCD	Acute gluten exposure	Blocking immune activation	Inducing tolerance to gluten	Preventing development of CeD
Probiotics Glutenases Tight junction modulator-Larazotide Binding agents	Budesonide	tTG inhibitors T cell recruitment Anti IL-15	Vaccine Nanoparticles Bacteria with gliadin particles	?Probiotics ?Vaccine rotavirus Breastfeeding duration Timing of introduction of dietary gluten in infancy

destruction, which was later reversed after the exclusion of gluten, showing that IL-15 upregulation in both IECs and the lamina propria is required for the development of villous atrophy, Furthermore, as in patients with CeD, gluten-fed DQ8-Dd-villin-IL-15tg mice developed plasmacytosis in the lamina propria, and had circulating anti-gliadin IgG and IgA antibodies and anti-DGP IgG antibodies that recognize tTG-2 in the germline configuration [222].

HLA-DQ2 mouse model for CeD has also been prepared; in a study, C57BL/6 mice humanized with HLA-DR3-DQ2 gene, were subjected to an oral gluten challenge post-sensitization with gliadin. The mice were found to have subsequently, a low Vh:Cd ratio, higher IEL counts on flow cytometry, higher anti-tTG antibody and higher inflammatory gene expression in the intestinal epithelia. However, there was no increase in intestinal permeability. Overall, the new model could enhance preclinical testing as well as research on pathogenesis in patients with HLA-DQ2.5, the predominant haplotype found in CeD [223].

7.6 Summary

CeD is a prevalent enteropathy worldwide, with a known trigger/antigen (gluten immunogenic peptides). However, it is currently doomed with an imperfect therapy, the gluten-free diet. Several milestones have been reached toward the development of other therapies, notably a better understanding of the different checkpoints in the disease pathophysiology, the ability to identify gluten-specific T and B cells, as well as potential markers of an early immune response to gluten, such as IL-2 (see Table 7.5). CeD is nevertheless a unique, unprecedented model for drug testing. Although some therapeutic trials are directed toward individuals with active, non-responsive CeD, a significant proportion of clinical trials seek to stimulate active disease with a gluten challenge in individuals previously in remission to assess whether the agent prevents immune activation, clinical symptoms and mucosal damage. This is complicated by the various disease phenotypes, wide inter-individual heterogeneity in baseline histology, sensitivity to gluten and daily gluten exposure in their regular GFD. Consensus on a standardized gluten challenge, as well as which endpoints would be clinically meaningful for the CeD population are still awaited.

Nevertheless, a first phase 3 trial is currently ongoing and several lessons have been learned from previous trials, notably the Hawthorne effect, the lack of specificity of the PRO for active CeD, and the possibility of non-compliance to the gluten challenge by study participants. On the other hand, it has also contributed to the development of quantitative histology, methods to assess immune activation (such as ELISPOT), as well as the identification of cytokines that could signify an early immune response to gluten as well as potentially differentiate individuals with CeD from NCGS on a GFD.

The main objectives of novel therapeutics are to improve day-to-day symptoms and quality of life of individuals with CeD following a GFD or to prevent symptoms in case of accidental gluten exposure. The ultimate hope is to completely prevent immune activation and allow gluten consumption.

We now have in the pipeline approaches targeting gluten digestion, intestinal microbiota, intestinal permeability, as well as exciting new approaches that modulate downstream immune activation, by either blocking pathogenic effector pathways or stimulating immune tolerance. In the context of the latter, CeD with its known antigen and specific T- cell/B- cell response may be an ideal disease model for understanding and developing therapies for other auto-immune diseases. In today's research landscape, continuing discoveries of various molecular pathways in CeD pathogenesis offer hope for efficient development of safe and effective therapies.

References

[1] J. Tomal, D. McKiernan, S. Guandalini, C.E. Semrad, S. Kupfer, Celiac patients' attitudes regarding novel therapies, Minerva. Gastroenterol Dietol. 62 (2016) 275–280.

[2] V. Estévez, J. Ayala, C. Vespa, M. Araya, The gluten-free basic food basket: A problem of availability, cost and nutritional composition, Eur. J. Clin. Nutr. 70 (2016) 1215–1217 https://doi.org/10.1038/ejcn.2016.139.

[3] A. Rubio-Tapia, I.D. Hill, C.P. Kelly, A.H. Calderwood, JA. Murray, ACG clinical guidelines: Diagnosis and management of celiac disease, Am. J. Gastroenterol. 108 (2013) 656–676 https://doi.org/10.1038/ajg.2013.79.

[4] Celiac Disease – FDA-requested Listening Session FDA n.d. https://doi.org/10.1038/ajg.2013.79 (accessed March 11, 2020).

[5] S.M. Barratt, J.S. Leeds, DS. Sanders, Factors influencing the type, timing and severity of symptomatic responses to dietary gluten in patients with biopsy-proven coeliac disease, J. Gastrointest Liver Dis. 22 (2013) 391–396.

[6] J.A. Silvester, L.A. Graff, L. Rigaux, J.R. Walker, DR. Duerksen, Symptomatic suspected gluten

exposure is common among patients with coeliac disease on a gluten-free diet, Aliment. Pharmacol. Ther. 44 (2016) 612–619 https://doi.org/10.1111/apt.13725.

[7] G. Goel, A.J.M. Daveson, C.E. Hooi, J.A. Tye-Din, S. Wang, E. Szymczak, et al., Serum cytokines elevated during gluten-mediated cytokine release in coeliac disease, Clin. Exp. Immunol. (2019) cei.13369 https://doi.org/10.1111/cei.13369.

[8] A.K. Cartee, K.S. King, S. Wang, J.L. Dzuris, R.P. Anderson, C.T. van Dyke, et al., 825 – An Acute Rise in Serum Il-2 Levels But Not Symptoms Differentiates Celiac Disease Subjects from Non-Celiac Gluten Sensitivity and Healthy Subjects in a Single Dose Randomized Double Blind Placebo Controlled Gluten Challenge, Gastroenterology. 156 (2019) S177-S178 https://doi.org/10.1016/s0016-5085(19)37233-6.

[9] H.A. Penny, E.M.R. Baggus, A. Rej, J.A. Snowden, DS. Sanders, Non-Responsive Coeliac Disease: A Comprehensive Review from the NHS England National Centre for Refractory Coeliac Disease, Nutrients. 12 (2020) https://doi.org/10.3390/nu12010216.

[10] D.A. Leffler, M. Dennis, B. Hyett, E. Kelly, D. Schuppan, CP. Kelly, {A figure is presented}Etiologies and Predictors of Diagnosis in Nonresponsive Celiac Disease, Clin. Gastroenterol Hepatol. 5 (2007) 445–450 https://doi.org/10.1016/j.cgh.2006.12.006.

[11] C.A Catassi Prospective, Double-Blind, Placebo-Controlled Trial to Establish a Safe Gluten Threshold for Patients With Celiac Disease PubMed https://pubmed.ncbi.nlm.nih.gov/17209192-a-prospective-double-blind-placebo-controlled-trial-to-establish-a-safe-gluten-threshold-for-patients-with-celiac-disease/?from_term=%2210mg%22+AND+%22gluten%22+&from_size=100&from_pos=3 (accessed March 12, 2020).

[12] M. Leonard, P. Cureton, A. Fasano, Indications and Use of the Gluten Contamination Elimination Diet for Patients with Non-Responsive Celiac Disease, Nutrients. 9 (2017) 1129 https://doi.org/10.3390/nu9101129.

[13] A.F. Costa, E. Sugai, M. De La Paz, Temprano, S.I. Niveloni, H. Vázquez, M.L. Moreno, et al., Gluten immunogenic peptide excretion detects dietary transgressions in treated celiac disease patients, World. J. Gastroenterol. 25 (2019) 1409–1420 https://doi.org/10.3748/wjg.v25.i11.1409.

[14] K. Gerasimidis, K. Zafeiropoulou, M. Mackinder, U.Z. Ijaz, H. Duncan, E. Buchanan, et al., Comparison of clinical methods with the faecal gluten immunogenic peptide to assess gluten intake in coeliac disease, J. Pediatr. Gastroenterol Nutr. 67 (2018) 356–360 https://doi.org/10.1097/MPG.0000000000002062.

[15] A.F. Costa, E. Sugai, M. Temprano de la P, S.I. Niveloni, H. Vázquez, M.L. Moreno, et al., Gluten immunogenic peptide excretion detects dietary transgressions in treated celiac disease patients, World. J. Gastroenterol. 25 (2019) 1409–1420 https://doi.org/10.3748/wjg.v25.i11.1409.

[16] J.A. Silvester, I. Comino, C.P. Kelly, C. Sousa, D.R. Duerksen, C.N. Bernstein, et al., Most Patients With Celiac Disease on Gluten-Free Diets Consume Measurable Amounts of Gluten, Gastroenterology. 158 (2020) 1497–1499 e1. https://doi.org/10.1053/j.gastro.2019.12.016.

[17] J.P. Stefanolo, M. Tálamo, S. Dodds, M. Temprano de la P, A.F. Costa, M.L. Moreno, et al., Real-world Gluten Exposure in Patients With Celiac Disease on Gluten-Free Diets, Determined From Gliadin Immunogenic Peptides in Urine and Fecal Samples, Clin. Gastroenterol Hepatol. (2020) https://doi.org/10.1016/j.cgh.2020.03.038.

[18] M.L. Lähdeaho, M. Mäki, K. Laurila, H. Huhtala, K. Kaukinen, Small- bowel mucosal changes and antibody responses after low- and moderate-dose gluten challenge in celiac disease, BMC. Gastroenterol. 11 (2011) https://doi.org/10.1186/1471-230X-11-129.

[19] A. Roy, M. Minaya, M. Monegro, J. Fleming, R.K. Wong, S. Lewis, et al., Partner Burden: A Common Entity in Celiac Disease, Dig. Dis. Sci. 61 (2016) 3451–3459 https://doi.org/10.1007/s10620-016-4175-5.

[20] A. Lee, R. Wolf, B. Lebwohl, E. Ciaccio, P. Green, Persistent Economic Burden of the Gluten Free Diet, Nutrients. 11 (2019) 399 https://doi.org/10.3390/nu11020399.

[21] M. Gobbetti, E. Pontonio, P. Filannino, C.G. Rizzello, M. De Angelis, R. Di Cagno, How to improve the gluten-free diet: The state of the art from a food science perspective, Food. Res. Int. 110 (2018) 22–32 https://doi.org/10.1016/j.foodres.2017.04.010.

[22] C. Olsson, P. Lyon, A. Hörnell, A. Ivarsson, YM. Sydner, Food that makes you different: The stigma

experienced by adolescents with celiac disease, Qual. Health. Res. 19 (2009) 976–984 https://doi.org/10.1177/1049732309338722.

[23] A. Sarkhy, M.I. El Mouzan, E. Saeed, A. Alanazi, S. Alghamdi, S. Anil, et al., Socioeconomic impacts of gluten-free diet among Saudi children with celiac disease, Pediatr. Gastroenterol Hepatol. Nutr. 19 (2016) 162–167 https://doi.org/10.5223/pghn.2016.19.3.162.

[24] J.F. Ludvigsson, B. Lebwohl, Q. Chen, G. Bröms, R.L. Wolf, P.H.R. Green, et al., Anxiety after coeliac disease diagnosis predicts mucosal healing: a population-based study, Aliment. Pharmacol. Ther. 48 (2018) 1091–1098 https://doi.org/10.1111/apt.14991.

[25] R.L. Wolf, B. Lebwohl, A.R. Lee, P. Zybert, N.R. Reilly, J. Cadenhead, et al., Hypervigilance to a Gluten-Free Diet and Decreased Quality of Life in Teenagers and Adults with Celiac Disease, Dig. Dis. Sci. 63 (2018) 1438–1448 https://doi.org/10.1007/s10620-018-4936-4.

[26] H. Leinonen, L. Kivelä, M.L. Lähdeaho, H. Huhtala, K. Kaukinen, K. Kurppa, Daily life restrictions are common and associated with health concerns and dietary challenges in adult celiac disease patients diagnosed in childhood, Nutrients. 11 (2019) https://doi.org/10.3390/nu11081718.

[27] H.M. Staudacher, M. Kurien, K. Whelan, Nutritional implications of dietary interventions for managing gastrointestinal disorders, Curr. Opin. Gastroenterol. 34 (2018) 105–111 https://doi.org/10.1097/MOG.0000000000000421.

[28] J. Kikut, N. Konecka, M. Szczuko, Quantitative assessment of nutrition and nutritional status of patients with celiac disease aged 13–18, Rocz. Panstw. Zakl. Hig. 70 (2019) 359–367 https://doi.org/10.32394/rpzh.2019.0084.

[29] N.R. Reilly, B. Lebwohl, R. Hultcrantz, P.H.R. Green, J.F. Ludvigsson, Increased risk of non-alcoholic fatty liver disease after diagnosis of celiac disease, J. Hepatol. 62 (2015) 1405–1411 https://doi.org/10.1016/j.jhep.2015.01.013.

[30] J.F. Ludvigsson, P. Elfström, U. BroomÉ, A. Ekbom, S.M. Montgomery, Celiac Disease and Risk of Liver Disease: A General Population-Based Study, Clin. Gastroenterol Hepatol. 5 (2007) https://doi.org/10.1016/j.cgh.2006.09.034.

[31] N. Sansotta, K. Amirikian, S. Guandalini, H. Jericho, Celiac Disease Symptom Resolution, J. Pediatr. Gastroenterol Nutr. 66 (2018) 48–52 https://doi.org/10.1097/MPG.0000000000001634.

[32] S. Shah, M. Akbari, R. Vanga, C.P. Kelly, J. Hansen, T. Theethira, et al., Patient perception of treatment burden is high in celiac disease compared with other common conditions, Am. J. Gastroenterol. 109 (2014) 1304–1311 https://doi.org/10.1038/ajg.2014.29.

[33] K. Gerasimidis, K. Zafeiropoulou, M. Mackinder, U.Z. Ijaz, H. Duncan, E. Buchanan, et al., Comparison of Clinical Methods With the Faecal Gluten Immunogenic Peptide to Assess Gluten Intake in Coeliac Disease, J. Pediatr. Gastroenterol Nutr. 67 (2018) 356–360 https://doi.org/10.1097/MPG.0000000000002062.

[34] J.A. Silvester, I. Comino, C.P. Kelly, C. Sousa, D.R. Duerksen, C.N. Bernstein, et al., Most Patients With Celiac Disease on Gluten-Free Diets Consume Measurable Amounts of Gluten, Gastroenterology. 158 (2019) https://doi.org/10.1053/j.gastro.2019.12.016.

[35] A. Ravelli, V. Villanacci, C. Monfredini, S. Martinazzi, V. Grassi, S. Manenti, How patchy is patchy villous atrophy?: Distribution pattern of histological lesions in the duodenum of children with celiac disease, Am. J. Gastroenterol. 105 (2010) 2103–2110 https://doi.org/10.1038/ajg.2010.153.

[36] D.C. Adelman, J. Murray, T.-.T. Wu, M. Mäki, P.H. Green, C.P. Kelly, Measuring Change In Small Intestinal Histology In Patients With Celiac Disease, Am. J. Gastroenterol. 113 (2018) 339–347 https://doi.org/10.1038/ajg.2017.480.

[37] G. Goel, T. King, A.J. Daveson, J.M. Andrews, J. Krishnarajah, R. Krause, et al., Epitope-specific immunotherapy targeting CD4-positive T cells in coeliac disease: two randomised, double-blind, placebo-controlled phase 1 studies, Lancet. Gastroenterol Hepatol. 2 (2017) 479–493 https://doi.org/10.1016/S2468-1253(17)30110-3.

[38] A.J.M. Daveson, A. Popp, J. Taavela, K.E. Goldstein, J. Isola, K.E. Truitt, et al., Baseline quantitative histology in therapeutics trials reveals villus atrophy in most patients with coeliac disease who appear well controlled on gluten-free diet, GastroHep. 2 (2020) 22–30 https://doi.org/10.1002/ygh2.380.

[39] B. Lebwohl, F. Granath, A. Ekbom, S.M. Montgomery, J.A. Murray, A. Rubio-Tapia, et al., Mucosal healing and mortality in coeliac disease, Aliment. Pharmacol. Ther. 37 (2013) 332–339 https://doi.org/10.1111/apt.12164.

[40] M.T. Bethune, P. Strop, Y. Tang, L.M. Sollid, C. Khosla, Heterologous Expression, Purification, Refolding, and Structural-Functional Characterization of EP-B2, a Self-Activating Barley Cysteine Endoprotease, Chem. Biol. 13 (2006) 637–647 https://doi.org/10.1016/j.chembiol.2006.04.008.

[41] L. Shan, T. Marti, L.M. Sollid, G.M. Gray, C. Khosla, Comparative biochemical analysis of three bacterial prolyl endopeptidases: Implications for coeliac sprue, Biochem. J. 383 (2004) 311–318 https://doi.org/10.1042/BJ20040907.

[42] J. Gass, C. Khosla, Prolyl endopeptidases, Cell. Mol. Life. Sci. 64 (2007) 345–355 https://doi.org/10.1007/s00018-006-6317-y.

[43] E. J, G. S, M. B, M. J, K. C, Protein Engineering of Improved Prolyl Endopeptidases for Celiac Sprue Therapy, Protein. Eng. Des. Sel. 21 (2008) https://doi.org/10.1093/PROTEIN/GZN050.

[44] J.A. Tye-Din, R.P. Anderson, R.A. Ffrench, G.J. Brown, P. Hodsman, M. Siegel, et al., The effects of ALV003 pre-digestion of gluten on immune response and symptoms in celiac disease in vivo, Clin. Immunol. 134 (2010) 289–295 https://doi.org/10.1016/j.clim.2009.11.001.

[45] M.L. Lähdeaho, K. Kaukinen, K. Laurila, P. Vuotikka, O.P. Koivurova, T. Kärjä-Lahdensuu, et al., Glutenase ALV003 attenuates gluten-induced mucosal injury in patients with celiac disease, Gastroenterology. 146 (2014) 1649–1658 https://doi.org/10.1053/j.gastro.2014.02.031.

[46] J.A. Murray, C.P. Kelly, P.H.R. Green, A. Marcantonio, T.T. Wu, M. Mäki, et al., No Difference Between Latiglutenase and Placebo in Reducing Villous Atrophy or Improving Symptoms in Patients With Symptomatic Celiac Disease, Gastroenterology. 152 (2017) 787–798 e2 https://doi.org/10.1053/j.gastro.2016.11.004.

[47] J.A. Syage, J.A. Murray, P.H.R. Green, C. Khosla, Latiglutenase Improves Symptoms in Seropositive Celiac Disease Patients While on a Gluten-Free Diet, Dig. Dis. Sci. 62 (2017) 2428–2432 https://doi.org/10.1007/s10620-017-4687-7.

[48] Solutions for Celiac https://www.solutionsforceliac.com/(accessed May 6, 2020).

[49] K. Lindfors, J. Lin, H.-.S. Lee, H. Hyöty, M. Nykter, K. Kurppa, et al., Metagenomics of the faecal virome indicate a cumulative effect of enterovirus and gluten amount on the risk of coeliac disease autoimmunity in genetically at risk children: the TEDDY study, Gut. (2019) gutjnl-2019-319809 https://doi.org/10.1136/gutjnl-2019-319809.

[50] C. Wolf, J.B. Siegel, C. Tinberg, A. Camarca, C. Gianfrani, S. Paski, et al., Engineering of Kuma030: A Gliadin Peptidase That Rapidly Degrades Immunogenic Gliadin Peptides in Gastric Conditions, J. Am. Chem. Soc. 137 (2015) 13106–13113 https://doi.org/10.1021/jacs.5b08325.

[51] C. Wolf, J.B. Siegel, C. Tinberg, A. Camarca, C. Gianfrani, S. Paski, et al., Engineering of Kuma030: A Gliadin Peptidase That Rapidly Degrades Immunogenic Gliadin Peptides in Gastric Conditions, J. Am. Chem. Soc. 137 (2015) 13106–13113 https://doi.org/10.1021/jacs.5b08325.

[52] I. Pultz, D. Leffler, T. Liu, P. Winkle, V. Jm, M. Hill TAK-062 effectively digests gluten in the human stomach: results of a phase 1 study. n.d.

[53] N. Gujral, R. Löbenberg, M. Suresh, H. Sunwoo, In-vitro and in-vivo binding activity of chicken egg yolk immunoglobulin γ (IgY) against gliadin in food matrix, J. Agric. Food. Chem. 60 (2012) 3166–3172 https://doi.org/10.1021/jf205319s.

[54] D.A. Sample, H.H. Sunwoo, H.Q. Huynh, H.L. Rylance, C.L. Robert, B.W. Xu, et al., AGY, a Novel Egg Yolk-Derived Anti-gliadin Antibody, Is Safe for Patients with Celiac Disease, Dig. Dis. Sci. 62 (2017) 1277–1285 https://doi.org/10.1007/s10620-016-4426-5.

[55] R.D. Hills, B.A. Pontefract, H.R. Mishcon, C.A. Black, S.C. Sutton, CR. Theberge, Gut microbiome: Profound implications for diet and disease, Nutrients. 11 (2019) https://doi.org/10.3390/nu11071613.

[56] Z.Y. Kho, SK. Lal, The human gut microbiome - A potential controller of wellness and disease, Front. Microbiol. 9 (2018) https://doi.org/10.3389/fmicb.2018.01835.

[57] A. Caminero, J.L. McCarville, H.J. Galipeau, C. Deraison, S.P. Bernier, M. Constante, et al., Duodenal bacterial proteolytic activity determines sensitivity to dietary antigen through protease-activated receptor-2, Nat. Commun. 10 (2019) 1198 https://doi.org/10.1038/s41467-019-09037-9.

[58] M. Olivares, A. Neef, G. Castillejo, G. De Palma, V. Varea, A. Capilla, et al., The HLA-DQ2 genotype selects for early intestinal microbiota composition in infants at high risk of developing coeliac disease, Gut. 64 (2015) 406–417 https://doi.org/10.1136/gutjnl-2014-306931.

[59] M.M. Leonard, S. Camhi, T.B. Huedo-Medina, A. Fasano, Celiac disease genomic, environmental, microbiome, and metabolomic (CDGEMM) study design: Approach to the future of personalized prevention of celiac disease, Nutrients. 7 (2015) 9325–9336 https://doi.org/10.3390/nu7115470.

[60] R. D'Arienzo, R. Stefanile, F. Maurano, G. Mazzarella, E. Ricca, R. Troncone, et al., Immunomodulatory effects of Lactobacillus casei administration in a mouse model of gliadin-sensitive enteropathy, Scand. J. Immunol. 74 (2011) 335–341 https://doi.org/10.1111/j.1365-3083.2011.02582.x.

[61] K. Lindfors, T. Blomqvist, K. Juuti-Uusitalo, S. Stenman, J. Venäläinen, M. Mäki, et al., Live probiotic Bifidobacterium lactis bacteria inhibit the toxic effects induced by wheat gliadin in epithelial cell culture, Clin. Exp. Immunol. 152 (2008) 552–558 https://doi.org/10.1111/j.1365-2249.2008.03635.x.

[62] C. Papista, V. Gerakopoulos, A. Kourelis, M. Sounidaki, A. Kontana, L. Berthelot, et al., Gluten induces coeliac-like disease in sensitised mice involving IgA, CD71 and transglutaminase 2 interactions that are prevented by probiotics, Lab. Investig. 92 (2012) 625–635 https://doi.org/10.1038/labinvest.2012.13.

[63] R. Francavilla, M. Piccolo, A. Francavilla, L. Polimeno, F. Semeraro, F. Cristofori, et al., Clinical and Microbiological Effect of a Multispecies Probiotic Supplementation in Celiac Patients with Persistent IBS-type Symptoms: A Randomized, Double-Blind, Placebo-controlled, Multicenter Trial, J. Clin. Gastroenterol. 53 (2019) E117–E125 https://doi.org/10.1097/MCG.0000000000001023.

[64] E. Smecuol, H.J. Hwang, E. Sugai, L. Corso, A.C. Cherñavsky, F.P. Bellavite, et al., Exploratory, randomized, double-blind, placebo-controlled study on the effects of bifidobacterium infantis natren life start strain super strain in active celiac disease, J. Clin. Gastroenterol. 47 (2013) 139–147 https://doi.org/10.1097/MCG.0b013e31827759ac.

[65] M.I. Pinto-Sánchez, E.C. Smecuol, M.P. Temprano, E. Sugai, A. González, M.L. Moreno, et al., Bifidobacterium infantis NLS Super Strain Reduces the Expression of α-Defensin-5, a Marker of Innate Immunity, in the Mucosa of Active Celiac Disease Patients, J. Clin. Gastroenterol. 51 (2017) 814–817 https://doi.org/10.1097/MCG.0000000000000687.

[66] J. Harnett, S.P. Myers, M. Rolfe, Significantly higher faecal counts of the yeasts candida and saccharomyces identified in people with coeliac disease, Gut. Pathog. 9 (2017) 26 https://doi.org/10.1186/s13099-017-0173-1.

[67] D.A. Leffler, C.P. Kelly, P.H.R. Green, R.N. Fedorak, A. DiMarino, W. Perrow, et al., Larazotide acetate for persistent symptoms of celiac disease despite a gluten-free diet: a randomized controlled trial, Gastroenterology. 148 (2015) 1311–1319 e6 https://doi.org/10.1053/j.gastro.2015.02.008.

[68] A. Fasano, Zonulin and its regulation of intestinal barrier function: The biological door to inflammation, autoimmunity, and cancer, Physiol. Rev. 91 (2011) 151–175 https://doi.org/10.1152/physrev.00003.2008.

[69] S. Gopalakrishnan, M. Durai, K. Kitchens, A.P. Tamiz, R. Somerville, M. Ginski, et al., Larazotide acetate regulates epithelial tight junctions in vitro and in vivo, Peptides. 35 (2012) 86–94 https://doi.org/10.1016/j.peptides.2012.02.015.

[70] K. Boger, A.T. Blikslager, C.P. Prior, J. Madan, S. Laumas, B.R. Krishnan, et al., Su1017 – Establishment and Characterization of a Leaky Porcine Jejunal Cell Line Grown As a 2-Dimensional Monolayer Using Crypt Culture Media and Their Response to the Tight Junction Agent Larazotide Acetate, Gastroenterology. 156 (2019) S–486 https://doi.org/10.1016/s0016-5085(19)38076-x.

[71] L. Hernandez, A. Carlson, T. Pridgen, K. Messenger, C.P. Prior, S. Laumas, et al., Sa1183 – Larazotide Stimulates Recovery of Ischemic-Injured Intestine in a Dose-Dependent Manner Associated with Restoration of Tight Junctions. Gastroenterology https://doi.org/10.1016/s0016-5085.

[72] Y. Jin, C.P. Prior, J. Madan, S. Laumas, B.R. Krishnan, AT. Blikslager, Su1019 – Larazotide Protects the Intestinal Tight Junction Barrier During Anoxia/Reoxygenation Injury Via Inhibition of Myosin Light Chain Kinase, Gastroenterology. 156 (2019) S-487 https://doi.org/10.1016/s0016-5085(19)38078-3.

[73] K.E. Black, J.A. Murray, CS. David, HLA-DQ Determines the Response to Exogenous Wheat Proteins: A Model of Gluten Sensitivity in Transgenic Knockout Mice, J. Immunol. 169 (2002) 5595–5600 https://doi.org/10.4049/jimmunol.169.10.5595.

[74] B.M. Paterson, K.M. Lammers, M.C. Arrieta, A. Fasano, JB. Meddings, The safety, tolerance, pharmacokinetic and pharmacodynamic effects of single doses of AT-1001 in coeliac disease subjects: A proof of concept study, Aliment. Pharmacol. Ther. 26 (2007) 757–766 https://doi.org/10.1111/j.1365-2036.2007.03413.x.

[75] D.A. Leffler, C.P. Kelly, H.Z. Abdallah, A.M. Colatrella, L.A. Harris, F. Leon, et al., A randomized, double-blind study of larazotide acetate to prevent the activation of celiac disease during gluten challenge, Am. J. Gastroenterol. 107 (2012) 1554–1562 https://doi.org/10.1038/ajg.2012.211.

[76] C.P. Kelly, P.H.R. Green, J.A. Murray, A. Dimarino, A. Colatrella, D.A. Leffler, et al., Larazotide acetate in patients with coeliac disease undergoing a gluten challenge: A randomised placebo-controlled study, Aliment. Pharmacol. Ther. 37 (2013) 252–262 https://doi.org/10.1111/apt.12147.

[77] A. Rubio-Tapia, I.D. Hill, C.P. Kelly, American College of Gastroenterology. ACG clinical guidelines: diagnosis and management of celiac disease, Am. J. Gastroenterol. 108 (2013) 656–676.

[78] J.S. Leeds, A.D. Hopper, D.P. Hurlstone, S.J. Edwards, M.E. Mcalindon, A.J. Lobo, et al., Is exocrine pancreatic insufficiency in adult coeliac disease a cause of persisting symptoms? Aliment. Pharmacol. Ther. 25 (2007) 265–271 https://doi.org/10.1111/j.1365-2036.2006.03206.x.

[79] Sodium Polystyrene Sulfonate. National Library of Medicine (US) 2006.

[80] M. Pinier, E.F. Verdu, M. Nasser-Eddine, C.S. David, A. Vézina, N. Rivard, et al., Polymeric Binders Suppress Gliadin-Induced Toxicity in the Intestinal Epithelium, Gastroenterology. 136 (2009) 288–298 https://doi.org/10.1053/j.gastro.2008.09.016.

[81] M. Pinier, G. Fuhrmann, H.J. Galipeau, N. Rivard, J.A. Murray, C.S. David, et al., The copolymer P(HEMA-co-SS) binds gluten and reduces immune response in gluten-sensitized mice and human tissues, Gastroenterology. 142 (316-25) (2012) e1–12 https://doi.org/10.1053/j.gastro.2011.10.038.

[82] J.L. McCarville, Y. Nisemblat, H.J. Galipeau, J. Jury, R. Tabakman, A. Cohen, et al., BL-7010 demonstrates specific binding to gliadin and reduces gluten-associated pathology in a chronic mouse model of gliadin sensitivity, PLoS. One. 9 (2014) https://doi.org/10.1371/journal.pone.0109972.

[83] E. B, S. E, W. DJ, F. F, S. O, The Production of a Recombinant Tandem Single Chain Fragment Variable Capable of Binding Prolamins Triggering Celiac Disease, BMC Biotechnol. 18 (2018) https://doi.org/10.1186/S12896-018-0443-0.

[84] V. Stadlmann, H. Harant, I. Korschineck, M. Hermann, F. Forster, A. Missbichler, Novel avian single-chain fragment variable (scFv) targets dietary gluten and related natural grain prolamins, toxic entities of celiac disease, BMC Biotechnol. 15 (2015) https://doi.org/10.1186/s12896-015-0223-z.

[85] R. Borghini, G. Donato, M. Marino, R. Casale, A. Picarelli, Culture of gastric biopsies in celiac disease and its relationship with gastritis and Helicobacter pylori infection, Dig. Liver. Dis. 50 (2018) 97–100 https://doi.org/10.1016/j.dld.2017.10.011.

[86] M. Heyman, J. Abed, C. Lebreton, N. Cerf-Bensussan, Intestinal permeability in coeliac disease: Insight into mechanisms and relevance to pathogenesis, Gut. 61 (2012) 1355–1364 https://doi.org/10.1136/gutjnl-2011-300327.

[87] M. Bruewer, M. Utech, A.I. Ivanov, A.M. Hopkins, C.A. Parkos, A. Nusrat, Interferon-γ induces internalization of epithelial tight junction proteins via a macropinocytosis-like process, FASEB J. 19 (2005) 923–933 https://doi.org/10.1096/fj.04-3260com.

[88] B. RN, B. PJ, E. BM, W. PL, G. M, Alterations in the Distribution and Activity of Transglutaminase During Tumour Growth and Metastasis, Carcinogenesis. 6 (1985) https://doi.org/10.1093/CARCIN/6.3.459.

[89] E. Junn, R.D. Ronchetti, M.M. Quezado, S.Y. Kim, MM. Mouradian, Tissue transglutaminase-induced aggregation of α-synuclein: Implications for Lewy body formation in Parkinson's disease and dementia with Lewy bodies, Proc. Natl. Acad. Sci. U. S. A. 100 (2003) 2047–2052 https://doi.org/10.1073/pnas.0438021100.

[90] D.M. Appelt, G.C. Kopen, L.J. Boyne, BJ. Balin, Localization of transglutaminase in hippocampal neurons: Implications for Alzheimer's disease, J. Histochem. Cytochem. 44 (1996) 1421–1427 https://doi.org/10.1177/44.12.8985134.

[91] P. Kahlem, H. Green, P. Djian, Transglutaminase action imitates Huntington's disease: Selective polymerization of huntingtin containing expanded polyglutamine, Mol. Cell. 1 (1998) 595–601 https://doi.org/10.1016/S1097-2765(00)80059-3.

[92] C. Marrano, P. de Macédo, P. Gagnon, D. Lapierre, C. Gravel, JW. Keillor, Synthesis and evaluation of novel dipeptide-bound 1,2,4-thiadiazoles as irreversible inhibitors of guinea pig liver transglutaminase, Bioorg. Med. Chem. 9 (2001) 3231–3241 https://doi.org/10.1016/S0968-0896(01)00228-0.

[93] K. Choi, M. Siegel, J.L. Piper, L. Yuan, E. Cho, P. Strnad, et al., Chemistry and biology of dihydroisoxazole derivatives: Selective inhibitors of human transglutaminase 2, Chem. Biol. 12 (2005) 469–475 https://doi.org/10.1016/j.chembiol.2005.02.007.

[94] V. Abadie, S.M. Kim, T. Lejeune, B.A. Palanski, J.D. Ernest, O. Tastet, et al., IL-15, gluten and HLA-DQ8 drive tissue destruction in coeliac disease, Nature. (2020) https://doi.org/10.1038/s41586-020-2003-8.

[95] D.M. Pinkas, P. Strop, A.T. Brunger, C. Khosla, Transglutaminase 2 undergoes a large conformational change upon activation, PLoS. Biol. 5 (2007) 2788–2796 https://doi.org/10.1371/journal.pbio.0050327.

[96] Z. Szondy, Z. Sarang, P. Molnár, T. Németh, M. Piacentini, P.G. Mastroberardino, et al., Transglutaminase 2-/- mice reveal a phagocytosis-associated crosstalk between macrophages and apoptotic cells, Proc. Natl. Acad. Sci. U. S. A. 100 (2003) 7812–7817 https://doi.org/10.1073/pnas.0832466100.

[97] Press release: Dr. Falk Pharma and Zedira announce completion of phase 1b clinical trial of ZED1227 for the treatment of celiac disease and move on to proof of concept study 2017 News ZEDIRA GmbH, Darmstadt, Germany https://zedira.com/News/Press-release-Dr-Falk-Pharma-and-Zedira-announce-completion-of-phase-1b-clinical-trial-of-ZED1227-for-the-treatment-of-celiac-disease-and-move-on-to-proof-of-concept-study_97?oswsid=b5e45e4b75ea60bce78c28aaf 705687a (accessed April 6, 2020).

[98] M.A.E. Ventura, K. Sajko, M. Hils, R. Pasternack, R. Greinwald, B. Tewes, et al., Su1161 - The Oral Transglutaminase 2 (TG2) Inhibitor Zed1227 Blocks TG2 Activity in a Mouse Model of Intestinal Inflammation, Gastroenterology. 154 (2018) S–490 https://doi.org/10.1016/s0016-5085.1831861-4.

[99] B. Jabri, V. Abadie, IL-15 functions as a danger signal to regulate tissue-resident T cells and tissue destruction, Nat. Rev. Immunol. 15 (2015) 771–783 https://doi.org/10.1038/nri3919.

[100] M.-.L. Lähdeaho, M. Scheinin, M. Pesu, L. Kivelä, Z. Lovró, J. Keisala, et al., 618 - AMG 714 (Anti-IL-15 MAB) Ameliorates the Effects of Gluten Consumption in Celiac Disease: A Phase 2A, Randomized, Double-Blind, Placebo-Controlled Study Evaluating AMG 714 in Adult Patients with Celiac Disease Exposed to a High-Dose Gluten Challenge, Gastroenterology. 154 (18) (2018) S–130 https://doi.org/10.1016/s0016-5085. 30861-8.

[101] M.-.L. Lähdeaho, M. Scheinin, P. Vuotikka, J. Taavela, A. Popp, J. Laukkarinen, et al., Safety and efficacy of AMG 714 in adults with coeliac disease exposed to gluten challenge: a phase 2a, randomised, double-blind, placebo-controlled study, Lancet. Gastroenterol Hepatol. 4 (2019) 948–959 https://doi.org/10.1016/S2468-1253(19)30264-X.

[102] (No Title) n.d. https://www.sec.gov/Archives/edgar/data/1695357/000149315219003585/ex99-1.htm (accessed April 28, 2020).

[103] K. Sestak, J.P. Dufour, D.X. Liu, N. Rout, X. Alvarez, J. Blanchard, et al., Beneficial effects of human anti-interleukin-15 antibody in gluten-sensitive rhesus macaques with celiac disease, Front. Immunol. 9 (2018) https://doi.org/10.3389/fimmu.2018.01603.

[104] C. Ciszewski, V. Discepolo, A. Pacis, N. Doerr, O. Tastet, T. Mayassi, et al., Identification of a γc Receptor Antagonist that Prevents Reprogramming of Human Tissue-resident Cytotoxic T Cells by IL15 and IL21, Gastroenterology. (2019) https://doi.org/10.1053/j.gastro.2019.10.006.

[105] R.D.A. Wilkinson, R. Williams, C.J. Scott, R.E. Burden, Cathepsin S: Therapeutic, diagnostic, and prognostic potential, Biol. Chem. 396 (2015) 867–882 https://doi.org/10.1515/hsz-2015-0114.

[106] M. Theron, D. Bentley, S. Nagel, M. Manchester, M. Gerg, T. Schindler, et al., Pharmacodynamic monitoring of RO5459072, a small molecule inhibitor of cathepsin S, Front. Immunol. 8 (2017) https://doi.org/10.3389/fimmu.2017.00806.

[107] M.J. Walters, Y. Wang, N. Lai, T. Baumgart, B.N. Zhao, D.J. Dairaghi, et al., Characterization of CCX282-B, an orally bioavailable antagonist of the CCR9 chemokine receptor, for treatment of inflammatory bowel disease, J. Pharmacol. Exp. Ther. 335 (2010) 61–69 https://doi.org/10.1124/jpet.110.169714.

[108] N.J. Tubo, M.A. Wurbel, T.T. Charvat, T.J. Schall, M.J. Walters, JJ. Campbell, A Systemically-Administered Small Molecule Antagonist of CCR9 Acts as a Tissue-Selective Inhibitor of Lymphocyte Trafficking, PLoS One. 7 (2012) https://doi.org/10.1371/journal.pone.0050498.

[109] T.B. Smallwood, P.R. Giacomin, A. Loukas, J.P. Mulvenna, R.J. Clark, JJ. Miles, Helminth immunomodulation in autoimmune disease, Front. Immunol. 8 (2017) https://doi.org/10.3389/fimmu.2017.00453.

[110] J.F. Bach, The effect of infections on susceptibility to autoimmune and allergic diseases, N. Engl. J. Med. 347 (2002) 911–920 https://doi.org/10.1056/NEJMra020100.

[111] P. Giacomin, M. Zakrzewski, T.P. Jenkins, X. Su, R. Al-Hallaf, J. Croese, et al., Changes in duodenal tissue-associated microbiota following hookworm infection and consecutive gluten challenges in humans with coeliac disease, Sci. Rep. 6 (2016) https://doi.org/10.1038/srep36797.

[112] J. Croese, P. Giacomin, S. Navarro, A. Clouston, L. McCann, A. Dougall, et al., Experimental hookworm infection and gluten microchallenge promote tolerance in celiac disease, J. Allergy. Clin. Immunol. 135 (2015) 508–516 e5 https://doi.org/10.1016/j.jaci.2014.07.022.

[113] A. Therrien, J.A. Silvester, D.A. Leffler, C.P. Kelly, Efficacy of Enteric-Release Oral Budesonide in Treatment of Acute Reactions to Gluten in Patients With Celiac Disease, Clin. Gastroenterol Hepatol. (2019) https://doi.org/10.1016/j.cgh.2019.03.029.

[114] C. Ciacci, L. Maiuri, I. Russo, R. Tortora, C. Bucci, C. Cappello, et al., Efficacy of budesonide therapy in the early phase of treatment of adult coeliac disease patients with malabsorption: An *in vivo / in vitro* pilot study, Clin. Exp. Pharmacol. Physiol. 36 (2009) 1170–1176 https://doi.org/10.1111/j.1440-1681.2009.05211.x.

[115] G. Pelaia, A. Vatrella, M.T. Busceti, F. Fabiano, R. Terracciano, R.M. Maria Gabriella Matera, Molecular and Cellular Mechanisms Underlying the Therapeutic Effects of Budesonide in Asthma, Pulm. Pharmacol. Ther. (2016) 15–21 https://doi.org/10.1016/j.pupt.2016.07.001.

[116] S. Jamma, A. Rubio-Tapia, C.P. Kelly, J. Murray, R. Najarian, S. Sheth, et al., Celiac Crisis Is a Rare but Serious Complication of Celiac Disease in Adults, Clin. Gastroenterol Hepatol. 8 (2010) 587–590 https://doi.org/10.1016/j.cgh.2010.04.009.

[117] S.S. Mukewar, A. Sharma, A. Rubio-Tapia, T.T. Wu, B. Jabri, JA. Murray, Open-Capsule Budesonide for Refractory Celiac Disease, Am. J. Gastroenterol. 112 (2017) 959–967 https://doi.org/10.1038/ajg.2017.71.

[118] M. Setty, V. Discepolo, V. Abadie, S. Kamhawi, T. Mayassi, A. Kent, et al., Distinct and Synergistic Contributions of Epithelial Stress and Adaptive Immunity to Functions of Intraepithelial Killer Cells and Active Celiac Disease, Gastroenterology. 149 (2015) 681e10-691.e10 https://doi.org/10.1053/j.gastro.2015.05.013.

[119] C. Ciszewski, V. Discepolo, A. Pacis, N. Doerr, O. Tastet, T. Mayassi, et al., Identification of a γc Receptor Antagonist That Prevents Reprogramming of Human Tissue-resident Cytotoxic T Cells by IL15 and IL21, Gastroenterology. 158 (2020) 625–637 e13 https://doi.org/10.1053/j.gastro.2019.10.006.

[120] T. Nata, A. Basheer, F. Cocchi, R. Van Besien, R. Massoud, S. Jacobson, et al., Targeting the binding interface on a shared receptor subunit of a cytokine family enables the inhibition of multiple member cytokines with selectable target spectrum, J. Biol. Chem. 290 (2015) 22338–22351 https://doi.org/10.1074/jbc.M115.661074.

[121] V.R. Villella, A. Venerando, G. Cozza, S. Esposito, E. Ferrari, R. Monzani, et al., A pathogenic role for cystic fibrosis transmembrane conductance regulator in celiac disease, EMBO J. 38 (2019) https://doi.org/10.15252/embj.2018100101.

[122] L. Maiuri, V.R. Villella, V. Raia, G. Kroemer, The gliadin-CFTR connection: New perspectives for the treatment of celiac disease, Ital. J. Pediatr. 45 (2019) https://doi.org/10.1186/s13052-019-0627-9.

[123] S. Esposito, V.R. Villella, E. Ferrari, R. Monzani, A. Tosco, F. Rossin, et al., Genistein antagonizes gliadin-induced CFTR malfunction in models of celiac disease, Aging (Albany NY). 11 (2019) 2003–2019 https://doi.org/10.18632/aging.101888.

[124] S. Yokoyama, P.Y. Perera, T.A. Waldmann, T. Hiroi, LP. Perera, Tofacitinib, a Janus Kinase inhibitor demonstrates efficacy in an IL-15 transgenic mouse model that recapitulates pathologic manifestations of celiac disease, J. Clin. Immunol. 33 (2013) 586–594 https://doi.org/10.1007/s10875-012-9849-y.

[125] S.E. Williams, T.I. Brown, A. Roghanian, JM. Sallenave, SLPI and elafin: One glove, many fingers, Clin. Sci. 110 (2006) 21–35 https://doi.org/10.1042/CS20050115.

[126] J.P. Motta, L.G. Bermúdez-Humarán, C. Deraison, L. Martin, C. Rolland, P. Rousset, et al., Food-grade bacteria expressing elafin protect against inflammation and restore colon homeostasis, Sci. Transl. Med. 4 (2012) https://doi.org/10.1126/scitranslmed.3004212.

[127] H.J. Galipeau, M. Wiepjes, J.P. Motta, J.D. Schulz, J. Jury, J.M. Natividad, et al., Novel role of the serine protease inhibitor elafin in gluten-related disorders, Am. J. Gastroenterol. 109 (2014) 748–756 https://doi.org/10.1038/ajg.2014.48.

[128] J.E. Ollague, CH. Nousari, Expression of elafin in dermatitis herpetiformis, Am. J. Dermatopathol. 40 (2018) 1–6 https://doi.org/10.1097/DAD.0000000000000915.

[129] J. Xia, E. Bergseng, B. Fleckenstein, M. Siegel, C.Y. Kim, C. Khosla, et al., Cyclic and dimeric gluten peptide analogues inhibiting DQ2-mediated antigen presentation in celiac disease, Bioorganic. Med. Chem. 15 (2007) 6565–6573 https://doi.org/10.1016/j.bmc.2007.07.001.

[130] N. Ji, A. Somanaboeina, A. Dixit, K. Kawamura, N.J. Hayward, C. Self, et al., Small Molecule Inhibitor of Antigen Binding and Presentation by HLA-DR2b as a Therapeutic Strategy for the Treatment of Multiple Sclerosis, J. Immunol. 191 (2013) 5074–5084 https://doi.org/10.4049/jimmunol.1300407.

[131] B. Jabri, LM. Sollid, T Cells in Celiac Disease, J. Immunol. 198 (2017) 3005–3014 https://doi.org/10.4049/jimmunol.1601693.

[132] L.S. Høydahl, R. Frick, I. Sandlie, GÅ. Løset, Targeting the MHC Ligandome by Use of TCR-Like Antibodies, Antibodies. 8 (2019) 32 https://doi.org/10.3390/antib8020032.

[133] G. Serena, S. Yan, S. Camhi, S. Patel, R.S. Lima, A. Sapone, et al., Proinflammatory cytokine interferon-γ and microbiome-derived metabolites dictate epigenetic switch between forkhead box protein 3 isoforms in coeliac disease, Clin. Exp. Immunol. 187 (2017) 490–506 https://doi.org/10.1111/cei.12911.

[134] B.R. Burton, G.J. Britton, H. Fang, J. Verhagen, B. Smithers, C.A. Sabatos-Peyton, et al., Sequential transcriptional changes dictate safe and effective antigen-specific immunotherapy, Nat. Commun. 5 (2014) https://doi.org/10.1038/ncomms5741.

[135] J.A. Tye-Din, A.J.M. Daveson, H.C. Ee, G. Goel, J. MacDougall, S. Acaster, et al., Elevated serum interleukin-2 after gluten correlates with symptoms and is a potential diagnostic biomarker for coeliac disease, Aliment. Pharmacol. Ther. 50 (2019) 901–910 https://doi.org/10.1111/apt.15477.

[136] P. Serra, P. Santamaria, Antigen-specific therapeutic approaches for autoimmunity, Nat. Biotechnol. 37 (2019) 238–251 https://doi.org/10.1038/s41587-019-0015-4.

[137] R. Kuo, E. Saito, S.D. Miller, LD. Shea, Peptide-Conjugated Nanoparticles Reduce Positive Co-stimulatory Expression and T Cell Activity to Induce Tolerance, Mol. Ther. 25 (2017) 1676–1685 https://doi.org/10.1016/j.ymthe.2017.03.032.

[138] D.R. Getts, D.M. Turley, C.E. Smith, C.T. Harp, D. McCarthy, E.M. Feeney, et al., Tolerance Induced by Apoptotic Antigen-Coupled Leukocytes Is Induced by PD-L1 + and IL-10–Producing Splenic Macrophages and Maintained by T Regulatory Cells, J. Immunol. 187 (2011) 2405–2417 https://doi.org/10.4049/jimmunol.1004175.

[139] S. Prasad, T. Neef, D. Xu, J.R. Podojil, D.R. Getts, L.D. Shea, et al., Tolerogenic Ag-PLG nanoparticles induce tregs to suppress activated diabetogenic CD4 and CD8 T cells, J. Autoimmun. 89 (2018) 112–124 https://doi.org/10.1016/j.jaut.2017.12.010.

[140] T.L. Freitag, J.R. Podojil, R.M. Pearson, F.J. Fokta, C. Sahl, M. Messing, et al., Gliadin Nanoparticles Induce Immune Tolerance to Gliadin in Mouse Models of Celiac Disease, Gastroenterology. 158 (2020) 1667–1681 e12 https://doi.org/10.1053/j.gastro.2020.01.045.

[141] C. Kelly, J. Murray, D.A Leffler, A. Bledsoe, G. Smithson, J. Podojil, R. First, A. Morris, M. Boyne, A. Elhofy, T.T. Wu, S. Miller, CNP-101 PREVENTS GLUTEN CHALLENGE INDUCED IMMUNE ACTIVATION IN ADULTS WITH CELIAC DISEASE, in: United Eur. Gastroenterol Week., Barcelona, 2019.

[142] X. Clemente-Casares, J. Blanco, P. Ambalavanan, J. Yamanouchi, S. Singha, C. Fandos, et al., Expanding antigen-specific regulatory networks to treat autoimmunity, Nature. 530 (2016) 434–440 https://doi.org/10.1038/nature16962.

[143] D.P. Cook, C. Gysemans, C. Mathieu, Lactococcus lactis as a versatile vehicle for tolerogenic immunotherapy, Front Immunol. 8 (2018) https://doi.org/10.3389/fimmu.2017.01961.

[144] I.L. Huibregtse, E.V. Marietta, S. Rashtak, F. Koning, P. Rottiers, C.S. David, et al., Induction of Antigen-Specific Tolerance by Oral Administration of Lactococcus lactis Delivered Immunodominant DQ8-Restricted Gliadin Peptide in Sensitized Nonobese Diabetic Ab° Dq8 Transgenic Mice, J. Immunol. 183 (2009) 2390–2396 https://doi.org/10.4049/jimmunol.0802891.

[145] L.F. Risnes, A. Christophersen, S. Dahal-Koirala, R.S. Neumann, G.K. Sandve, V.K. Sarna, et al., Disease-driving CD4+ T cell clonotypes persist for decades in celiac disease, J. Clin. Invest. 128 (2018) 2642–2650 https://doi.org/10.1172/JCI98819.

[146] T. Mayassi, K. Ladell, H. Gudjonson, J.E. McLaren, D.G. Shaw, M.T. Tran, et al., Chronic Inflammation Permanently Reshapes Tissue-Resident Immunity in Celiac Disease, Cell. 176 (2019) 967–981 e19 https://doi.org/10.1016/j.cell.2018.12.039.

[147] R. Bouziat, R. Hinterleitner, J.J. Brown, J.E. Stencel-Baerenwald, M. Ikizler, T. Mayassi, et al., Reovirus infection triggers inflammatory responses to dietary antigens and development of celiac disease, Science. 356 (2017) 44–50 80- https://doi.org/10.1126/science.aah5298.

[148] R. Bouziat, S.B. Biering, E. Kouame, K.A. Sangani, S. Kang, J.D. Ernest, et al., Murine Norovirus Infection Induces TH1 Inflammatory Responses to Dietary Antigens, Cell Host Microbe. 24 (2018) 677–688 e5 https://doi.org/10.1016/j.chom.2018.10.004.

[149] R. Bouziat, R. Hinterleitner, J.J. Brown, J.E. Stencel-Baerenwald, M. Ikizler, T. Mayassi, et al., Reovirus infection triggers inflammatory responses to dietary antigens and development of celiac disease, Science. 356 (2017) 44–50 80- https://doi.org/10.1126/science.aah5298.

[150] L.C. Stene, M.C. Honeyman, E.J. Hoffenberg, J.E. Haas, R.J. Sokol, L. Emery, et al., Rotavirus infection frequency and risk of celiac disease autoimmunity in early childhood: A longitudinal study, Am. J. Gastroenterol. 101 (2006) 2333–2340 https://doi.org/10.1111/j.1572-0241.2006.00741.x.

[151] K.M. Kemppainen, K.F. Lynch, E. Liu, M. Lönnrot, V. Simell, T. Briese, et al., Factors That Increase Risk of Celiac Disease Autoimmunity After a Gastrointestinal Infection in Early Life, Clin. Gastroenterol Hepatol. 15 (2017) 694–702 e5 https://doi.org/10.1016/j.cgh.2016.10.033.

[152] K. Lindfors, J. Lin, H.S. Lee, H. Hyöty, M. Nykter, K. Kurppa, et al., Metagenomics of the faecal virome indicate a cumulative effect of enterovirus and gluten amount on the risk of coeliac disease autoimmunity in genetically at risk children: The TEDDY study, Gut. (2019) https://doi.org/10.1136/gutjnl-2019-319809.

[153] N. Radlovic, Z. Lekovic, V. Radlovic, D. Simic, B. Vuletic, S. Ducic, et al., Celiac crisis in children in Serbia, Ital. J. Pediatr. 42 (2016) https://doi.org/10.1186/s13052-016-0233-z.

[154] L.S. Høydahl, L. Richter, R. Frick, O. Snir, K.S. Gunnarsen, O.J.B. Landsverk, et al., Plasma Cells Are the Most Abundant Gluten Peptide MHC-expressing Cells in Inflamed Intestinal Tissues From Patients With Celiac Disease, Gastroenterology. 156 (2019) 1428–1439 e10 https://doi.org/10.1053/j.gastro.2018.12.013.

[155] A. Christophersen, E.G. Lund, O. Snir, E. Solà, C. Kanduri, S. Dahal-Koirala, et al., Distinct phenotype of CD4+ T cells driving celiac disease identified in multiple autoimmune conditions, Nat. Med. 25 (2019) 734–737 https://doi.org/10.1038/s41591-019-0403-9.

[156] S.A. Jenks, K.S. Cashman, E. Zumaquero, U.M. Marigorta, A.V. Patel, X. Wang, et al., Distinct Effector B Cells Induced by Unregulated Toll-like Receptor 7 Contribute to Pathogenic Responses in Systemic Lupus Erythematosus, Immunity. 49 (2018) 725–739 e6 https://doi.org/10.1016/j.immuni.2018.08.015.

[157] R. Iversen, L.M. Sollid, Autoimmunity provoked by foreign antigens: In celiac disease, exogenous gluten drives T cell-B cell interactions that cause autoimmunity, Science. 368 (2020) 132–133 80- https://doi.org/10.1126/science.aay3037.

[158] J. Petersen, L. Ciacchi, M.T. Tran, K.L. Loh, Y. Kooy-Winkelaar, N.P. Croft, et al., T cell receptor cross-reactivity between gliadin and bacterial peptides in celiac disease, Nat. Struct. Mol. Biol. 27 (2020) 49–61 https://doi.org/10.1038/s41594-019-0353-4.

[159] P. Brar, G.Y. Kwon, I.I. Egbuna, S. Holleran, R. Ramakrishnan, G. Bhagat, et al., Lack of correlation of degree of villous atrophy with severity of clinical presentation of coeliac disease, Dig. Liver. Dis. 39 (2007) 26–29 https://doi.org/10.1016/j.dld.2006.07.014.

[160] H.S. Kim, D.M. Courtney, D.M. McCarthy, D. Cella, Patient-reported Outcome Measures in Emergency Care Research: A Primer for Researchers, Peer Reviewers, and Readers, Acad. Emerg. Med. (2020) https://doi.org/10.1111/acem.13918.

[161] D.A. Leffler, C.P. Kelly, H.Z. Abdallah, A.M. Colatrella, L.A. Harris, F. Leon, et al., A randomized, double-blind study of larazotide acetate to prevent the activation of celiac disease during gluten challenge, Am. J. Gastroenterol. 107 (2012) 1554–1562 https://doi.org/10.1038/ajg.2012.211.

[162] J. Svedlund, I. Sjödin, G. Dotevall, GSRS-A clinical rating scale for gastrointestinal symptoms in patients with irritable bowel syndrome and peptic ulcer disease, Dig. Dis. Sci. 33 (1988) 129–134 https://doi.org/10.1007/BF01535722.

[163] L.D. McLeod, C.D. Coon, S.A. Martin, S.E. Fehnel, R.D. Hays, Interpreting patient-reported outcome results: US FDA guidance and emerging methods, Expert Rev. Pharmacoeconomics Outcomes Res. 11 (2011) 163–169 https://doi.org/10.1586/erp.11.12.

[164] J.A. Murray, C.P. Kelly, P.H.R. Green, A. Marcantonio, T.-.T. Wu, M. Mäki, et al., No Difference Between Latiglutenase and Placebo in Reducing Villous Atrophy or Improving Symptoms in Patients With Symptomatic Celiac Disease, Gastroenterology. 152 (2017) 787–798 e2 https://doi.org/10.1053/j.gastro.2016.11.004.

[165] H. Crocker, C. Jenkinson, M. Peters, Quality of life in coeliac disease: item reduction, scale development and psychometric evaluation of the Coeliac Disease Assessment Questionnaire (CDAQ), Aliment. Pharmacol. Ther. 48 (2018) 852–862 https://doi.org/10.1111/apt.14942.

[166] MJ. Bruins, The clinical response to gluten challenge: A review of the literature, Nutrients. 5 (2013) 4614–4641 https://doi.org/10.3390/nu5114614.

[167] E. Smecuol, H.J. Hwang, E. Sugai, L. Corso, A.C. Cherñavsky, F.P. Bellavite, et al., Exploratory, randomized, double-blind, placebo-controlled study on the effects of Bifidobacterium infantis natren life start strain super strain in active celiac disease, J. Clin. Gastroenterol. 47 (2013) 139–147 https://doi.org/10.1097/MCG.0b013e31827759ac.

[168] C.P. Kelly, P.H.R. Green, J.A. Murray, A. Dimarino, A. Colatrella, D.A. Leffler, et al., Larazotide acetate in patients with coeliac disease undergoing a gluten challenge: a randomised placebo-controlled study, Aliment. Pharmacol. Ther. 37 (2013) 252–262 https://doi.org/10.1111/apt.12147.

[169] M.-.L. Lähdeaho, M. Scheinin, P. Vuotikka, J. Taavela, A. Popp, J. Laukkarinen, et al., Safety and efficacy of AMG 714 in adults with coeliac disease exposed to gluten challenge: a phase 2a, randomised, double-blind, placebo-controlled study, Lancet. Gastroenterol Hepatol. 4 (2019) 948–959 https://doi.org/10.1016/S2468-1253(19)30264-X.

[170] J.P.W. Burger, B. de Brouwer, J. IntHout, P.J. Wahab, M. Tummers, JPH. Drenth, Systematic review with meta-analysis: Dietary adherence influences normalization of health-related quality of life in coeliac disease, Clin. Nutr. 36 (2017) 399–406 https://doi.org/10.1016/j.clnu.2016.04.021.

[171] D.A. Leffler, S. Acaster, K. Gallop, M. Dennis, C.P. Kelly, DC. Adelman, A Novel Patient-Derived Conceptual Model of the Impact of Celiac Disease in Adults: Implications for Patient-Reported Outcome and Health-Related Quality-of-Life Instrument Development, Value. Heal. 20 (2017) 637–643 https://doi.org/10.1016/j.jval.2016.12.016.

[172] S. Ford, R. Howard, J. Oyebode, Psychosocial aspects of coeliac disease: A cross-sectional survey of a UK population, Br. J. Health. Psychol. 17 (2012) 743–757 https://doi.org/10.1111/j.2044-8287.2012.02069.x.

[173] A. Paavola, K. Kurppa, A. Ukkola, P. Collin, M.L. Lähdeaho, H. Huhtala, et al., Gastrointestinal symptoms and quality of life in screen-detected celiac disease, Dig. Liver. Dis. 44 (2012) 814–818 https://doi.org/10.1016/j.dld.2012.04.019.

[174] S.F. Bakker, F. Pouwer, M.E. Tushuizen, R.P. Hoogma, C.J. Mulder, S. Simsek, Compromised quality of life in patients with both Type 1 diabetes mellitus and coeliac disease, Diabet. Med. 30 (2013) 835–839 https://doi.org/10.1111/dme.12205.

[175] W. Häuser, A. Stallmach, W.F. Caspary, J. Stein, Predictors of reduced health-related quality of life in adults with coeliac disease, Aliment. Pharmacol. Ther. 25 (2007) 569–578 https://doi.org/10.1111/j.1365-2036.2006.03227.x.

[176] A.R. Lee, R. Wolf, I. Contento, H. Verdeli, PHR. Green, Coeliac disease: The association between quality of life and social support network participation, J. Hum. Nutr. Diet. 29 (2016) 383–390 https://doi.org/10.1111/jhn.12319.

[177] F. Casellas, L. Rodrigo, A.J. Lucendo, F. Fernández-Bañares, J. Molina-Infante, S. Vivas, et al., Benefit on health-related quality of life of adherence to gluten-free diet in adult patients with celiac disease, Rev. Esp. Enfermedades. Dig. 107 (2015) 196–201.

[178] J.F. Ludvigsson, C. Ciacci, P.H.R. Green, K. Kaukinen, I.R. Korponay-Szabo, K. Kurppa, et al., Outcome measures in coeliac disease trials: The Tampere recommendations, Gut. 67 (2018) 1410–1424 https://doi.org/10.1136/gutjnl-2017-314853.

[179] S.D. Dorn, L. Hernandez, M.T. Minaya, C.B. Morris, Y. Hu, J. Leserman, et al., The development and validation of a new coeliac disease quality of life survey (CD-QOL), Aliment. Pharmacol. Ther. 31 (2010) 666–675 https://doi.org/10.1111/j.1365-2036.2009.04220.x.

[180] W. Häser, J. Gold, A. Stallmach, W.F. Caspary, J. Stein, Development and validation of the Celiac Disease Questionnaire (CDQ), a disease-specific health-related quality of life measure for adult patients with celiac disease, J. Clin. Gastroenterol. 41 (2007) 157–166 https://doi.org/10.1097/01.mcg.0000225516.05666.4e.

[181] D. Adelman Patient Reported Outcome Instrument. 14/425,932, 2013.

[182] N.E. Jordan, Y. Li, D. Magrini, S. Simpson, N.R. Reilly, A.R. Defelice, et al., Development and validation of a celiac disease quality of life instrument for North American children, J. Pediatr. Gastroenterol Nutr. 57 (2013) 477–486 https://doi.org/10.1097/MPG.0b013e31829b68a1.

[183] D. Leffler, D. Schuppan, K. Pallav, R. Najarian, J.D. Goldsmith, J. Hansen, et al., Kinetics of the histological, serological and symptomatic responses to gluten challenge in adults with coeliac disease, Gut. 62 (2013) 996–1004 https://doi.org/10.1136/gutjnl-2012-302196.

[184] MN. Marsh, Gluten, major histocompatibility complex, and the small intestine. A molecular and immunobiologic approach to the spectrum of gluten sensitivity ('celiac sprue'), Gastroenterology. 102 (1992) 330–354 https://doi.org/10.1016/0016-5085(92)91819-P.

[185] G. Oberhuber, G. Granditsch, H. Vogelsang, The histopathology of coeliac disease: Time for a standardized report scheme for pathologists, Eur. J. Gastroenterol Hepatol. 11 (1999) 1185–1194 https://doi.org/10.1097/00042737-199910000-00019.

[186] G.R. Corazza, VV.C disease, J Clin Pathol 58 (2005) 573–574 https://doi.org/10.1136/jcp.2004.023978.

[187] A. Ensari, Gluten-sensitive enteropathy (celiac disease): Controversies in diagnosis and classification, Arch. Pathol. Lab. Med. 134 (2010) 826–836 https://doi.org/10.1043/1543-2165-134.6.826.

[188] J. Taavela, O. Koskinen, H. Huhtala, M.L. Lähdeaho, A. Popp, K. Laurila, et al., Validation of Morphometric Analyses of Small-Intestinal Biopsy Readouts in Celiac Disease, PLoS One. 8 (2013) https://doi.org/10.1371/journal.pone.0076163.

[189] B.S. Höroldt, M.E. McAlindon, T.J. Stephenson, M. Hadjivassiliou, DS. Sanders, Making the diagnosis of coeliac disease: Is there a role for push enteroscopy? Eur. J. Gastroenterol Hepatol. 16 (2004) 1143–1146 https://doi.org/10.1097/00042737-200411000-00010.

[190] J.A. Murray, A. Rubio-tapia, C.T. Van Dyke, D.L. Brogan, M.A. Knipschield, B. Lahr, et al., Mucosal Atrophy in Celiac Disease: Extent of Involvement, Correlation With Clinical Presentation, and Response to Treatment, Clin. Gastroenterol Hepatol. 6 (2008) 186–193 https://doi.org/10.1016/j.cgh.2007.10.012.

[191] E.J. Ciaccio, G. Bhagat, S.K. Lewis, PH. Green, Recommendations to quantify villous atrophy in video capsule endoscopy images of celiac disease patients, World. J. Gastrointest. Endosc. 8 (2016) 653 https://doi.org/10.4253/wjge.v8.i18.653.

[192] J. Vicnesh, J.K.E. Wei, E.J. Ciaccio, S.L. Oh, G. Bhagat, S.K. Lewis, et al., Automated diagnosis of celiac disease by video capsule endoscopy using DAISY Descriptors, J. Med. Syst. 43 (2019) 157 https://doi.org/10.1007/s10916-019-1285-6.

[193] R.P. Anderson, P. Degano, A.J. Godkin, D.P. Jewell, AVS. Hill, In vivo antigen challenge in celiac disease identifies a single transglutaminase-modified peptide as the dominant A-gliadin T-cell epitope, Nat. Med. 6 (2000) 337–342 https://doi.org/10.1038/73200.

[194] M. Ráki, L.E. Fallang, M. Brottveit, E. Bergseng, H. Quarsten, K.E.A. Lundin, et al., Tetramer visualization of gut-homing gluten-specific T cells in the peripheral blood of celiac disease patients, Proc. Natl. Acad. Sci. U. S. A. 104 (2007) 2831–2836 https://doi.org/10.1073/pnas.0608610104.

[195] G. Goel, J.A. Tye-Din, S.-.W. Qiao, A.K. Russell, T. Mayassi, C. Ciszewski, et al., Cytokine release and gastrointestinal symptoms after gluten challenge in celiac disease, Sci. Adv. 5 (2019) eaaw7756 https://doi.org/10.1126/sciadv.aaw7756.

[196] G. Goel, A.J.M. Daveson, C.E. Hooi, J.A. Tye-Din, S. Wang, E. Szymczak, et al., Serum cytokines elevated during gluten-mediated cytokine release in coeliac disease, Clin. Exp. Immunol. 199 (2020) 68–78 https://doi.org/10.1111/cei.13369.

[197] L.F. Risnes, A. Christophersen, S. Dahal-Koirala, R.S. Neumann, G.K. Sandve, V.K. Sarna, et al., Disease-driving CD4+ T cell clonotypes persist for decades in celiac disease, J. Clin. Invest. 128 (2018) 2642–2650 https://doi.org/10.1172/JCI98819.

[198] D.A. Leffler, C.P. Kelly, P.H.R. Green, R.N. Fedorak, A. Dimarino, W. Perrow, et al., Larazotide acetate for persistent symptoms of celiac disease despite a gluten-free diet: A randomized controlled trial, Gastroenterology. 148 (2015) 1311–1319.e6 https://doi.org/10.1053/j.gastro.2015.02.008.

[199] M. Linsalata, G. Riezzo, B. D'Attoma, C. Clemente, A. Orlando, F. Russo, Noninvasive biomarkers of gut barrier function identify two subtypes of patients suffering from diarrhoea predominant-IBS: a case-control study, BMC Gastroenterol. 18 (2018) 167 https://doi.org/10.1186/s12876-018-0888-6.

[200] C. Sturgeon, A. Fasano, Zonulin, a regulator of epithelial and endothelial barrier functions, and its involvement in chronic inflammatory diseases, Tissue. Barriers. 4 (2016) https://doi.org/10.1080/21688370.2016.1251384.

[201] B. Morón, A.K. Verma, P. Das, J. Taavela, L. Dafik, T.R. Diraimondo, et al., CYP3A4-catalyzed simvastatin metabolism as a non-invasive marker of small intestinal health in celiac disease, Am. J. Gastroenterol. 108 (2013) 1344–1351 https://doi.org/10.1038/ajg.2013.151.

[202] M. Adriaanse, DA. Leffler, Serum markers in the clinical management of celiac disease, Dig. Dis. 33 (2015) 236–243 https://doi.org/10.1159/000371405.

[203] G.J. Tack, J.M.W. van de Water, M.J. Bruins, E.M.C. Kooy-Winkelaar, J. van Bergen, P. Bonnet, et al., Consumption of gluten with gluten-degrading enzyme by celiac patients: A pilot-study, World. J. Gastroenterol. 19 (2013) 5837–5847 https://doi.org/10.3748/wjg.v19.i35.5837.

[204] B.N. Salden, V. Monserrat, F.J. Troost, M.J. Bruins, L. Edens, R. Bartholomé, et al., Randomised clinical study: Aspergillus niger-derived enzyme digests gluten in the stomach of healthy volunteers, Aliment. Pharmacol. Ther. 42 (2015) 273–285 https://doi.org/10.1111/apt.13266.

[205] A.J.M. Daveson, J.A. Tye-Din, G. Goel, K.E. Goldstein, H.L. Hand, K.M. Neff, et al., Masked bolus gluten challenge low in FODMAPs implicates nausea and vomiting as key symptoms associated with immune activation in treated coeliac disease, Aliment. Pharmacol. Ther. 51 (2020) 244–252 https://doi.org/10.1111/apt.15551.

[206] M.L. Lähdeaho, M. Scheinin, P. Vuotikka, J. Taavela, A. Popp, J. Laukkarinen, et al., Safety and efficacy of AMG 714 in adults with coeliac disease exposed to gluten challenge: a phase 2a, randomised, double-blind, placebo-controlled study, Lancet. Gastroenterol Hepatol. 4 (2019) 948–959 https://doi.org/10.1016/S2468-1253(19)30264-X.

[207] V. Villanacci, L. Lorenzi, F. Donato, R. Auricchio, P. Dziechciarz, J. Gyimesi, et al., Histopathological evaluation of duodenal biopsy in the PreventCD project. An observational interobserver agreement study, APMIS. 126 (2018) 208–214 https://doi.org/10.1111/apm.12812.

[208] K. Gottlieb, S. Travis, B. Feagan, F. Hussain, W.J. Sandborn, P. Rutgeerts, Central Reading of Endoscopy Endpoints in Inflammatory Bowel Disease Trials, Inflamm. Bowel. Dis. 21 (2015) 2475–2482 https://doi.org/10.1097/MIB.0000000000000470.

[209] B.G. Feagan, W.J. Sandborn, G. D'Haens, S. Pola, J.W.D. McDonald, P. Rutgeerts, et al., The role of centralized reading of endoscopy in a randomized controlled trial of mesalamine for ulcerative colitis, Gastroenterology. 145 (2013) https://doi.org/10.1053/j.gastro.2013.03.025.

[210] B.C. Dickson, C.J. Streutker, R. Chetty, Coeliac disease: An update for pathologists, J. Clin. Pathol. 59 (2006) 1008–1016 https://doi.org/10.1136/jcp.2005.035345.

[211] P.J. Wahab, J.W.R. Meijer, C.J.J. Mulder, Histologic follow-up of people with celiac disease on a gluten-free diet: Slow and incomplete recovery, Am. J. Clin. Pathol. 118 (2002) 459–463 https://doi.org/10.1309/EVXT-851X-WHLC-RLX9.

[212] J. Taavela, K. Kurppa, P. Collin, M.L. Lähdeaho, T. Salmi, P. Saavalainen, et al., Degree of Damage to the Small Bowel and Serum Antibody Titers Correlate With Clinical Presentation of Patients With Celiac Disease, Clin. Gastroenterol Hepatol. 11 (2013) https://doi.org/10.1016/j.cgh.2012.09.030.

[213] A. Rubio-Tapia, M.W. Rahim, J.A. See, B.D. Lahr, T.T. Wu, JA. Murray, Mucosal recovery and mortality in adults with celiac disease after treatment with a gluten-free diet, Am. J. Gastroenterol. 105 (2010) 1412–1420 https://doi.org/10.1038/ajg.2010.10.

[214] B. Lebwohl, R.C. Kapel, A.I. Neugut, P.H.R. Green, R.M. Genta, Adherence to biopsy guidelines increases celiac disease diagnosis, Gastrointest. Endosc. 74 (2011) 103–109 https://doi.org/10.1016/j.gie.2011.03.1236.

[215] T. McCarty, C. O'Brien, A. Gremida, C. Ling, T. Rustagi, Efficacy of duodenal bulb biopsy for diagnosis of celiac disease: a systematic review and meta-analysis, Endosc. Int. Open. 06 (2018) E1369–E1378 https://doi.org/10.1055/a-0732-5060.

[216] M. Kurien, K.E. Evans, A.D. Hopper, M.F. Hale, S.S. Cross, DS. Sanders, Duodenal bulb biopsies for diagnosing adult celiac disease: Is there an optimal biopsy site? Gastrointest. Endosc. 75 (2012) 1190–1196 https://doi.org/10.1016/j.gie.2012.02.025.

[217] A. Rostom, J.A. Murray, M.F. Kagnoff, American Gastroenterological Association (AGA) Institute Technical Review on the Diagnosis and Management of Celiac Disease, Gastroenterology. 131 (2006) 1981–2002 https://doi.org/10.1053/j.gastro.2006.10.004.

[218] E.F. Verdu, X. Huang, J. Natividad, J. Lu, P.A. Blennerhassett, C.S. David, et al., Gliadin-dependent neuromuscular and epithelial secretory responses in gluten-sensitive HLA-DQ8 transgenic mice, Am. J. Physiol – Gastrointest Liver Physiol. 294 (2007) https://doi.org/10.1152/ajpgi.00225.2007.

[219] E. Marietta, K. Black, M. Camilleri, P. Krause, R.S. Rogers, C. David, et al., A new model for dermatitis herpetiformis that uses HLA-DQ8 transgenic NOD mice, J. Clin. Invest. 114 (2004) 1090–1097 https://doi.org/10.1172/jci21055.

[220] D. Sblattero, F. Maurano, G. Mazzarella, M. Rossi, S. Auricchio, F. Florian, et al., Characterization of the Anti-Tissue Transglutaminase Antibody Response in Nonobese Diabetic Mice, J. Immunol. 174 (2005) 5830–5836 https://doi.org/10.4049/jimmunol.174.9.5830.

[221] A. Caminero, E.F. Verdu, Metabolism of wheat proteins by intestinal microbes: Implications for wheat related disorders, Gastroenterol. Hepatol. 42 (2019) 449–457 https://doi.org/10.1016/j.gastrohep.2019.04.001.

[222] V. Abadie, S.M. Kim, T. Lejeune, B.A. Palanski, J.D. Ernest, O. Tastet, et al., IL-15, gluten and HLA-DQ8 drive tissue destruction in coeliac disease, Nature. 578 (2020) 600–604 https://doi.org/10.1038/s41586-020-2003-8.

[223] A.V. Clarizio, H.J. Galipeau, J. Jury, L. Rondeau, J. Godbout, L. Williams, et al., A19 NOVEL HLA-DQ2 TRANSGENIC MICE DEVELOP GLUTEN-IMMUNOPATHOLOGY FOLLOWING GLUTEN SENSITIZATION, J. Can. Assoc. Gastroenterol. 3 (2020) 22–23 https://doi.org/10.1093/JCAG/GWZ047.018.

CHAPTER 8

Dermatitis herpetiformis – a cutaneous manifestation of coeliac disease

Teea Salmi[a,b], Kaisa Hervonen[a,b], Timo Reunala[a]
[a]Celiac Disease Research Center, Faculty of Medicine and Health Technology, Tampere University, Tampere, Finland
[b]Department of Dermatology, Tampere University Hospital, Tampere, Finland

Dermatitis herpetiformis (DH) was described as a dermatological entity by Louis Duhring in 1884, four years before Samuel Gee defined the symptoms of coeliac disease [1,2]. Though patients with DH rarely presented with overt gastrointestinal symptoms as present in coeliac disease, small bowel biopsies taken in the 1960s showed unexpectedly villous atrophy in most patients [3,4] (Table 8.1). Use of granular IgA deposits in papillary dermis as a diagnostic criterion of DH [5] confirmed that a quarter of the patients had normal small bowel villous architecture with increased density of intraepithelial lymphocytes. Later it became evident that this finding was a representative of coeliac-type minor enteropathy [6]. In the 1970s further observations linking DH and coeliac disease together (Table 8.1) were identical human leucocyte antigen (HLA) pattern [7], response of rash to a gluten-free diet (GFD) in the patients with and without villous atrophy [8,9] and occurrence of both diseases in the same families [10]. Today, genetic studies have shown that virtually every patient with DH and coeliac disease has the alleles contributing to the HLA-DQ2 or HLA-DQ8 haplotype [11], and even monozygotic twin pairs can be affected by both phenotypes of the disease [12].

Risk of lymphoma is well documented in coeliac disease [13]. Lymphoma occurs also in DH (Table 8.1) and importantly, GFD treatment seems to reduce the risk [14,15]. Moreover, DH as coeliac disease is known for associated autoimmune disorders such as thyroid disease and type 1 diabetes [16,17]. A breakthrough in the coeliac disease research was the discovery by Dieterich et al. [18] in 1997 that tissue transglutaminase (TG2) was the target autoantigen for IgA antibody responses. A few years later, Sárdy et al. [19] showed that epidermal transglutaminase (TG3) was the autoantigen in DH (Table 8.1). At present, TG3 enzyme is known to coexists with IgA antibodies in the dermal deposits forming tightly bound immune complexes [19,20].

Coeliac Disease and Gluten-Related Disorders.
DOI: https://doi.org/10.1016/B978-0-12-821571-5.00009-X

Table 8.1 Major findings linking Dermatitis herpetiformis to coeliac disease.

	Year	Authors
Small bowel villous atrophy	1966, 1967	Marks et al. [3], Fry et al. [4]
Same HLA pattern	1972	Katz et al. [7]
Rash heals with a gluten-free diet	1973, 1978	Fry et al. [8], Reunala et al. [9]
Familial occurrence	1976	Reunala et al. [10]
Risk of lymphoma	1983, 1996	Leonard et al. [14], Lewis et al. [15]
Occurrence of autoimmune disorders	1985, 1997	Cunningham & Zone [16], Reunala & Collin [17]
Transglutaminase as autoantigen	2002	Sárdy et al. [19]

8.1 Epidemiology of dermatitis herpetiformis

8.1.1 Prevalence and incidence of dermatitis herpetiformis

Prevalence figures of DH are available from a few European countries and Utah, USA [21–23]. In the Finnish study prevalence was 75 per 100,000 and in the U.K. study it was 30 per 100,000, eight times lower than that of coeliac disease [21,22] (Table 8.2). These figures indicate that DH is the most common extraintestinal manifestation of coeliac disease [24]. In agreement with this, DH has shown to occur in up to 14% of the coeliac disease patient cohort or register studies from USA, Ireland, Wales and Germany [25–28]. Coeliac disease occurs also in Asia [29] and small DH patient series have been described from China and India [30,31]. In contrast, DH patients reported from Japan seem not to represent a gluten-sensitive disorder like the classical DH occurring in the caucasoid people [32].

A nationwide register in Finland including a total of 31,385 adult patients with coeliac disease and 3671 with DH in 1980-2014 showed that the incidence of these diseases was about the same in the 1980s [33–35] (see Fig. 8.1). Thereafter, the incidence of coeliac disease has sharply increased whereas the incidence of DH is slowly decreasing. A Finnish cohort study [21] and a U.K. register study [22] covering time periods of 30 and 20 years, respectively, confirmed that the annual incidence rate of DH has decreased significantly from 5.2 to 2.7 per 100,000 in the Finnish and from 1.8 to 0.8 per 100,000 in the U.K. study (see Fig. 8.1)

The decreasing incidence rate of DH in Finland and U.K., along with a simultaneous rapid increase in coeliac disease, fits to hypothesis that subclinical, undiagnosed coeliac disease is a prerequisite for the development of DH. When increasing amount of patients with coeliac disease are at present found by better knowledge of the disorder and search by serology, the large pool of hidden cases from which DH develops is diminishing.

Table 8.2 Differences between Dermatitis herpetiformis and coeliac disease.

	Dermatitis herpetiformis	**Coeliac disease**
Gender	Slightly more males	Females predominate
Age at onset	Mainly adults	Children and adults
Diagnostic delay [42,43]	1.6 – 3.2 years	3 – 11 years
Prevalence in Finland and United Kingdom [21,22,34]	75 and 30 per 100,000	660 and 240 per 100,000
Incidence [21,22]	Decreasing	Increasing
Rash with IgA-TG3 deposits	100%	0%
Small bowel villous atrophy	75%*	100%**
Response to a gluten-free diet [8,9,44]	Slow; months, at onset most patients need dapsone to control the rash	Rapid; days or weeks until gastrointestinal symptoms resolve
Health-related quality of life on a gluten-free diet [45,46]	Normal	Decreased
Long-term prognosis on a gluten-free diet [13,47,48]	Excellent	All-cause and lymphoma mortality may be increased

TG=Transglutaminase
* The remainder has coeliac-type inflammation in the small bowel mucosa.
** Potential coeliac disease with normal mucosa architecture, inflammation and positive TG2 serology also exist.

In support of subsequent development of DH from pre-existing coeliac disease are the patients initially diagnosed with coeliac disease who did not follow or only partially followed a GFD and subsequently developed DH [36]. Moreover, adult DH patients frequently have coeliac-type dental enamel defects, which developed in childhood as a result of malabsorption or immune alteration caused by undiagnosed coeliac disease [37].

8.1.2 Gender and age at onset in dermatitis herpetiformis

Earlier studies in DH have shown male to female ratios ranging up to 2:1 whereas two recent large DH studies found the ratio to be close to 1:1 [21,22]. Even this is in sharp contrast to coeliac disease in which females outnumber males (Table 8.2). The reason for the gender differences between DH and coeliac disease remains unknown.

Coeliac disease can be diagnosed at any age with the peak incidences being in childhood and between 40 and 60 years of age [38]. In contrast to coeliac disease, DH in childhood seems to be rare occurring in only 4% in the Finnish but a higher percentage in an Italian series [39,40] (Table 8.2).

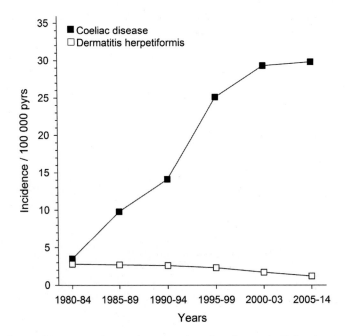

Figure 8.1 *Incidence of Dermatitis herpetiformis and coeliac disease in Finland, 1980–2014.* The data includes 3671 adult patients with Dermatitis herpetiformis and 31,385 with coeliac disease registered with the Social Insurance Institution of Finland. Reprinted from Reunala et al. Nutrients.

Interestingly, a significant increase in the mean age at diagnosis has occurred in a cohort of 477 DH patients collected from 1970 onwards in Finland [21]. The increase was from 35 to 51 years in men and from 36 to 46 years in women. A similar increasing trend in the mean age at diagnosis has also been observed in adult coeliac disease [33,41]. One explanation for this trend may be a lower lifetime gluten load. For example, in Finland the annual consumption of wheat, rye, and other cereals per person has decreased from 150 kg to 71 kg over the past 50 years [35].

8.2 Pathogenesis of dermatitis herpetiformis: from gut to skin

In DH, pathognomic granular IgA deposits occur in papillary dermis of perilesional and uninvolved skin [49], and it has long been suspected that these IgA deposits derive from the gut. In 2002, Sárdy et al. [19] demonstrated that the autoantigen for deposited cutaneous IgA is TG3. This enzyme is typically expressed in the epidermis and closely related, but not identical, to TG2 autoantigen of coeliac disease [18]. At present, it is known that TG3 enzyme in IgA aggregates is active possibly resulting in covalent cross-linking of the complex to dermal structures [20]. These IgA–TG3 complexes disappear from the papillary dermis first after many years on GFD treatment but reappear within months on gluten challenge [50,51].

As in coeliac disease, untreated DH patients have frequently circulating IgA autoantibodies to TG2 but also high avidity antibodies to TG3 both of which disappear during GFD treatment [19,52,53]. The TG3 enzyme has not been detected in the small bowel similarly to the TG2. However, DH patients have been shown to secrete high levels of IgA class TG3 antibodies into the gut organ culture medium, and also have TG3-antibody-positive cells in the small bowel mucosa [54,55]. These findings suggest that TG3 antibody production occurs in the small bowel of DH patients similarly to TG2 antibody production in coeliac disease [56]. At present, a valid hypothesis is that the immunopathogenesis of DH starts from hidden coeliac disease in the gut with TG2 and also TG3 autoantibody responses possibly due to epitope spreading. The next step is immune complex deposition of high avidity IgA class TG3 autoantibodies together with the TG3 enzyme in the papillary dermis [19,20]. Whether TG3 derives as immune complexes from the circulation is not known but a possibility [57,58]. TG3 may also be shedding from the epidermis and be aggregated by IgA autoantibodies in the upper papillary dermis as suggested by passive transfer of DH sera or TG3 antibodies to human skin-grafted mice [59].

An unanswered question is why the DH lesions have a predilection for areas in the knees, elbows, and buttocks, although IgA–TG3 aggregates are also deposited at sites never involved in the formation of blisters [49]. The most likely explanation for this unique distribution of the skin lesions involves the influence of local factors, such as pressure and stretching. Direct activation of TG3 in dermal aggregates by mechanical force similarly to TG2 activation in vascular walls [60], could be an initiator. This would result in release of fibrinogen from the aggregates, which, besides being a clotting factor, is an inflammatory protein capable of attracting T cells, neutrophils and macrophages, all of which have been shown to influx into the developing DH blisters [20,61]. Early appearance of urokinase-type plasminogen activator mRNA in basal keratinocytes further supports the role of fibrinogen release in blister formation [61,62].

8.3 Clinical features of dermatitis herpetiformis

The clinical picture of DH consists of an itchy rash occurring in the predilection sites of elbows, extensor surfaces of forearms, knees and buttocks including the sacral area. The rash is polymorphic, consisting of small blisters, papules and erythema, but due to intense pruritus and scratching erosions, crusts and post-inflammatory hyperpigmentation often predominate the clinical picture (Figure 8.2 A-B). The localization of the rash is so typical for DH that itchy rash in these skin sites usually rises suspicion of DH. However, the severity of the rash varies individually, and in more intense disease also other sites such

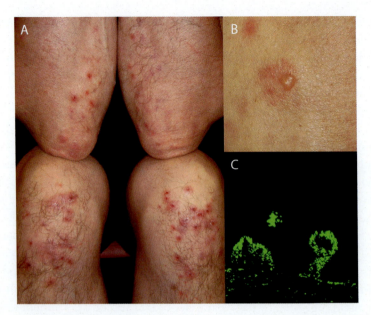

Figure 8.2 *Dermatitis herpetiformis (DH).* (A) A patient with dermatitis herpetiformis: typical rash with excoriated blisters and papules on elbows and knees, (B) Intact blister and small papules in forearm, (C) Direct immunofluorescence finding: granular IgA deposits in papillary dermis.

as scalp, face and upper back may be affected. A rare cutaneous finding in DH is acral purpura, which may be seen either with the typical DH rash or as the only manifestation of the disease [63–65]. Interestingly, the clinical picture and severity of DH rash seems to have remained quite unchanged during recent decades [42], and not become milder or otherwise non–classical as seen in coeliac disease symptoms [66–68].

In addition to typical rash DH patients may sometimes have gastrointestinal symptoms or signs of malabsorption, but despite the similar gluten-sensitive enteropathy they are considerably less frequent than in coeliac disease [69,70]. In recent studies up to one third of DH patients have reported some abdominal complaints, but these are typically mild and include bloating, loose stools, transient or chronic diarrhea and abdominal pain [71,72].

Coeliac-type dental enamel defects of permanent teeth are common in children and adults with DH [37,73]. Oral lesions are uncommon in DH, although erosions with burning or pain have sporadically been described [74,75].

The differential diagnosis of DH includes various itchy, excoriated and blistering skin diseases. Linear IgA disease and sometimes bullous pemphigoid may clinically resemble DH but direct immunofluorescence (IF) biopsy easily differentiates these disorders [76,77]. Other itchy skin diseases, such as scabies, atopic and nummular dermatitis, lichen planus and urticaria may also clinically resemble DH, but the most obvious difference is the predilection site of these diseases [78].

Despite the highly characteristic clinical picture of DH familiar for dermatologists, the suspicion of the correct diagnosis is not always self-evident in primary care. The mean delay in DH diagnosis has been reported to be as long as 3.2 years [42,79] but still notably shorter as in coeliac disease [43] (Table 8.2). In recent studies from Finland one-third of the DH and coeliac patients had diagnostic delays of at least two and ten years, respectively, and in both studies female sex was significantly associated with the delayed diagnosis [42,80]. Fortunately, the diagnostic delay both in DH and coeliac disease has shortened during the last decades [42,81].

8.4 Diagnosis of dermatitis herpetiformis

8.4.1 Skin and small bowel biopsy findings

In DH, granular IgA deposits at the dermal papillae were detected by direct IF already in 1969 [5], and since then, the finding has proven to be highly specific for DH [82]. Skin biopsy should be taken from normal-appearing skin adjacent to the rash, i.e. perilesional skin, since even though IgA deposits are distributed widely in the skin of DH patients, they occur in greater amounts near the active lesions [49]. The granular deposition of IgA mostly occurs at the dermoepidermal junction at the tips of the dermal papillae (Figure 8.2 C) and occasionally along the entire junction and also vessel walls. In addition to characteristic granular IgA deposition, also DH cases with fibrillar IgA deposits have been presented especially in Japan [83]. However, it seems that Japanese DH differs from Caucasian cases in several aspects and is probably not even gluten-dependent [32].

IgA deposits in DH skin are directed against TG3, and it has been demonstrated, that the IgA and TG3 not only co-localize in active DH, but they also disappear in parallel during GFD treatment [50]. Interestingly, there are two reports demonstrating granular IgA deposits in the skin of coeliac disease patients [84,85]. However, the number of investigated coeliac disease patients with IgA deposition has been thus far very limited, the IgA deposits were weak and best identified by confocal microscopy and co-localization with TG3 was seen only in the first study [85]. Hence, granular IgA deposition can still be used as the best diagnostic criterion for DH.

Histopathological findings in DH lesional skin, however, are not entirely specific for DH as similar findings are detected in other blistering skin diseases as well. If an intact vesicle or inflamed skin is biopsied, the typical findings in DH include neutrophilic microabscesses in the dermal papillae with or without subepidermal blister [86]. A recent study documented that although the specificity of histologic examination for DH was high (0.95), the sensitivity was only 0.75 [87]. Thus, if IF finding is compatible with DH, a biopsy for histological analysis is not necessary [88].

Small bowel biopsies obtained during upper gastrointestinal endoscopy at diagnosis reveal coeliac-type small bowel mucosal villous atrophy and crypt hyperplasia in approximately ¾ of the DH patients [71]. However, similarly to coeliac disease, it has recently been shown that the severity of small bowel mucosal lesion has become less severe in DH [89,90]. Correspondingly with coeliac disease, typical for DH is intraepithelial lymphocytosis and increased densities of $\gamma\delta+$ T-cells in frozen small bowel mucosal samples, appearing even in those DH subjects without evident shortening of intestinal villi [91]. Moreover, TG2 targeted IgA deposits have been shown to exist in frozen small bowel mucosal samples of the majority of DH patients, but the prevalence of these deposits seems to be lower than in coeliac disease [92–94].

8.4.2 Serum coeliac antibody measurements

Circulating IgA class antibodies against TG2, the autoantigen of coeliac disease, was detected in DH in 1999 [95]. Since then, TG2 enzyme has proven to be the target for endomysial antibodies (EmA) as well [96], and both ELISA-based TG2-antibodies and indirect IF based EmA tests show high reliability in coeliac disease. However, performing EmA test requires experience and the test is more laborious, and hence IgA-class TG2 antibody ELISA test is suggested to be used as the first-line serological test. In DH, TG2 antibody titers have shown to correlate with the degree of villous atrophy and more reliably detect DH patients with villous atrophy than those with normal villous structures [90,97].

IgA-class antibodies against TG3, the autoantigen of DH, also exist in the serum of the majority of DH patients [19,52]. In DH, these antibodies have shown gluten–dependency disappearing from the serum in parallel with TG2 antibodies [53]. Interestingly, TG3 antibodies have also shown to occur in the serum of coeliac disease patients, but the prevalence, titers as well as the affinity of the antibodies is lower in coeliac disease compared to DH [19,52], and moreover, in coeliac disease the titers of TG3 antibodies increase with age [98]. At present, the exact specificity of TG3 antibody measurements for DH and the most applicable titers and test kits are to be clarified, and hence TG3 antibody testing is not routinely applied. Similarly, the role of deamidated gliadin peptide (DGP) antibodies in DH as well as in coeliac disease diagnostics is thus far obscure, even though they have shown to be rather reliable in small DH series [52,99].

8.4.3 Diagnostic work-up of dermatitis herpetiformis

Upon suspicion of DH, thorough patient history and clinical investigation are the first-line procedures. In addition to skin and possible abdominal symptoms (described in detail above), also signs of malabsorption and DH associated diseases should be observed. Moreover, family history of DH and coeliac disease should be inquired. If clinical picture is compatible with DH, all patients should undergo skin biopsy. The golden standard in DH diagnostics is the direct IF examination of perilesional skin biopsy and the detection of granular IgA deposits in the dermal papillae, and together with compatible clinical picture, this finding confirms the DH diagnosis. Repetition of IF examination should be considered if IgA is negative despite strong clinical suspicion. False negative IF results occur in about 5% of the patients and are possible especially if the biopsy has been taken from blisters or inflamed skin [49,82].

In addition to IF examination, serum IgA-class TG2-antibodies should be investigated when DH is suspected, and if positive, the result supports, but alone is not sufficient for, the diagnosis. It must be borne in mind that seronegativity occurs especially in DH patients with normal small bowel villous architecture [90,97]. Typical histopathological findings detected in lesional skin biopsy have supportive value in DH diagnosis. HLA determination should only be applied in obscure cases, and if performed, HLA DQ2 and DQ8 negative result practically excludes DH and coeliac disease.

Small bowel mucosal investigation is not necessary for DH diagnosis [88,100]. Evidence further shows that the severity of small bowel lesion in DH patients does not have any influence on the clinical recovery or long-term general health during GFD treatment [101].

8.5 Treatment of dermatitis herpetiformis

8.5.1 Gluten-free diet

The most crucial treatment for DH is a strict life-long GFD, which should be initiated to all patients after the diagnosis is confirmed. Maintaining a GFD means permanent avoidance of wheat, barley and rye and foods otherwise containing gluten. The safety of oats in DH has been demonstrated in two short-term challenge studies [102,103], and thus oats are nowadays allowed as part of GFD in most countries [44]. Oats may, however, easily get contaminated by gluten traces from other cereals for example during harvesting, transportation and production, and it is important to ensure that only non-contaminated oats are used [44,104]. Initiating and maintaining a strict GFD requires wide knowledge of tolerable foods and the diet, and therefore dietary guidance given preferably by a dietitian should be offered to all newly diagnosed DH patients. Moreover, education and support given by coeliac disease patient organizations to adhere on the GFD treatment and manage the costs are important for the patients [105].

In DH, maintaining a strict GFD leads to healing of the rash, recovery of the small bowel villous architecture and disappearance of circulating coeliac autoantibodies. Gastrointestinal symptoms, when present, usually alleviate within a few weeks, the coeliac autoantibodies disappear in one year, but healing of the small bowel mucosal changes may take longer, more than one year [106]. Resolution of the DH rash also happens slowly, taking several months or even a couple of years until total clearance is achieved on a strict GFD [8,9]. The disappearance of deposited IgA and TG3 from the papillary dermis takes even longer and cutaneous immune deposits persist even after patients have been asymptomatic for several years on a strict GFD [50,107,108].

8.5.2 Dapsone

The majority of the DH patients, especially those with widespread rash and intense itch, need additional dapsone treatment at the onset of GFD treatment. Dapsone is a sulfone antibiotic with potent anti-inflammatory properties, and it is the drug of choice in DH [100]. Dapsone relieves the itch and the rash in DH within a few days, but has no effect on the enteropathy or IgA deposits in the skin. The starting dose in adults is 25-50 mg/day which is usually enough to control the itch and development of new lesions, but if necessary, the dose can gradually be increased up to 100 mg/day. In children, the recommended starting dose of dapsone is 0.5 mg/kg/day. After the rash has disappeared, dapsone dose is slowly reduced and finally stopped as the GFD alone controls the rash. This will take an average of two years on a strict GFD [107,109]

Dapsone is usually well tolerated when recommended doses are used. However, hematological side-effects, the most common being dose dependent hemolysis and methemoglobinemia, may sometimes occur. Patients with glucose-6-phosphate dehydrogenase deficiency are more prone to these side effects, and hence should be treated with caution. Agranulocytosis is a rare event, occurring at the beginning of dapsone treatment. Other possible side-effects usually appearing within the first three months include elevation of transaminases, headache, dizziness, nausea and dapsone hypersensitivity syndrome. Peripheral motor neuropathy may occur after years of taking high doses of dapsone [110]. Clinical and laboratory monitoring, including complete blood cell count, liver and renal function tests, are necessary both at baseline and during the follow-up of dapsone treatment.

8.6 Long-term prognosis of dermatitis herpetiformis

Lymphoma is the most severe and well-documented complication also in DH (Table 8.1). The risk of lymphoma has shown to be increased up to 6 to 10 fold [15,111], but fortunately, increased lymphoma risk occurs only during the first five years after DH diagnosis and GFD adherence [48]. In DH, the risk of gastrointestinal carcinomas has not been reported to be increased, which is in contrast to coeliac disease [47,112]. Contrary to increased mortality rates associated with coeliac disease [13], the mortality rate of DH patients has shown to be decreased (Table 8.2). A Finnish study of 476 DH patients of whom 98% adhered to a GFD presented a significantly decreased standardized mortality rate (0.70) [48], and similarly, a study from UK found a slightly, albeit not significantly lowered mortality rate among DH patients (hazard ratio 0.93) [47]. Moreover, the bone fracture risk does not seem to be increased in DH as documented in coeliac disease [47,113,114].

In rare cases, DH rash persists despite long-term adherence to a GFD treatment. In such cases, dietary transgressions are the most common reason and dietary consultation should be considered. DH non-responsive to GFD, i.e. refractory DH, has shown to exist in about 2% of DH patients [108]. However, in contrast to refractory coeliac disease, persistence of small bowel villous atrophy nor development of lymphoma has not been reported in refractory DH patients [108,115]. Persistent gastrointestinal symptoms known to occur in treated coeliac disease [116] seem not to affect DH patients [45]. Moreover, quality of life of GFD-treated DH patients seems to parallel to that of controls, which is in contrast to coeliac disease [45,46,72].

The associations between DH and other autoimmune disease must be kept in mind during long-term follow-up of DH patients. DH has been linked to autoimmune thyroid diseases and type 1 diabetes [16,17,117], and further, less reliable evidence suggests associations with for example Sjögren's syndrome, vitiligo and alopecia areata. Moreover, a register-based study recently associated DH with subsequent development of bullous pemphigoid, another autoimmune bullous skin disease [118].

8.7 Concluding remarks

DH presenting with intense itch and blistering rash typically in elbows, knees and buttocks is a cutaneous manifestation of coeliac disease. Though overt gastrointestinal symptoms are rare, ¾ of the DH patients have villous atrophy in the small bowel and the rest coeliac-type inflammatory changes. DH diagnosis relies on the detection of granular IgA deposits in skin biopsy samples. The DH autoantigen is TG3, which also is deposited in the papillary dermis. DH affects mostly adults and slightly more males. At present, DH prevalence is 1 to 8 compared to coeliac disease. The incidence of DH is decreasing, and the main cause is most likely improved diagnostics of coeliac disease. GFD adherence in DH should be strict and life-long, and DH patients following dietary treatment have an excellent long-term prognosis.

References

[1] LA. Duhring, Dermatitis herpetiformis. Landmark article, JAMA. 250 (2) (1983) 212–216.
[2] MS. Losowsky, A history of coeliac disease, Dig. Dis. 26 (2) (2008) 112–120.
[3] J. Marks, S. Shuster, AJ. Watson, Small-bowel changes in dermatitis herpetiformis, Lancet. 2 (1966) 1280–1282.
[4] L. Fry, P. Keir, R.M. McMinn, J.D. Cowan, A.V. Hoffbrand, Small-intestinal structure and function and haematological changes in dermatitis herpetiformis, Lancet. 2 (7519) (1967) 729–733.
[5] J.B. van der Meer, Granular deposits of immunoglobulins in the skin of patients with dermatitis herpetiformis. An immunofluorescent study, Br. J. Dermatol. 81 (7) (1969) 493–503.
[6] E. Savilahti, T. Reunala, M. Mäki, Increase of lymphocytes bearing the gamma/delta T cell receptor in the jejunum of patients with dermatitis herpetiformis, Gut. 33 (2) (1992) 206–211.
[7] S.I. Katz, Z.M. Falchuk, M.V. Dahl, G.N. Rogentine, W. Strober, HL-A8: a genetic link between dermatitis herpetiformis and gluten-sensitive enteropathy, J. Clin. Invest. 51 (11) (1972) 2977–2980.
[8] L. Fry, P.P. Seah, D.J. Riches, A.V. Hoffbrand, Clearance of skin lesions in dermatitis herpetiformis after gluten withdrawal, Lancet. 1 (7798) (1973) 288–291.
[9] T. Reunala, K. Blomqvist, S. Tarpila, H. Halme, K. Kangas, Gluten-free diet in dermatitis herpetiformis. I. Clinical response of skin lesions in 81 patients, Br. J. Dermatol. 97 (5) (1977) 473–480.

[10] T. Reunala, O.P. Salo, A. Tiilikainen, O. Selroos, P. Kuitunen, Family studies in dermatitis herpetiformis, Ann. Clin. Res. 8 (4) (1976) 254–261.

[11] A. Balas, J.L. Vicario, A. Zambrano, D. Acuna, D. Garcıa-Novo, Absolute linkage of celiac disease and dermatitis herpetiformis to HLA-DQ, Tissue Antigens. 50 (1) (1997) 52–56.

[12] K. Hervonen, H.P. Karell, P. Collin, J. Partanen, T. Reunala, Concordance of dermatitis herpetiformis and celiac disease in monozygous twins, J. Invest. Dermatol. 115 (6) (2000) 990–993.

[13] M. Tio, M.R. Cox, G.D. Eslick, Meta-analysis: coeliac disease and the risk of all-cause mortality, any malignancy and lymphoid malignancy, Aliment. Pharmacol. Ther. 35 (5) (2012) 540–551.

[14] J.N. Leonard, W.F. Tucker, J.S. Fry, C.A. Coulter, A.W. Boylston, R.M. McMinn, G.P. Haffenden, A.F. Swain, L. Fry, Increased incidence of malignancy in dermatitis herpetiformis, Br. Med. J. 286 (6358) (1983) 16–18.

[15] H.M. Lewis, T.L. Reunala, J.J. Garioch, J.N. Leonard, J.S. Fry, P. Collin, D. Evans, L. Fry, Protective effect of gluten-free diet against development of lymphoma in dermatitis herpetiformis, Br. J. Dermatol. 135 (3) (1996) 363–367.

[16] M.J. Cunningham, J.J. Zone, Thyroid abnormalities in dermatitis herpetiformis. Prevalence of clinical thyroid disease and thyroid autoantibodies, Ann. Intern Med. 102 (2) (1985) 194–196.

[17] T. Reunala, P. Collin, Diseases associated with dermatitis herpetiformis, Br. J. Dermatol. 136 (3) (1997) 315–318.

[18] W. Dieterich, T. Ehnis, M. Bauer, P. Donner, U. Volta, E.O. Riecken, Schuppan D Identification of tissue transglutaminase as the autoantigen of celiac disease, Nat. Med. 3 (7) (1997) 797–801.

[19] M. Sárdy, S. Kárpáti, B. Merkl, M. Paulsson, N. Smyth, Epidermal transglutaminase (TGase 3) is the autoantigen of dermatitis herpetiformis, J. Exp. Med. 195 (6) (2002) 747–757.

[20] T.B. Taylor, L.A. Schmidt, L.J. Meyer, J.J. Zone, Transglutaminase 3 present in the IgA aggregates in dermatitis herpetiformis skin is enzymatically active and binds soluble fibrinogen, J. Invest. Dermatol. 135 (2) (2015) 623–625.

[21] T.T. Salmi, K. Hervonen, H. Kautiainen, P. Collin, T. Reunala, Prevalence and incidence of dermatitis herpetiformis: A 40-year prospective study from Finland, Br. J. Dermatol. 165 (2) (2011) 354–359.

[22] J. West, K.M. Fleming, L.J. Tata, T.R. Card, CJ. Crooks, Incidence and prevalence of celiac disease and dermatitis herpetiformis in the UK over two decades: population-based study, Am. J. Gastroenterol. 109 (5) (2014) 757–768.

[23] J.B. Smith, J.E. Tulloch, L.J. Meyer, J.J. Zone, The incidence and prevalence of dermatitis herpetiformis in Utah, Arch. Dermatol. 128 (12) (1992) 1608–1610.

[24] D.A. Leffler, P.H. Green, A. Fasano, Extraintestinal manifestations of coeliac disease, Nat. Rev. Gastroenterol. Hepatol. 12 (10) (2015) 561–571.

[25] S. Krishnareddy, S.K. Lewis, PH. Green, Dermatitis herpetiformis: clinical presentations are independent of manifestations of celiac disease, Am. J. Clin. Dermatol. 15 (1) (2014) 51–56.

[26] P. Dominguez Castro, G. Harkin, M. Hussey, B. Christopher, C. Kiat, J.L. Chin, et al., Changes in presentation of celiac disease in Ireland from the 1960s to 2015, Clin. Gastroenterol. Hepatol. 15 (6) (2017) 864–871.

[27] N.D. Hawkes, G.L. Swift, P.M. Smith, H.R. Jenkins, Incidence and presentation of coeliac disease in South Glamorgan, Eur. J. Gastroenterol. Hepatol. 12 (3) (2000) 345–349.

[28] W. Häuser, J. Gold, J. Stein, W.F. Caspary, A. Stallmach, Health-related quality of life in adult coeliac disease in Germany: results of a national survey, Eur. J. Gastroenterol. Hepatol. 18 (7) (2006) 747–754.

[29] S. Ashtari, M.A. Pourhoseingholi, K. Rostami, H.A. Aghdaei, M. Rostami-Nejad, L. Busani, et al., Prevalence of gluten-related disorders in Asia-Pacific region: a systematic review, J. Gastrointestin. Liver. Dis. 28 (1) (2019) 95–105.

[30] F. Zhang, B. Yang, Y. Lin, S. Chen, G. Zhou, G. Wang, et al., Dermatitis herpetiformis in China: a report of 22 cases, J. Eur. Acad. Dermatol. Venereol. 26 (7) (2012) 903–907.

[31] S. Handa, G. Dabas, D. De, R. Mahajan, D. Chatterjee, U.N. Saika, BD. Radotra, A retrospective study of dermatitis herpetiformis from an immunobullous disease clinic in north India, Int. J. Dermatol. 57 (8) (2018) 959–964.

[32] C. Ohata, N. Ishii, T. Hamada, Y. Shimomura, H. Niizeki, T. Dainichi, et al., Distinct characteristics in Japanese dermatitis herpetiformis: a review of all 91 Japanese patients over the last 35 years, Clin. Dev. Immunol. 2012 (2012) 562168.

[33] P. Collin, H. Huhtala, L. Virta, L. Kekkonen, T. Reunala, Diagnosis of celiac disease in clinical practice: Physician's alertness to the condition essential, J. Clin. Gastroenterol. 41 (2) (2007) 152–156.

[34] L.J. Virta, M.M. Saarinen, K.-L. Kolho, Declining trend in the incidence of biopsy-verified coeliac disease in the adult population of Finland, 2005-2014, Aliment. Pharmacol. Ther. 46 (11-12) (2017) 1085–1093.

[35] T. Reunala, T.T. Salmi, K. Hervonen, K. Kaukinen, P. Collin, Dermatitis herpetiformis: a common extraintestinal manifestation of coeliac disease, Nutrients. 10 (5) (2018) 602.

[36] T.T. Salmi, K. Hervonen, K. Kurppa, P. Collin, K. Kaukinen, T. Reunala, Coeliac disease evolving into dermatitis herpetiformis in patients adhering to normal or gluten-free diet, Scand. J. Gastroenterol. 50 (4) (2015) 387–392.

[37] L. Aine, M. Mäki, T. Reunala, Coeliac-type dental enamel defects in patients with dermatitis herpetiformis, Acta Derm. Venereol. 72 (1) (1992) 25–27.

[38] P.H. Green, C. Cellier, Celiac disease, N. Engl. J Med. 357 (17) (2007) 1731–1743.

[39] K. Hervonen, T.T. Salmi, K. Kurppa, K. Kaukinen, P. Collin, T. Reunala, Dermatitis herpetiformis in children: A long-term follow-up study, Br. J. Dermatol. 171 (5) (2014) 1242–1243.

[40] E. Antiga, A. Verdelli, A. Calabri, P. Fabbri, M. Caproni, Clinical and immunopathological features of 159 patients with dermatitis herpetiformis: An Italian experience, G. Ital. Dermatol. Venereol. 148 (2) (2013) 163–169.

[41] D. Rampertab, N. Pooran, P. Brar, P. Singh, P.H. Green, Trends in the presentation of celiac disease, Am. J. Med. 119 (4) (2006) 355e9–355e14.

[42] E. Mansikka, T. Salmi, K. Kaukinen, P. Collin, H. Huhtala, T. Reunala, et al., Diagnostic delay in dermatitis herpetiformis in a high-prevalence area, Acta Derm. Venereol. 98 (2) (2018) 195–199.

[43] P.H.R. Green, S.N. Stavropoulos, S.G. Panagi, S.L. Goldstein, D.J. Mcmahon, S.G. Absan H, et al., Characteristics of adult celiac disease in the USA: results of a national survey, Am. J. Gastroenterol. 96 (1) (2001) 126–131.

[44] C. Ciacci, P. Ciclitira, M. Hadjivassiliou, K. Kaukinen, J.F. Ludvigsson, N. McGough, et al., The gluten-free diet and its current application in coeliac disease and dermatitis herpetiformis, United European Gastroenterol. J. 3 (2) (2015) 121–135.

[45] C. Pasternack, K. Kaukinen, K. Kurppa, M. Mäki, P. Collin, T. Reunala, et al., Quality of life and gastrointestinal symptoms in long-term treated Dermatitis herpetiformis patients: A cross sectional study in Finland, Am. J. Clin. Dermatol. 16 (6) (2015) 545–552.

[46] J.P.W. Burger, B. de Brouwer, J. IntHout, P.J. Wahab, M. Tummers, J.P.H. Drenth, Systematic review with meta-analysis: Dietary adherence influences normalization of health-related quality of life in coeliac disease, Clin. Nutr. 36 (2) (2017) 399–406.

[47] N.R. Lewis, R.F. Logan, R.B. Hubbard, J. West, No increase in risk of fracture, malignancy or mortality in dermatitis herpetiformis: A cohort study, Aliment. Pharmacol. Ther. 27 (11) (2008) 1140–1147.

[48] K. Hervonen, A. Alakoski, T.T. Salmi, S. Helakorpi, H. Kautiainen, K. Kaukinen, E. Pukkala, P. Collin, T. Reunala, Reduced mortality in dermatitis herpetiformis: A population-based study of 476 patients, Br. J. Dermatol 167 (6) (2012) 1331–1337.

[49] J.J. Zone, L.J. Meyer, MJ. Petersen, Deposition of granular IgA relative to clinical lesions in dermatitis herpetiformis, Arch. Dermatol. 132 (8) (1996) 912–918.

[50] M. Hietikko, K. Hervonen, T. Salmi, T. Ilus, J.J. Zone, K. Kaukinen, et al., Disappearance of epidermal transglutaminase and IgA deposits from the papillary dermis of patients with dermatitis herpetiformis after a long-term gluten-free diet, Br. J. Dermatol. 178 (3) (2018) e198–e201.

[51] E. Mansikka, K. Hervonen, K. Kaukinen, T. Ilus, P. Oksanen, K. Lindfors, et al., Gluten challenge induces skin and small bowel relapse in long-term gluten-free diet-treated dermatitis herpetiformis, J. Invest. Dermatol. 139 (10) (2019) 2108–2114.

[52] C.M. Hull, M. Liddle, N. Hansen, L.J. Meyer, L. Schmidt, T. Taylor, et al., Elevation of IgA anti-epidermal transglutaminase antibodies in dermatitis herpetiformis, Br. J. Dermatol. 159 (1) (2008) 120–124.

[53] T. Reunala, T.T. Salmi, K. Hervonen, K. Laurila, H. Kautiainen, P. Collin, K. Kaukinen, IgA antiepidermal transglutaminase antibodies in dermatitis herpetiformis: a significant but not complete response to a gluten-free diet treatment, Br. J. Dermatol. 172 (4) (2015) 1139–1141.

[54] M. Hietikko, K. Hervonen, T. Ilus, T. Salmi, H. Huhtala, K. Laurila, et al., Ex vivo culture of duodenal biopsies from patients with dermatitis herpetiformis indicates that ransglutaminase 3 antibody production occurs in the gut, Acta Derm. Venereol. 98 (3) (2018) 366–372.

[55] H. Sankari, M. Hietikko, K. Kurppa, K. Kaukinen, E. Mansikka, H. Huhtala, et al., Intestinal TG3- and TG2-specific plasma cell responses in dermatitis herpetiformis patients undergoing a gluten challenge, Nutrients. 13 (12) (2020) 467 2.

[56] M. Hietikko, O. Koskinen, K. Kurppa, K. Laurila, P. Saavalainen, T. Salmi, et al., Small-intestinal TG2-specific plasma cells at different stages of coeliac disease, BMC Immunol. 19 (1) (2018) 36.

[57] K. Preisz, M. Sárdy, A. Horváth, KS. Immunoglobulin, complement and epidermal transglutaminase deposition in the cutaneous vessels in dermatitis herpetiformis, J. Eur. Acad. Dermatol. Venereol. 19 (1) (2005) 74–79.

[58] A. Görög, K. Németh, K. Kolev, J.J. Zone, B. Mayer, P. Silló, et al., Circulating transglutaminase 3-Immunoglobulin A immune complexes in dermatitis herpetiformis, J. Invest. Dermatol. 136 (8) (2016) 1729–1731.

[59] J.J. Zone, L.A. Schmidt, T.B. Taylor, C.M. Hull, M.C. Sotiriou, T.D. Jaskowski, et al., Dermatitis herpetiformis sera or goat anti-transglutaminase-3 transferred to human skin-grafted mice mimics dermatitis herpetiformis immunopathology, J. Immunol. 186 (7) (2011) 4474–4480.

[60] G. Huelsz-Prince, A.M. Belkin, E. VanBavel, EN. Bakker, Activation of extracellular transglutaminase 2 by mechanical force in the arterial wall, J. Vasc. Res. 50 (5) (2013) 383–395.

[61] S. Reitamo, T. Reunala, Y.T. Konttinen, O. Saksela, OP. Salo, Inflammatory cells, IgA, C3, fibrin and fibronectin in skin lesions in dermatitis herpetiformis, Br. J. Dermatol. 105 (2) (1981) 167–177.

[62] K. Airola, M. Vaalamo, T. Reunala, U.K. Saarialho-Kere, Enhanced expression of interstitial collagenase, stromelysin-1, and urokinase plasminogen activator in lesions of dermatitis herpetiformis, J. Invest. Dermatol. 105 (2) (1995) 184–189.

[63] S. Karpati, E. Torok, I. Kosnai, Discrete palmar and plantar symptoms in children with dermatitis herpetiformis Duhring, Cutis. 37 (3) (1986) 184–187.

[64] H. Tu, L. Parmentier, M. Stieger, Z. Spanou, M. Horn, H. Beltraminelli, et al., Acral purpura as leading clinical manifestation of dermatitis herpetiformis: report of two adult cases with a review of the literature, Dermatology. 227 (1) (2013) 1–4.

[65] D. Zaghi, D. Witheiler, A.M. Menter, Petechial eruption on fingers. Dermatitis herpetiformis, JAMA Dermatol. 150 (12) (2014) 1353–1354.

[66] J. Visakorpi, M. Mäki, Changing clinical features of coeliac disease, Acta. Pediatr. 83 (395) (1994) 10–13 Suppl.

[67] D. Tapsas, E. Hollen, L. Stenhammar, K. Fält-Magnusson, The clinical presentation of coeliac disease in 1030 Swedish children: changing features over the past four decades, Dig. Liver. Dis. 48 (1) (2016) 16–22.

[68] M. Spijkerman, I.J. Tan, J.J. Kolkman, S. Withoff, C. Wijmenga, M.C. Wisschedijk, et al., A large variety of clinical features and concomitant disorders in coeliac disease-a cohort study in the Netherlands, Dig. Liver. Dis. 48 (5) (2016) 499–505.

[69] T. Reunala, I. Kosnai, S. Karpati, P. Kuitunen, E. Török, E. Savilahti, Dermatitis herpetiformis: jejunal findings and skin response to gluten free diet, Arch. Dis. Child. 59 (6) (1984) 517–522.

[70] D.J. Gawkrodger, A. Ferguson, R.S. Barnetson, Nutritional status in patients with dermatitis herpetiformis, Am. J. Clin. Nutr. 48 (2) (1988) 355–360.

[71] A. Alakoski, T. Salmi, K. Hervonen, H. Kautiainen, M. Salo, K. Kaukinen, et al., Chronic gastritis in dermatitis herpetiformis: a controlled study, Clin. Dev. Immunol. 2012 (2012) 640630.

[72] C. Pasternack, K. Kaukinen, K. Kurppa, M. Mäki, P. Collin, K. Hervonen, et al., Gastrointestinal symptoms increase the burden of illness in dermatitis herpetiformis: a prospective study, Acta Derm. Venereol. 97 (1) (2017) 58–62.

[73] L. Aine, T. Reunala, M. Mäki, Dental enamel defects in children with dermatitis herpetiformis, J. Pediatr. 118 (1991) 572–574.

[74] H. Lähteenoja, K. Irjala, M. Viander, E. Vainio, A. Toivanen, S. Syrjänen, Oral mucosa is frequently affected in patients with dermatitis herpetiformis, Arch. Dermatol. 134 (6) (1998) 756–758.

[75] P. Patinen, J. Hietanen, M. Malmström, T. Reunala, E. Savilahti, Iodine and gliadin challenge on oral mucosa in dermatitis herpetiformis, Acta Derm. Venereol. 82 (2) (2002) 86–89.

[76] K.T. Amber, D.F. Murrell, E. Schmidt, P. Joly, L. Borradori, Autoimmune subepidermal bullous diseases of the skin and mucosae: clinical features, diagnosis and management, Clin. Rev. Allergy. Immunol. 54 (1) (2018) 26–51.

[77] V. Venning, Linear IgA disease: clinical presentation, diagnosis, and pathogenesis, Dermatol. Clin. 29 (3) (2011) 453–458.

[78] D. Bolotin, V. Petronic-Rosic, Dermatitis herpetiformis. Part I. Epidemiology, pathogenesis, and clinical presentation, J. Am. Acad. Dermatol. 64 (6) (2011) 1017–1024.

[79] C. Rose, E.-.B. Bröcker, Z. Zillicens, Clinical, histological and immunopathological findings in 32 patients with dermatitis herpetiformis Duhring, J. Dtsch. Dermatol. Ges 8 (4) (2010) 265–270.

[80] V. Fuchs, K. Kurppa, H. Huhtala, P. Collin, M. Mäki, K. Kaukinen, Factors associated with long diagnostic delay in celiac disease, Scand. J. Gastroenterol. 49 (11) (2014) 1304–1310.

[81] S.D. Rampertab, N. Pooran, P. Brar, P. Singh, P.H. Green, Trends in the presentation of celiac disease, Am. J. Med. 119 (4) (2006) 9–14.

[82] P.P. Seah, L. Fry, Immunoglobulins in the skin in dermatitis herpetiformis and their relevance in diagnosis, Br. J. Dermatol. 92 (2) (1975) 157–166.

[83] T. Makino, T. Shimizu, Fibrillar-type Dermatitis Herpetiformis, Eur. J. Dermatol. 29 (2) (2019) 115–120.

[84] V. Bonciolini, E. Antiga, B. Bianchi, E. DelBianco, A. Ninci, V. Maio, et al., Granular IgA deposits in the skin of patients with coeliac disease: is it always dermatitis herpetiformis? Acta Derm. Venereol. 99 (1) (2019) 78–83.

[85] C. Cannistraci, I. Lesnoni La Parola, G. Cardinali, G. Bolasco, N. Aspite, V. Stigliano, et al., Co-localization of IgA and TG3 in healthy skin of coeliac patients, J. Eur. Acad. Dermatol. Venereol. 21 (4) (2007) 509–514.

[86] J. Pierard, I. Whimster, The histological diagnosis of dermatitis herpetiformis, bullous pemphigoid and erythema multiforme, Br. J. Dermatol. 73 (1961) 253–266.

[87] S.C. Bresler, S.R. Granter, Utility of direct immunofluorescence testing for IgA in patients with high and low clinical suspicion for dermatitis herpetiformis, Am. J. Clin. Pathol. 144 (6) (2015) 880–884.

[88] M. Caproni, E. Antiga, L. Melani, P. Fabbri, Guidelines for the diagnosis and treatment of dermatitis herpetiformis, J. Eur. Acad. Dermatol. Venereol. 23 (6) (2009) 633–638.

[89] P. Brar, G.Y. Kwon, I.I. Egbuna, S. Holleran, R. Ramakrishnan, G. Bhagat, P.H. Green, Lack of correlation of degree of villous atrophy with severity of clinical presentation of coeliac disease, Dig. Liver. Dis. 39 (1) (2007) 26–29.

[90] E. Mansikka, K. Hervonen, T.T. Salmi, H. Kautiainen, K. Kaukinen, P. Collin, T. Reunala, The decreasing prevalence of severe villous atrophy in dermatitis herpetiformis: A 45-year experience in 393 patients, J. Clin. Gastroenterol. 51 (2017) 235–239.

[91] T.T. Järvinen, K. Kaukinen, K. Laurila, S. Kyrönpalo, M. Rasmussen, M. Mäki, et al., Intraepithelial lymphocytes in celiac disease, Am. J. Gastroenterol. 98 (6) (2003) 1332–1337.

[92] I.R. Korponay-Szabó, T. Halttunen, Z. Szalai, K. Laurila, R. Kiraly, J. Kovacs, et al., In vivo targeting of intestinal and extraintestinal transglutaminase 2 by coeliac autoantibodies, Gut. 53 (5) (2004) 641–648.

[93] O. Koskinen, P. Collin, I. Korponay-Szabo, T. Salmi, S. Iltanen, K. Haimila, et al., Gluten-dependent small bowel mucosal transglutaminase 2–specific IgA deposits in overt and mild enteropathy coeliac disease, J. Pediatr. Gastroenterol. Nutr. 47 (4) (2008) 436–442.

[94] T. Salmi, K. Hervonen, K. Laurila, P. Collin, M. Mäki, O. Koskinen, et al., Small bowel transglutaminase 2-specific IgA deposits in dermatitis herpetiformis, Acta Derm. Venereol. 94 (4) (2014) 393–397.

[95] W. Dieterich, D. Schuppan, E. Laag, L. Bruckner-Tuderman, T. Reunala, S. Kárpáti, et al., Antibodies to tissue transglutaminase as serologic markers in patients with dermatitis herpetiformis, J. Invest. Dermatol. 113 (1) (1999) 133–136.

[96] I.R. Korponay-Szabó, K. Laurila, Z. Szondy, T. Halttunen, Z. Szalai, I. Dahlbom, et al., Missing endomysial and reticulin binding of coeliac antibodies in transglutaminase 2 knockout tissues, Gut. 52 (2) (2003) 199–204.

[97]   I. Dahlbom, I.R. Korponay-Szabo, J.B. Kovács, Z. Szalai, M. Mäki, T. Hansson, Prediction of clinical and mucosal severity of coeliac disease and dermatitis herpetiformis by quantification of IgA/IgG serum antibodies to tissue transglutaminase, J. Pediatr. Gastroenterol Nutr. 50 (2) (2010) 140–146.

[98]   T. Salmi, K. Kurppa, K. Hervonen, K. Laurila, P. Collin, H. Huhtala, et al., Serum transglutaminase 3 antibodies correlate with age at celiac disease diagnosis, Dig. Liver. Dis. 48 (6) (2016) 632–637.

[99]   M. Kasperkiewicz, C. Dähnrich, C. Probst, L. Komorowski, W. Stöcker, W. Schlumberger, et al., Novel assay for detecting celiac disease-associated autoantibodies in dermatitis herpetiformis using deamidated gliadin-analogous fusion peptides, J. Am. Acad. Dermatol. 66 (4) (2012) 583–588.

[100] D. Bolotin, V. Petronic-Rosic, Dermatitis herpetiformis: part II. Diagnosis, management, and prognosis, J. Am. Acad. Dermatol. 64 (6) (2011) 1027–1033.

[101] E. Mansikka, K. Hervonen, K. Kaukinen, P. Collin, H. Huhtala, T. Reunala, Prognosis of dermatitis herpetiformis patients with and without villous atrophy at diagnosis, Nutrients. 10 (5) (2018) 641.

[102] C.M. Hardman, J.J. Garioch, J.N. Leonard, HJ. Thomas, Absence of toxicity of oats in patients with dermatitis herpetiformis, N. Engl. J. Med. 3378 (26) (1997) 1884–1887.

[103] T. Reunala, P. Collin, K. Holm, P. Pikkarainen, A. Miettinen, N. Vuolteenaho, M. Mäki, Tolerance to oats in dermatitis herpetiformis, Gut. 43 (1998) 490–493.

[104] I. Hoffmanova, D. Sanchez, A. Szczepankova, Tlaskalova-Hogenova H. The pros and cons of using oat in celiac disease, Nutrients. 11 (10) (2019) 2345.

[105] J.F. Ludvigsson, T. Card, P.J. Ciclitira, G.L. Swift, I. Nasr, D.S. Sanders, C. Ciacci, Support for patients with celiac disease: A literature review, United European Gastroenterol. J. 3 (2) (2015) 146–159.

[106] L. Elli, F. Ferretti, S. Orlando, M. Vecchi, E. Monguzzi, L. Roncoroni, et al., Management of celiac disease in daily clinical practice, Eur. J. Intern. Med. 61 (2019) 15–24.

[107] J.J. Garioch, H.M. Lewis, S.A. Gargent, J.N. Leonard, L. Fry, 25 years' experience of a gluten-free diet in the treatment of dermatitis herpetiformis, Br. J. Dermatol. 131 (4) (1994) 541–545.

[108] K. Hervonen, T. Salmi, T. Ilus, K. Paasikivi, M. Vornanen, K. Laurila, et al., Dermatitis herpetiformis refractory to gluten-free dietary treatment, Acta Derm. Venereol. 96 (1) (2016) 82–86.

[109] J.N. Leonard, L. Fry, Treatment and management of dermatitis herpetiformis, Clin. Dermatol. 9 (3) (1991) 403–408.

[110] Y.I. Zhu, MJ. Stiller, Dapsone and sulfones in dermatology. Overview and update, J. Am. Acad. Dermat. 45 (3) (2001) 420–434.

[111] K. Hervonen, M. Vornanen, H. Kautiainen, P. Collin, T. Reunala, Lymphoma in patients with dermatitis herpetiformis and their first-degree relatives, Br. J. Dermatol. 152 (1) (2005) 82–86 152.

[112] M. Viljamaa, K. Kaukinen, E. Pukkala, K. Hervonen, T. Reunala, P. Collin, Malignancies and mortality in patients with coeliac disease and dermatitis herpetiformis: 30-year population-based study, Dig. Liver. Dis. 38 (6) (2006) 374–380.

[113] C. Pasternack, I. Koskinen, K. Hervonen, K. Kaukinen, J. Järvelin, T. Reunala, et al., Risk of fractures in dermatitis herpetiformis and coeliac disease: a register-based study, Scand. J. Gastroenterol. 54 (7) (2019) 843–848.

[114] K. Heikkilä, J. Pearce, M. Mäki, K. Kaukinen, Celiac disease and bone fractures: a systematic review and meta-analysis, J. Clin. Endocrinol. Metab. 100 (1) (2015) 25–34.

[115] G. Malamut, C. Cellier, Refractory celiac disease: epidemiology and clinical manifestations, Dig. Dis. 33 (2) (2015) 221–226.

[116] P. Laurikka, T. Salmi, P. Collin, H. Huhtala, M. Mäki, K. Kaukinen, K. Kurppa, Gastrointestinal symptoms in celiac disease patients on a long-term gluten-free diet, Nutrients. 8 (7) (2016) 429.

[117] K. Hervonen, M. Viljamaa, P. Collin, M. Knip, T. Reunala, The occurrence of type 1 diabetes in patients with dermatitis herpetiformis and their first-degree relatives, Br. J. Dermatol. 150 (1) (2004) 136–138.

[118] O. Varpuluoma, J. Jokelainen, A.K. Försti, M. Timonen, L. Huilaja, K. Tasanen, Dermatitis herpetiformis and celiac disease increase the risk of bullous pemphigoid, J. Invest. Dermatol. 139 (3) (2019) 600–604.

CHAPTER 9

Noncoeliac gluten sensitivity

Knut E.A. Lundin[a], Margit Brottveit[b], Gry Skodje[c]
[a]KG Jebsen Coeliac Disease Research Centre, Institute of Clinical Medicine, Faculty of Medicine, University of Oslo, Oslo, Norway and Dept of gastroenterology, Division of Surgery, inflammationa and transplantation, Oslo University Hospital Rikshospitalet, N-0372 Oslo, Norway
[b]Healthy Life Centre, Municipality of Nes, N-2150 Årnes, Norway
[c]Department of gastroenterology, Division of Medicine, Oslo University Hospital Ullevål, N-0450 Oslo, Norway

9.1 Introduction

The role of gluten from wheat, rye and barley as the disease driver of coeliac disease is undisputed, and the treatment for coeliac disease is a gluten-free diet. However, the community consumption of gluten-free food by far exceeds that consumed by people with a diagnosis of coeliac disease. Furthermore, the vast majority of people who avoid gluten do not have coeliac disease, even by the most vigorous clinical examinations. How can this be? We here review the current knowledge on the clinical entity "noncoeliac gluten sensitivity" (NCGS), the theories of action of gluten and possibly other components of gluten containing cereals, proposed clinical work-up and some cautious remarks. The discussion on NCGS today revolves around two major hypothesis; 1) NCGS is caused by immune activation by gluten or "gluten-like" proteins in the cereals, 2) NCGS is caused by triggering of abdominal symptoms caused by Fermentable Oligo-, Di-, Monosaccharides And Polyols (FODMAP) [1].

The discussion on the triggers of the clinical picture has even led to nomenclature change. The condition started out as «non-celiac gluten sensitivity» [2,3] but has later been renamed to noncoeliac wheat sensitivity (NCWS) [4]. For simplicity we will here use the term NCGS (see Fig. 9.1).

9.2 The clinical picture

One of the first clinical descriptions of the condition was published by Holmes et al., in 1980 [5]. Since then, a substantial amount of research and corresponding literature has been performed. The proportion of papers on NCGS, compared to those on coeliac disease has steadily increased during the last few decades [2]. Furthermore, NCGS has been the focus of repeated reviews by expert groups [2–4] and is well recognized in expert reviews and guidelines on coeliac disease and other gluten-related disorders [6–9].

Our experience in adults indicate that most people that are on a gluten free diet have started this without any formal evaluation by health care professionals, but rather by advice from friends, from social media including Facebook, magazines etc. Although they

Coeliac Disease and Gluten-Related Disorders.
DOI: https://doi.org/10.1016/B978-0-12-821571-5.00004-0

Figure 9.1 *Some examples of FODMAP containing food.* Photo by Øyvind Horgmo, University of Oslo.

did not get any professional guidance, they still adhere rather strictly to the diet, at least in our experience [10]. We often also notice children that are on a gluten-free diet based on parent's preferences, also in those cases without professional evaluation. Such individuals are often assumed to be prone to somatization of their clinical status, however, we could not find any signs of that in a case series of subjects with self-instituted gluten-free diet [11].

The condition is rather frequent in the general population, as for instance reported by a UK survey [12]. A population-based questionnaire received 1002 responses, the self-reported prevalence of "gluten sensitivity" was 13%, whereas 3.7% consumed a gluten-free diet and 0.8% reported they had a diagnosis of coeliac disease. Similar figures in the range around 10% have been presented from other countries, so the public perception that gluten avoidance is very frequent, is supported [13–15].

It is, of course, essential to rule out coeliac disease as a cause of perceived gluten intolerance. Wahnschaffe et al., made interesting observations in a cohort of "gluten sensitive" individuals [16, 17]. They observed that although they could not find serological signs of coeliac disease by analyzing their blood samples, they did find anti-TG2 antibodies in gut lavage and also correlation with expression of HLA-DQ2. This closely resembles previous findings from Ann Fergusons group that intestinal efflux of antiendomysium antibodies is seen in coeliac disease patients also in the case of "sero-negativity" [18]. It further correlates with publications where duodenal anti-TG2 antibodies, not detectable in serum, can be shown using immunohistochemistry on mucosal samples [19]. With today's understanding, a possible explanation for the finding of Wahnschaffe and colleagues would be that these individuals actually can have "evolving coeliac disease," meaning they belong to the clinical spectrum of "potential coeliac disease" [6, 9].

We investigated a cohort of possible NCGS patients from the general public [20]. In this "presocial media" time, they were invited by newspaper advertisement. We received response from 136 individuals that had commenced a strict or semi-strict gluten-free diet without clinical work-up [10]. They were asked to donate a blood sample for HLA typing. We found that 43 of them carried HLA-DQ2.5, thus could possibly have coeliac disease and were suitable for an "HLA-DQ:gluten tetramer" test that is known to identify coeliac disease patients after a 3-day gluten challenge [21]. The proportion of DQ2.5+ respondents was approximately as expected from the general population in Norway. We proceeded with 35 of these 43 indivduals. After a 3 day gluten challenge, with gastroduodenoscopy and an HLA-DQ:gluten tetramer test, we concluded that 3 of the 35, out of the original 136 respondents, received a diagnosis of coeliac disease [20]. Thus, actual coeliac disease seems to be rare in a NCGS population and a clinical response to "gluten" is not a reliable marker for coeliac disease. This closely resembles a previous report from the Tampere group where they did endoscopy and "allergy investigation" in 93 consecutive adults that reported abdominal symptoms after ingestion of cereals [22]. Eight of these 93 were found to have coeliac disease, and another seven were labeled as "latent coeliac disease".

Much focus has been paid to the IBS-like picture of NCGS, but there are other aspects of NCGS that deserves attention. The role of gluten in neurological disorders like "gluten ataxia" is dealt with elsewhere in this book. In subjects with NCGS it has been reported that they suffer from lack of concentration – a phenomenon usually referred to as "foggy brain" [23], and has received support by recent functional MRI studies [24]. Participants in a randomized controlled study, that could not find any signs of abdominal symptoms caused by gluten, still showed higher overall depression sores after gluten challenge [25]. Signs of gluten sensitivity (to be discussed later) can also be seen in chronic fatigue syndrome [26]. And furthermore, a strong association between psychological function of children with Down syndrome and both urine levels of and antibodies against gluten was found [27].

9.3 IBS – NCGS – or both? What can be learned from RCTs?

While many of us have broad impressions from clinical practice, it is generally recognized that data from randomized controlled trials more accurately expresses the "real truth". The results from such trials do not give a really clear picture – this is not the least because of lack of uniform trial designs, lack of uniform criteria for selection of patients, and lack of standardization of the gluten challenge vehicle. This is only a selection, for a systematic review and meta-analysis of re-challenge studies see Lionetti et al. [28].

Gluten challenge RCTs	Sample size	Type of challenge	Vehicle	Outcome measure	Critical remarks	Outcome result
Skodje et al. [29]	59	Double-blinded, triple arm cross-over trial	Müsli bars vs spiked with fructan vs spiked with pure gluten (5.7 g/day)	Gastrointestinal symptoms	The patients were not diag-nosed with NCGS in adv-ance, only self-instituted GFD	Only fructan induced symptoms (gluten similar to placebo)

(continued on next page)

Gluten challenge RCTs	Sample size	Type of challenge	Vehicle	Outcome measure	Critical remarks	Outcome result
Biesiekierski et al. [30]	34	Double-blinded rechallenge trial	Bread and muffins 16 g gluten/day	Overall and gastrointestinal symptoms	Self- instituted GFD	*Symptoms not adequately controlled:* 13/19 (68%) in gluten group vs 6/15 (40%) in placebo group, p=0.001
Biesiekierski et al. [31]	37	Double-blinded, triple arm cross-over trial	Gluten (16 g/day), whey and placebo spiked in natural food delivered to the participants	Overall and gastrointestinal symptoms	The patients were not diagnosed with NCGS in advance, only self-instituted GFD	No specific or dose-dependent effects of gluten in patients with NCGS placed diets low in FODMAPs
Carroccio et al. [32]	920 IBS	Double-blinded placebo-controlled challenge	Wheat capsules	Gastrointestinal and overall symptoms	Suprisingly little placebo response	276 (30%) sensitive to wheat; asymptomatic on diet, symptoms during challenge
Elli et al. [33]	98	Multicenter double-blind placebo-controlled trial with crossover	5.6 g of gluten capsules per day	Gastrointestinal and overall symptoms	No evaluation of FODMAPS during study	14% symptomatic relapse during gluten challenge
Di Sabatino et al. [34]	59	Prospective, randomized, placebo-controlled, cross-over trial	4.375 g gluten capsules per day	Gastrointestinal and overall symptoms	Self-instituted GFD	Significantly increase in overall and gastrointestinal symptoms in gluten compared to placebo

(continued on next page)

Gluten challenge RCTs	Sample size	Type of challenge	Vehicle	Outcome measure	Critical remarks	Outcome result
Zanini et al. [35]	35	Double-blind, cross over challenge study	Gluten-containing flour 7.89 g gluten/day	Identifying gluten from placebo and increase in gastrointestinal symptoms	FODMAPs in gluten free flour	12 (34%) identified gluten and diagnosed with NCGS
Skodje et al. [36]	56	Open challenge	Bread 8 g gluten/day	Gastrointestinal or overall symptoms	Open challenge Self instituted GFD	47 (85%) diagnosed with NCGS
Shahbazkhani et al. [37]	72	Elimination and then Placebo controlled challenge	Powder 52 g gluten/day-(n=35) or placebo (n=37)	Gastrointestinal or overall symptoms		26/35 symptom worsening in gluten group vs 6/37 in placebo group, p<0.001
Dale et al. [38]	20	Double-blinded placebo-controlled food challenge	Muffins 11 g gluten/day	Gastrointestinal symptoms		Response to placebo > gluten (p=0.012) 4/20 NCGS due to correct identification of periods

The most striking observation from this plethora of clinical studies is the diversity of outcomes, ranging from no effect of gluten to an overwhelming signal versus placebo. Taken together, however, one must conclude that the RCTs do not generally support by the clinical impression of the care givers and the NCGS subjects that gluten ingestion usually givers strong symptoms.

Another striking observation may also be the difference between blinded and open challenges. This is not the least evident in our own research. In a blinded RCT with cross-over design, no effect by gluten was observed [29]. However, when the same individuals were challenged in an open manner with regular sandwich bread, 85% responded to the challenge and were deemed as fulfilling the criteria for NCGS [39]. Of course, an open challenge is subject to several flaws including expectations by the participants and care givers, and the fact that regular bread always will contain both gluten and other possible culprits of NCGS.

A third observation is further that the studies use a range of different scoring systems, and no agreement has been made. An expert group suggested some few years back how to score the symptoms based on their clinical experience, but only a single publication has addressed the performance of different scoring systems [3, 36].

The fourth observation is that different vehicles for challenge have been used. That would otherwise mimic the situation in drug testing where the same formulation is used in different countries. We attempted to at least use the same source for gluten in our RCTs in collaboration with the Australian group, but still the formulations were different [29, 31]. A well-known manufacturer of gluten-free food made an attempt to produce a cookie product that could be used by several investigators, but as far as we know, the production has been halted and the product was never really tested.

9.4 The gluten-free market

People who avoid gluten can, of course, rely on natural gluten-free food that by origin does not contain any gluten; vegetables, meat, fish. Consumption of these foodstuffs would never be registered as part of gluten-free food. Rather, with this label we consider special products for the gluten-free market, typically bread, bagels, buns etc. The price of this diet exceeds by far that of a regular diet [40, 41]. A report from US summarized that USD 15 500 000 000 were spent on retail sales of gluten-free foods in 2016 [42]. That is remarkable, especially since the nutritional value of gluten-free replacement products have been questioned by many authors [41, 43–45]. In particular, the gluten-free products generally contain less fiber and protein, more salt and saturated fat than their gluten-containing counterparts. Thus, people avoiding gluten has a need for dietary education and nutritional follow-up, a need that most probably rarely is followed by proper care [39].

9.5 Immunobiology of gluten in NCGS (and CD)

There is, in fact, a substantial amount of evidence that "gluten" has effects in both cell culture and animal models. Some of these effects are of interest for coeliac disease pathogenesis, of course. Some of these effects have been claimed to be purely part of how the innate immune system recognizes gluten. But it is still somewhat unclear why these effects on the innate immune system is only seen in patients that have *bona fide* coeliac disease and not in all individuals.

Animal models would be ideal to study NCGS but has not been really useful [46]. Special interest has been focused on the findings that gluten-free diet seems to prevent diabetes in NOD mice, an animal model for spontaneous Type 1 diabetes [47]. Such mice further develop signs of small intestinal enteropathy when fed gluten containing diet [48]. It was also shown that the lower incidence of spontaneous diabetes in these mice could be attributed to changes in the intestinal microbiome [49]. The effects of gluten have been studied in detail by the group of Karsten Buschard, as summarized in a recent review [50].

However, it is clear, as stated by the authors, that human interventional trials are needed to confirm proposed mechanisms. The effects of gluten are not limited to laboratory animals but is also seen in domestic dogs, including a well-known coeliac disease like condition in Irish setters [51, 52] but also other conditions without enteropathy [53]. This has led to widespread avoidance of gluten in industry produced pet food. The significance for human NCGS remains obscure.

The effects of gluten in cell culture systems have been extensively studied. The effects of gluten derived peptides on agglutination of so-called K563 cells was shown as early as 1984 [54]. This phenomenon has later been dissected further and also used to predict possible toxicity of gluten for coeliac and other gluten-sensitive patients, although one cannot say it has been widely accepted and adapted [55,56]. Gluten has further been shown to induce maturation of antigen presenting cells [57]. A non-gluten protein, amylase trypsin inhibitor directly activates monocytes, macrophages and dendritic cells via the Toll-like Receptor 4 [58]. This protein co-precipitates with gluten in separation of proteins and may in fact have contributed to some of the effects of gluten in other experiments. The role of the ATI will be discussed in more details below.

One hallmark of coeliac is the reactivation of mucosal pathology and immune activation upon re-challenge with gluten [9,59]. Very recently, it was demonstrated that gluten challenge of coeliac disease patients induces a rapid flux of the cytokines Interleukin-2, Interleukin- 8 and Interleukin-10, among others [60,61]. Interleukin-2 is a *bona fide* T helper cell cytokine, whereas the others are linked to innate immune activation and could therefore be more expected to be increased in subjects suffering from NCGS. However, this was not the case when serum from such gluten challenges were investigated [62]. This calls into question the notion that NCGS is caused by activation of the innate immune system by gluten.

9.6 The possible culprits

While it is clear to the patient and often to the clinician that the clinical picture of NCGS is caused by ingestion of gluten-containing food, the scientific picture of the culprit of NCGS is much more uncertain.

a) Gluten is, of course, a very likely candidate. By gluten one usually refers to a large set of proteins from wheat, rye, and barley of which the wheat proteins have been extensively studied. The gluten proteins of wheat can be divided into α, β, γ and ω gliadins and the low- and high molecular glutenins. In coeliac disease, these proteins drive the disease when they are presented by the coeliac disease associated HLA-DQ2.5, -DQ2.2, and -DQ8 molecules, this has been mapped in detail although still more epitopes are likely to exist [63,64]. It can, however, bluntly be stated that limited amounts of this knowledge is of value for our considerations of NCGS. There are reasons for this. Firstly, T cell recognition of gluten derived peptides is unlikely to play a role in NCGS. Secondly, in the lack of a defined molecular understanding of the possible pathogenesis of NCGS, a search for defined peptides is difficult. Thirdly, no attempts have been made to perform challenge experiments with gluten peptides or recombinant gluten proteins in well-defined NCGS population.

b) Peptide 31-43 of alpha-gliadin stands out as a possible, but unproven, candidate as a culprit in NCGS. The history of this peptide epitope started with a seminal paper by the Kasarda and Kagnoff research groups [65]. They performed short *ex vivo* culture experiments with biopsies from coeliac disease patients. Such experiments were widely employed in the past. They were done on biopsies from untreated coeliac disease patients, where the readout would be improvement of morphology in the absence of toxic gluten fragments. Alternatively, biopsies from treated patients were used, where the readout would be induction of villous atrophy in the presence of toxic fragments. At any rate, a peptide from alpha-gliadin was found to be active after repeated fractionation by high performance liquid chromatography (HPLC).

Inspired by these findings, we performed T cell cloning experiments looking for recognition of this peptide presented by the HLA-DQ2.5 molecule in coeliac disease. It was successful, in the sense that T cells were isolated [66]. However, it turned out that the peptide 31-43 did not bind well to HLA-DQ2.5 [67] and was not found as an epitope for small intestinal T cells in coeliac disease [63,68,69]. With today's knowledge, it is even possible that the effects observed by Kasarda and Kagnoff were due not to the peptide 31-43 itself, but rather due to other fragments in gliadin preparations [68].

At any rate, the 31-43 peptide of α-gliadin has a very long history showing its activity in diverse cell systems, although frankly, the receptor mediating the effect and the signaling pathways remain uncertain. It was shown that both total gliadin, the P31-43 as well as the longer 33-mer peptide could activate murine macrophages via the MyD88 pathway, an affect also shown for amylase trypsin inhibitors as discussed below [70]. The 31-43 peptide interferes with lysosome activation [71] and induces activation of the innate immune system in mice [72]. The same peptide also has functions, including activation of TLR-7, in both human cells like Caco-2 and in biopsies from coeliac disease patients [73]. The significance of these findings for NCGS are not really clear.

c) ATI

The amylase trypsin inhibitors (ATI) is an interesting group of proteins that are expressed by the cereals when they are under pressure of parasites [74]. It has been known for years to be involved in food allergy and in exercise-induced anaphylactic reactions [75]. When gluten is purified from the wheat flour, the ATIs will co-precipitate with gluten so the action of the two sets of proteins can be hard to dissociate. The possible importance of the ATIs was highlighted by the important publication by the Detlef Schuppan group in 2012 [58] and a subsequent report in 2017 [76]. In a series of elegant experiments they showed that the ATIs activate mouse toll like receptor (TLR) causing innate immune activation. The effect of purified ATI was also shown with recombinant ATI, ruling out possible contamination. While most of their work was done on mouse cells, data from human coeliac disease tissues supported the animal model. Since then, several other papers from the same group has been published where the role of the ATIs have been investigated in other disease models [77,78]. There are some loose ends; 1) can the effect of the recombinant proteins be tracked down to specific peptides?, 2) will human macrophages be triggered via TLR4?, 3) will the ATIs activate the mucosal immune system in humans with or without coeliac disease, 4) will the ATIs trigger NCGS? There seems to be several knowledge gaps [79].

d) Fructans are oligo- and polysaccharides of short chains of fructose units, and the major dietary source of fermentable carbohydrates (FODMAP) [80,81]. Fructans with a short chain length are generally referred to as fructooligosaccharides or oligofructose, and the longer chain are called inulins [82]. Fructans are storage carbohydrates in plants, mostly obtained from wheat and onion, which are relatively low in fructans but are consumed in large quantities [81,83]. Minimal digestion of fructans occurs in the small intestine due to the inability of the mammalian intestine to hydrolyze glycosidic linkages. Due to their small molecular size, fructans are likely to exert an osmotic effect leading to increased delivery of water to the colon [84]. Byproducts of the fermentation of fructans in colon include gases such as carbon dioxide, methane and hydrogen which may induce abdominal pain and bloating in subjects with visceral hypersensitivity/IBS [84,85]. Recently, a randomized, double-blind place-controlled challenge aimed to investigate the effect of gluten (without fructan) and fructan (without gluten), on gastrointestinal symptoms in individuals with self-reported NCGS [29]. No significant effect of gluten was found as compared to placebo and fructan. In contrast, a small daily dose of 2.1 g of fructans induced greater symptoms on multiple criteria, including the Gastronitestinal symptom rating scale-IBS version, after a 7-day challenge [29]. The study indicated that fructans are more likely to induce symptoms in those reporting sensitivities to wheat, rye and barley.

9.7 Serological and genetic markers

It goes without saying that the use of serology and/or genetic markers are highly desired in the investigation of possible NCGS. The problem is, of course, that the clinical diagnosis lacks objective markers – this has implications for precise delineations of patient materials. Very little is known about genetic markers, apart from a signal relating to HLA-DQ but this may in fact be related to "contamination" of the patient materials by coeliac disease patients – or by the common belief among clinicians that HLA-DQ2 and HLA-DQ8 are markers for gluten intolerance in the broader sense [17]. A large Italian, multi center study could not find any association between NCGS and HLA haplotypes [23].

Antibodies to gliadin, either of the IgA or the IgG class, are of special interest, of course. Volta and co-workers first studied a cohort of NCGS subjects compared to a matched cohort of coeliac disease patients, this was soon after expanded to a larger cohort from an Italian multi-center study [23,86]. They found in their first study that 56% of the NCGS subjects and 81% of the coeliac disease patients had increased levels of IgG to gluten, whereas only the latter had IgA antibodies to TG2 and IgG antibodies to DGP [86]. The later multi-center study on a total of 486 patients that were classified with NCGS did, however, showed that only 25% of the individuals tested, had increased levels of IgG to gluten [23]. It should be noted that the diagnosis NCGS in this study was done by ruling out coeliac disease, then a withdrawal of gluten-containing food from the diet with clinical improvement, followed by deterioration upon re-challenge with regular food. Thus, the diagnosis cannot be said to be "definite" in all cases, as previously discussed.

These observations were taken further in collaboration with the group of Alaedini, when serum samples from individuals with "clinical wheat sensitivity," as judged by open withdrawal and reinstitution of wheat, rye, and barley from their diet, were analyzed for mucosal damage as well as immune reactivity against both gluten to microbial agents [87]. They found increased levels of both soluble CD14 and lipopolysaccharide (LPS) binding protein, and of antibodies to LPS and flagellin, and circulating levels of fatty acid-binding protein 2, indicating intestinal damage. Unfortunately, a control group of IBS patients was not included. Most recently, the same group of investigators reported that NCGS subjects and coeliac disease patients differed in their IgG subclass reactivity against gluten [88]. Not only antibodies to gluten and microbial agents are of potential value other serum markers involved in intestinal integrity have been proposed from an Italian multi-center study [89]. They investigated serum zonulin in different patient groups, and found it to be increased in both NCGS and CD. A diagnostic algoritm was proposed, it gave a 89% diagnostic accuracy for NCGS. The levels were further followed after diet treatment and dropped, but in NCGS only in patients carrying the coeliac disease associated HLA-DQ2 and -DQ8 genes. This is not easy to understand since there is very little support of adaptive T cell reactivity to gluten in NCGS. All these aspects support the notion of an organic substrate for NCGS – but the mechanism still remains elusive.

9.8 Clinical investigation – the view of the clinical dietician

The role of a clinical dietician in suspicion of any adverse reaction to food is to confirm the diagnosis by a detailed clinical history through a thorough diet and symptom assessment. A clinical dietician is qualified to educate on structured elimination and re-challenge of suspected offending foods and to construct a nutritional adequate diet avoiding unnecessary restrictions. Open food challenges are still accepted in clinical settings [90, 91] and are the first step diagnostic tool to explore adverse reactions to food when individuals are self-instituted on an elimination diet. Open food challenge may be followed by blinded challenge where there is a need for confirmation, but it is unclear how this shall be done in the clinical setting [9, 92]. In cases of self-reported NCGS where coeliac disease is not adequately excluded before gluten has been eliminated from the diet, gluten-containing food needs to be re-introduced in sufficient amount to get reliable coeliac disease diagnostics. A gluten exposure equivalent to four slices of white wheat bread daily for minimum two weeks followed by duodenal biopsy has been a clinical practice [9].

Individuals on a self-instituted gluten-free diet where coeliac disease is properly excluded usually have a strong belief that gluten is the culprit. Their observation of symptom improvement in response to a wheat-free diet may be correct.

However, if gluten is the cause of NCGS, then a gluten-free diet should lead to resolution of symptoms to a level consistent with the healthy population. Two surveys of individuals who reported they were gluten sensitive, even those strictly on a gluten-free diet, remained moderately symptomatic [36,93]. This was confirmed in the setting of a clinical trial with symptoms assessed by daily entries into a diary using a visual analogue scale [31]. In fact, this group had considerable further improvement of symptoms when taught how to reduce FODMAP intake.

Evidence to date suggests that 50-80% of individuals with IBS report symptomatic benefit from low FODMAP diet [94]. A speculation is that individuals with NCGS belong to the group of 20–50% that will not respond to low-FODMAP diet. In absence of reliable diagnostic tools to identify NCGS, the role of a dietician is to guide the individual to identify symptom triggering foods, to clarify what is a normal symptom and to interpret the change in symptoms. When the low FODMAP diet does not work, and other conditions with IBS-like symptoms are not present, the role of a dietician is to work with the individual to find other solutions and aid with appropriate delivery of dietary therapy while ensuring nutritional and psychological health [95]. That could include the use of gluten-free diet in absence of coeliac disease.

9.9 The FODMAP approach

Dietary restriction of FODMAP is effective in the management of functional gastrointestinal symptoms that occur in IBS [96–98]. The diet involves restriction of high-FODMAP foods for 4-6 weeks to achieve symptom relief followed by systematic reintroduction to identify tolerance threshold for individual FODMAP, which enables long-term self-management of symptoms and nutritional adequacy [99]. The low-FODMAP diet is a complex intervention and should be implemented with counseling from a clinical dietician [81, 100].

The gluten-free diet is necessarily a diet low in fructans and is light version of the low FODMAP diet, since all wheat, rye and barley products are excluded. Therefore, symptom relief in self-reported NCGS could be explained by the reduction of fructans in the diet. Individuals avoiding food based on gluten content, will be eating other FODMAP-rich foods, such as the fructan-rich onion and garlic, resulting in persistent gastrointestinal symptoms. A low-FODMAP diet is not a gluten-free diet. Products with small amounts of wheat or gluten are allowed in a low-FODMAP diet. Spelt sourdough and oat products are low-FODMAP alternatives to wheat that contain gluten. Likewise, naturally gluten-free foods such as honey, apple and cauliflower are allowed in a gluten-free diet, but they contain high amounts of the FODMAPs fructose and polyols [80].

9.10 Summary

We have here summarized the history of NCGS and described some of the research supporting its existence, but also highlighted several of the controversies in this field. There is no doubt that avoidance of dietary gluten is a widespread phenomenon in the whole Western world, but only to some degree supported by science. The discussion about the role of gluten or gluten-like proteins like ATI as possible culprits, whether immune recognition is involved, as opposed to alternative view that it is the FODMAP moiety of cereals that cause the clinical effects, is still running.

References

[1] R. De Giorgio, U. Volta, P.R. Gibson, Sensitivity to wheat, gluten and FODMAPs in IBS: facts or fiction? Gut. 65 (2016) 169–178.

[2] C. Catassi, J.C. Bai, B. Bonaz, G. Bouma, A. Calabro, A. Carroccio, G. Castillejo, C. Ciacci, F. Cristofori, J. Dolinsek, R. Francavilla, L. Elli, P. Green, W. Holtmeier, P. Koehler, S. Koletzko, C. Meinhold, D. Sanders, M. Schumann, D. Schuppan, R. Ullrich, A. Vecsei, U. Volta, V. Zevallos, A. Sapone, A. Fasano, Non-Celiac Gluten sensitivity: the new frontier of gluten related disorders, Nutrients 5 (2013) 3839–3853.

[3] C. Catassi, L. Elli, B. Bonaz, G. Bouma, A. Carroccio, G. Castillejo, C. Cellier, F. Cristofori, L. de Magistris, J. Dolinsek, W. Dieterich, R. Francavilla, M. Hadjivassiliou, W. Holtmeier, U. Korner, D.A. Leffler, K.E. Lundin, G. Mazzarella, C.J. Mulder, N. Pellegrini, K. Rostami, D. Sanders, G.I. Skodje, D. Schuppan, R. Ullrich, U. Volta, M. Williams, V.F. Zevallos, Y. Zopf, A. Fasano, Diagnosis of Non-Celiac Gluten Sensitivity (NCGS): The Salerno Experts' Criteria, Nutrients 7 (2015) 4966–4977.

[4] C. Catassi, A. Alaedini, C. Bojarski, B. Bonaz, G. Bouma, A. Carroccio, G. Castillejo, L. De Magistris, W. Dieterich, D. Di Liberto, L. Elli, A. Fasano, M. Hadjivassiliou, M. Kurien, E. Lionetti, C.J. Mulder, K. Rostami, A. Sapone, K. Scherf, D. Schuppan, N. Trott, U. Volta, V. Zevallos, Y. Zopf, D.S. Sanders, The Overlapping Area of Non-Celiac Gluten Sensitivity (NCGS) and Wheat-Sensitive Irritable Bowel Syndrome (IBS): An Update, Nutrients (2017) 9.

[5] B.T. Cooper, G.K. Holmes, R. Ferguson, R.A. Thompson, R.N. Allan, W.T. Cooke, Gluten-sensitive diarrhea without evidence of celiac disease, Gastroenterology 79 (1980) 801–806.

[6] J.F. Ludvigsson, D.A. Leffler, J.C. Bai, F. Biagi, A. Fasano, P.H. Green, M. Hadjivassiliou, K. Kaukinen, C.P. Kelly, J.N. Leonard, K.E. Lundin, J.A. Murray, D.S. Sanders, M.M. Walker, F. Zingone, C. Ciacci, The Oslo definitions for coeliac disease and related terms, Gut. 62 (2013) 43–52.

[7] J.F. Ludvigsson, J.C. Bai, F. Biagi, T.R. Card, C. Ciacci, P.J. Ciclitira, P.H. Green, M. Hadjivassiliou, A. Holdoway, D.A. van Heel, K. Kaukinen, D.A. Leffler, J.N. Leonard, K.E. Lundin, N. McGough, M. Davidson, J.A. Murray, G.L. Swift, M.M. Walker, F. Zingone, D.S. Sanders, B.S.G. Coeliac, Disease Guidelines Development Group, and Gastroenterology British Society of, Diagnosis and management of adult coeliac disease: guidelines from the British Society of Gastroenterology, Gut. 63 (2014) 1210–1228.

[8] A. Rubio-Tapia, I.D. Hill, C.P. Kelly, A.H. Calderwood, J.A. Murray, G.A.C. of, ACG clinical guidelines: diagnosis and management of celiac disease, Am. J. Gastroenterol. 108 (2013) 656–676, quiz 77.

[9] A. Al-Toma, U. Volta, R. Auricchio, G. Castillejo, D.S. Sanders, C. Cellier, C.J. Mulder, K.E.A. Lundin, European Society for the Study of Coeliac Disease (ESsCD) guideline for coeliac disease and other gluten-related disorders, United European Gastroenterol. J. 7 (2019) 583–613.

[10] A. Lovik, G. Skodje, J. Bratlie, M. Brottveit, K.E. Lundin, Diet adherence and gluten exposure in coeliac disease and self-reported non-coeliac gluten sensitivity, Clin. Nutr. 36 (2017) 275–280.

[11] M. Brottveit, P.O. Vandvik, S. Wojniusz, A. Lovik, K.E. Lundin, B. Boye, Absence of somatization in non-coeliac gluten sensitivity, Scand. J. Gastroenterol. 47 (2012) 770–777.

[12] I. Aziz, N.R. Lewis, M. Hadjivassiliou, S.N. Winfield, N. Rugg, A. Kelsall, L. Newrick, D.S. Sanders, A UK study assessing the population prevalence of self-reported gluten sensitivity and referral characteristics to secondary care, Eur. J. Gastroenterol. Hepatol. 26 (2014) 33–39.

[13] F. Cabrera-Chavez, D.M. Granda-Restrepo, J.G. Aramburo-Galvez, A. Franco-Aguilar, D. Magana-Ordorica, J. Vergara-Jimenez Mde, N. Ontiveros, Self-Reported Prevalence of Gluten-Related Disorders and Adherence to Gluten-Free Diet in Colombian Adult Population, Gastroenterol. Res. Pract. 2016 (2016) 4704309.

[14] A. Carroccio, O. Giambalvo, F. Blasca, R. Iacobucci, A. D'Alcamo, P. Mansueto, Self-Reported Non-Celiac Wheat Sensitivity in High School Students: Demographic and Clinical Characteristics, Nutrients (2017) 9.

[15] T. van Gils, P. Nijeboer, I. Jssennagger CE, D.S. Sanders, C.J. Mulder, G. Bouma, Prevalence and Characterization of Self-Reported Gluten Sensitivity in The Netherlands, Nutrients (2016) 8.

[16] U. Wahnschaffe, R. Ullrich, E.O. Riecken, J.D. Schulzke, Celiac disease-like abnormalities in a subgroup of patients with irritable bowel syndrome, Gastroenterology 121 (2001) 1329–1338.

[17] U. Wahnschaffe, J.D. Schulzke, M. Zeitz, R. Ullrich, Predictors of clinical response to gluten-free diet in patients diagnosed with diarrhea-predominant irritable bowel syndrome, Clin. Gastroenterol. Hepatol. 5 (2007) 844–850, quiz 769.

[18] E. Arranz, J. Bode, K. Kingstone, A. Ferguson, Intestinal antibody pattern of coeliac disease: association with gamma/delta T cell receptor expression by intraepithelial lymphocytes, and other indices of potential coeliac disease, Gut. 35 (1994) 476–482.

[19] K. Kaukinen, M. Peraaho, P. Collin, J. Partanen, N. Woolley, T. Kaartinen, T. Nuutinen, T. Halttunen, M. Maki, I. Korponay-Szabo, Small-bowel mucosal transglutaminase 2-specific IgA deposits in coeliac disease without villous atrophy: a prospective and randomized clinical study, Scand. J. Gastroenterol. 40 (2005) 564–572.

[20] M. Brottveit, M. Raki, E. Bergseng, L.E. Fallang, B. Simonsen, A. Lovik, S. Larsen, E.M. Loberg, F.L. Jahnsen, L.M. Sollid, K.E. Lundin, Assessing possible celiac disease by an HLA-DQ2-gliadin Tetramer Test, Am. J. Gastroenterol. 106 (2011) 1318–1324.

[21] M. Raki, L.E. Fallang, M. Brottveit, E. Bergseng, H. Quarsten, K.E. Lundin, L.M. Sollid, Tetramer visualization of gut-homing gluten-specific T cells in the peripheral blood of celiac disease patients, Proc. Natl. Acad. Sci. U. S. A. 104 (2007) 2831–2836.

[22] K. Kaukinen, K. Turjanmaa, M. Maki, J. Partanen, R. Venalainen, T. Reunala, P. Collin, Intolerance to cereals is not specific for coeliac disease, Scand. J. Gastroenterol. 35 (2000) 942–946.

[23] U. Volta, M.T. Bardella, A. Calabro, R. Troncone, G.R. Corazza, Sensitivity Study Group for Non-Celiac Gluten, An Italian prospective multicenter survey on patients suspected of having non-celiac gluten sensitivity, BMC Med. 12 (2014) 85.

[24] I.D. Croall, N. Hoggard, I. Aziz, M. Hadjivassiliou, D.S. Sanders, Brain fog and non-coeliac gluten sensitivity: Proof of concept brain MRI pilot study, PLoS One 15 (2020) e0238283.

[25] S.L. Peters, J.R. Biesiekierski, G.W. Yelland, J.G. Muir, P.R. Gibson, Randomised clinical trial: gluten may cause depression in subjects with non-coeliac gluten sensitivity - an exploratory clinical study, Aliment. Pharmacol. Ther. 39 (2014) 1104–1112.

[26] M. Uhde, A.C. Indart, X.B. Yu, S.S. Jang, R. De Giorgio, P.H.R. Green, U. Volta, S.D. Vernon, A. Alaedini, Markers of non-coeliac wheat sensitivity in patients with myalgic encephalomyelitis/chronic fatigue syndrome, Gut. 68 (2019) 377–378.

[27] E. Nygaard, K.L. Reichelt, J.F. Fagan, The relation between the psychological functioning of children with Down syndrome and their urine peptide levels and levels of serum antibodies to food proteins, Downs. Syndr. Res. Pract. 6 (2001) 139–145.

[28] E. Lionetti, A. Pulvirenti, M. Vallorani, G. Catassi, A.K. Verma, S. Gatti, C. Catassi, Re-challenge Studies in Non-celiac Gluten Sensitivity: A Systematic Review and Meta-Analysis, Front. Physiol. 8 (2017) 621.

[29] G.I. Skodje, V.K. Sarna, I.H. Minelle, K.L. Rolfsen, J.G. Muir, P.R. Gibson, M.B. Veierod, C. Henriksen, K.E.A. Lundin, Fructan, Rather Than Gluten, Induces Symptoms in Patients With Self-Reported Non-Celiac Gluten Sensitivity, Gastroenterology 154 (2018) 529–539 e2.

[30] J.R. Biesiekierski, E.D. Newnham, P.M. Irving, J.S. Barrett, M. Haines, J.D. Doecke, S.J. Shepherd, J.G. Muir, P.R. Gibson, Gluten causes gastrointestinal symptoms in subjects without celiac disease: a double-blind randomized placebo-controlled trial, Am. J. Gastroenterol. 106 (2011) 508–514, quiz 15.

[31] J.R. Biesiekierski, S.L. Peters, E.D. Newnham, O. Rosella, J.G. Muir, P.R. Gibson, No effects of gluten in patients with self-reported non-celiac gluten sensitivity after dietary reduction of fermentable, poorly absorbed, short-chain carbohydrates, Gastroenterology 145 (2013) 320–328 e1-3.

[32] A. Carroccio, P. Mansueto, G. Iacono, M. Soresi, A. D'Alcamo, F. Cavataio, I. Brusca, A.M. Florena, G. Ambrosiano, A. Seidita, G. Pirrone, G.B. Rini, Non-celiac wheat sensitivity diagnosed by double-blind placebo-controlled challenge: exploring a new clinical entity, Am. J. Gastroenterol. 107 (2012) 1898–1906, quiz 907.

[33] L. Elli, C. Tomba, F. Branchi, L. Roncoroni, V. Lombardo, M.T. Bardella, F. Ferretti, D. Conte, F. Valiante, L. Fini, E. Forti, R. Cannizzaro, S. Maiero, C. Londoni, A. Lauri, G. Fornaciari, N. Lenoci, R. Spagnuolo, G. Basilisco, F. Somalvico, B. Borgatta, G. Leandro, S. Segato, D. Barisani, G. Morreale, E. Buscarini, Evidence for the Presence of Non-Celiac Gluten Sensitivity in Patients with Functional Gastrointestinal Symptoms: Results from a Multicenter Randomized Double-Blind Placebo-Controlled Gluten Challenge, Nutrients 8 (2016) 84.

[34] A. Di Sabatino, U. Volta, C. Salvatore, P. Biancheri, G. Caio, R. De Giorgio, M. Di Stefano, G.R. Corazza, Small Amounts of Gluten in Subjects With Suspected Nonceliac Gluten Sensitivity: A Randomized, Double-Blind, Placebo-Controlled, Cross-Over Trial, Clin. Gastroenterol. Hepatol. 13 (2015) 1604–1612 e3.

[35] B. Zanini, R. Basche, A. Ferraresi, C. Ricci, F. Lanzarotto, M. Marullo, V. Villanacci, A. Hidalgo, A. Lanzini, Randomised clinical study: gluten challenge induces symptom recurrence in only a minority of patients who meet clinical criteria for non-coeliac gluten sensitivity, Aliment. Pharmacol. Ther. 42 (2015) 968–976.

[36] G.I. Skodje, C. Henriksen, T. Salte, T. Drivenes, I. Toleikyte, A.M. Lovik, M.B. Veierod, K.E. Lundin, Wheat challenge in self-reported gluten sensitivity: a comparison of scoring methods, Scand. J. Gastroenterol. 52 (2017) 185–192.

[37] B. Shahbazkhani, A. Sadeghi, R. Malekzadeh, F. Khatavi, M. Etemadi, E. Kalantri, M. Rostami-Nejad, K. Rostami, Non-Celiac Gluten Sensitivity Has Narrowed the Spectrum of Irritable Bowel Syndrome: A Double-Blind Randomized Placebo-Controlled Trial, Nutrients 7 (2015) 4542–4554.

[38] H.F. Dale, J.G. Hatlebakk, N. Hovdenak, S.O. Ystad, G.A. Lied, The effect of a controlled gluten challenge in a group of patients with suspected non-coeliac gluten sensitivity: A randomized, double-blind placebo-controlled challenge, Neurogastroenterol. Motil. (2018).

[39] G.I. Skodje, I.H. Minelle, K.L. Rolfsen, M. Iacovou, K.E.A. Lundin, M.B. Veierod, C. Henriksen, Dietary and symptom assessment in adults with self-reported non-coeliac gluten sensitivity, Clin. Nutr. ESPEN 31 (2019) 88–94.

[40] A.R. Lee, D.L. Ng, J. Zivin, P.H. Green, Economic burden of a gluten-free diet, J. Hum. Nutr. Diet. 20 (2007) 423–430.

[41] L. Fry, A.M. Madden, R. Fallaize, An investigation into the nutritional composition and cost of gluten-free versus regular food products in the UK, J. Hum. Nutr. Diet. 31 (2018) 108–120.

[42] B. Niland, B.D. Cash, Health Benefits and Adverse Effects of a Gluten-Free Diet in Non-Celiac Disease Patients, Gastroenterol. Hepatol. (NY) 14 (2018) 82–91.

[43] V. Melini, F. Melini, Gluten-Free Diet: Gaps and Needs for a Healthier Diet, Nutrients (2019) 11.

[44] M. Cornicelli, M. Saba, N. Machello, M. Silano, S. Neuhold, Nutritional composition of gluten-free food versus regular food sold in the Italian market, Dig. Liver Dis. 50 (2018) 1305–1308.

[45] N. Pellegrini, C. Agostoni, Nutritional aspects of gluten-free products, J. Sci. Food Agric. 95 (2015) 2380–2385.

[46] P. Amnuaycheewa, J.A. Murray, E.V. Marietta, Animal models to study non-celiac gluten sensitivity, Minerva Gastroenterol. Dietol. 63 (2017) 22–31.

[47] D.P. Funda, A. Kaas, T. Bock, H. Tlaskalova-Hogenova, K. Buschard, Gluten-free diet prevents diabetes in NOD mice, Diabetes Metab. Res. Rev. 15 (1999) 323–327.

[48] F. Maurano, G. Mazzarella, D. Luongo, R. Stefanile, R. D'Arienzo, M. Rossi, S. Auricchio, R. Troncone, Small intestinal enteropathy in non-obese diabetic mice fed a diet containing wheat, Diabetologia 48 (2005) 931–937.

[49] E.V. Marietta, A.M. Gomez, C. Yeoman, A.Y. Tilahun, C.R. Clark, D.H. Luckey, J.A. Murray, B.A. White, Y.C. Kudva, G. Rajagopalan, Low incidence of spontaneous type 1 diabetes in non-obese diabetic mice raised on gluten-free diets is associated with changes in the intestinal microbiome, PLoS One 8 (2013) e78687.

[50] M. Haupt-Jorgensen, L.J. Holm, K. Josefsen, K. Buschard, Possible Prevention of Diabetes with a Gluten-Free Diet, Nutrients (2018) 10.

[51] F. Biagi, S. Maimaris, C.G. Vecchiato, M. Costetti, G. Biagi, Gluten-sensitive enteropathy of the Irish Setter and similarities with human celiac disease, Minerva Gastroenterol. Dietol. 66 (2020) 151–156.

[52] E.J. Hall, R.M. Batt, Abnormal permeability precedes the development of a gluten sensitive enteropathy in Irish setter dogs, Gut. 32 (1991) 749–753.

[53] M. Lowrie, O.A. Garden, M. Hadjivassiliou, R.J. Harvey, D.S. Sanders, R. Powell, L. Garosi, The Clinical and Serological Effect of a Gluten-Free Diet in Border Terriers with Epileptoid Cramping Syndrome, J. Vet. Intern. Med. 29 (2015) 1564–1568.

[54] S. Auricchio, G. De Ritis, M. De Vincenzi, E. Mancini, M. Minetti, O. Sapora, V. Silano, Agglutinating activity of gliadin-derived peptides from bread wheat: implications for coeliac disease pathogenesis, Biochem. Biophys. Res. Commun. 121 (1984) 428–433.

[55] M. Silano, M. De Vincenzi, In vitro screening of food peptides toxic for coeliac and other gluten-sensitive patients: a review, Toxicology 132 (1999) 99–110.

[56] M. Silano, M. Dessi, M. De Vincenzi, H. Cornell, In vitro tests indicate that certain varieties of oats may be harmful to patients with coeliac disease, J. Gastroenterol. Hepatol. 22 (2007) 528–531.

[57] G. De Palma, J. Kamanova, J. Cinova, M. Olivares, H. Drasarova, L. Tuckova, Y. Sanz, Modulation of phenotypic and functional maturation of dendritic cells by intestinal bacteria and gliadin: relevance for celiac disease, J. Leukoc. Biol. 92 (2012) 1043–1054.

[58] Y. Junker, S. Zeissig, S.J. Kim, D. Barisani, H. Wieser, D.A. Leffler, V. Zevallos, T.A. Libermann, S. Dillon, T.L. Freitag, C.P. Kelly, D. Schuppan, Wheat amylase trypsin inhibitors drive intestinal inflammation via activation of toll-like receptor 4, J. Exp. Med. 209 (2012) 2395–2408.

[59] J.F. Ludvigsson, C. Ciacci, P.H. Green, K. Kaukinen, I.R. Korponay-Szabo, K. Kurppa, J.A. Murray, K.E.A. Lundin, M.J. Maki, A. Popp, N.R. Reilly, A. Rodriguez-Herrera, D.S. Sanders, D. Schuppan, S. Sleet, J. Taavela, K. Voorhees, M.M. Walker, D.A. Leffler, Outcome measures in coeliac disease trials: the Tampere recommendations, Gut. 67 (2018) 1410–1424.

[60] G. Goel, T. King, A.J. Daveson, J.M. Andrews, J. Krishnarajah, R. Krause, G.J.E. Brown, R. Fogel, C.F. Barish, R. Epstein, T.P. Kinney, P.B. Miner Jr., J.A. Tye-Din, A. Girardin, J. Taavela, A. Popp, J. Sidney, M. Maki, K.E. Goldstein, P.H. Griffin, S. Wang, J.L. Dzuris, L.J. Williams, A. Sette, R.J. Xavier, L.M. Sollid, B. Jabri, R.P. Anderson, Epitope-specific immunotherapy targeting CD4-positive T cells in coeliac disease: two randomised, double-blind, placebo-controlled phase 1 studies, Lancet. Gastroenterol. Hepatol. 2 (2017) 479–493.

[61] V.K. Sarna, K.E.A. Lundin, L. Morkrid, S.W. Qiao, L.M. Sollid, A. Christophersen, HLA-DQ-Gluten Tetramer Blood Test Accurately Identifies Patients With and Without Celiac Disease in Absence of Gluten Consumption, Gastroenterology 154 (2018) e6 886-96.

[62] J.A. Tye-Din, G.I. Skodje, V.K. Sarna, J.L. Dzuris, A.K. Russell, G. Goel, S. Wang, K.E. Goldstein, L.J. Williams, L.M. Sollid, K.E. Lundin, R.P. Anderson, Cytokine release after gluten ingestion differentiates coeliac disease from self-reported gluten sensitivity, United European Gastroenterol. J. 8 (2020) 108–118.

[63] L.M. Sollid, J.A. Tye-Din, S.W. Qiao, R.P. Anderson, C. Gianfrani, F. Koning, Update 2020: nomenclature and listing of celiac disease-relevant gluten epitopes recognized by CD4(+) T cells, Immunogenetics 72 (2020) 85–88.

[64] L.M. Sollid, B. Jabri, Triggers and drivers of autoimmunity: lessons from coeliac disease, Nat. Rev. Immunol. 13 (2013) 294–302.

[65] M.F. Kagnoff, R.K. Austin, J.J. Hubert, J.E. Bernardin, D.D. Kasarda, Possible role for a human adenovirus in the pathogenesis of celiac disease, J. Exp. Med. 160 (1984) 1544–1557.

[66] H.A. Gjertsen, K.E. Lundin, L.M. Sollid, J.A. Eriksen, E. Thorsby, T cells recognize a peptide derived from alpha-gliadin presented by the celiac disease-associated HLA-DQ (alpha 1*0501, beta 1*0201) heterodimer, Hum. Immunol. 39 (1994) 243–252.

[67] B.H. Johansen, H.A. Gjertsen, F. Vartdal, S. Buus, E. Thorsby, K.E. Lundin, L.M. Sollid, Binding of peptides from the N-terminal region of alpha-gliadin to the celiac disease-associated HLA-DQ2 molecule assessed in biochemical and T cell assays, Clin. Immunol. Immunopathol. 79 (1996) 288–293.

[68] H. Sjostrom, K.E. Lundin, O. Molberg, R. Korner, S.N. McAdam, D. Anthonsen, H. Quarsten, O. Noren, P. Roepstorff, E. Thorsby, L.M. Sollid, Identification of a gliadin T-cell epitope in coeliac disease: general importance of gliadin deamidation for intestinal T-cell recognition, Scand. J. Immunol. 48 (1998) 111–115.

[69] O. Molberg, S.N. McAdam, R. Korner, H. Quarsten, C. Kristiansen, L. Madsen, L. Fugger, H. Scott, O. Noren, P. Roepstorff, K.E. Lundin, H. Sjostrom, L.M. Sollid, Tissue transglutaminase selectively modifies gliadin peptides that are recognized by gut-derived T cells in celiac disease, Nat. Med. 4 (1998) 713–717.

[70] K.E. Thomas, A. Sapone, A. Fasano, S.N. Vogel, Gliadin stimulation of murine macrophage inflammatory gene expression and intestinal permeability are MyD88-dependent: role of the innate immune response in Celiac disease, J. Immunol. 176 (2006) 2512–2521.

[71] M.V. Barone, M. Nanayakkara, G. Paolella, M. Maglio, V. Vitale, R. Troiano, M.T. Ribecco, G. Lania, D. Zanzi, S. Santagata, R. Auricchio, R. Troncone, S. Auricchio, Gliadin peptide P31-43 localises to endocytic vesicles and interferes with their maturation, PLoS One 5 (2010) e12246.

[72] R.E. Araya, M.F. Gomez Castro, P. Carasi, J.L. McCarville, J. Jury, A.M. Mowat, E.F. Verdu, F.G. Chirdo, Mechanisms of innate immune activation by gluten peptide p31-43 in mice, Am. J. Physiol. Gastrointest. Liver Physiol. 311 (2016) G40–G49.

[73] M. Nanayakkara, G. Lania, M. Maglio, R. Auricchio, C. De Musis, V. Discepolo, E. Miele, B. Jabri, R. Troncone, S. Auricchio, M.V. Barone, P31-43, an undigested gliadin peptide, mimics and enhances the innate immune response to viruses and interferes with endocytic trafficking: a role in celiac disease, Sci. Rep. 8 (2018) 10821.

[74] S.B. Altenbach, W.H. Vensel, F.M. Dupont, The spectrum of low molecular weight alpha-amylase/protease inhibitor genes expressed in the US bread wheat cultivar Butte 86, BMC Res. Notes 4 (2011) 242.

[75] E.A. Pastorello, L. Farioli, A. Conti, V. Pravettoni, S. Bonomi, S. Iametti, D. Fortunato, J. Scibilia, C. Bindslev-Jensen, B. Ballmer-Weber, A.M. Robino, C. Ortolani, Wheat IgE-mediated food allergy in European patients: alpha-amylase inhibitors, lipid transfer proteins and low-molecular-weight glutenins. Allergenic molecules recognized by double-blind, placebo-controlled food challenge, Int. Arch. Allergy Immunol. 144 (2007) 10–22.

[76] V.F. Zevallos, V. Raker, S. Tenzer, C. Jimenez-Calvente, M. Ashfaq-Khan, N. Russel, G. Pickert, H. Schild, K. Steinbrink, D. Schuppan, Nutritional Wheat Amylase-Trypsin Inhibitors Promote Intestinal Inflammation via Activation of Myeloid Cells, Gastroenterology 152 (2017) 1100–1113 e12.

[77] V.F. Zevallos, V.K. Raker, J. Maxeiner, P. Scholtes, K. Steinbrink, D. Schuppan, Dietary wheat amylase trypsin inhibitors exacerbate murine allergic airway inflammation, Eur. J. Nutr. 58 (2019) 1507–1514.

[78] I. Bellinghausen, B. Weigmann, V. Zevallos, J. Maxeiner, S. Reissig, A. Waisman, D. Schuppan, J. Saloga, Wheat amylase-trypsin inhibitors exacerbate intestinal and airway allergic immune responses in humanized mice, J. Allergy Clin. Immunol. 143 (2019) 201–212 e4.

[79] Y. Reig-Otero, J. Manes, L. Manyes, Amylase-Trypsin Inhibitors in Wheat and Other Cereals as Potential Activators of the Effects of Nonceliac Gluten Sensitivity, J. Med. Food 21 (2018) 207–214.

[80] J.G. Muir, R. Rose, O. Rosella, K. Liels, J.S. Barrett, S.J. Shepherd, P.R. Gibson, Measurement of short-chain carbohydrates in common Australian vegetables and fruits by high-performance liquid chromatography (HPLC), J. Agric. Food Chem. 57 (2009) 554–565.

[81] H.M. Staudacher, P.M. Irving, M.C. Lomer, K. Whelan, Mechanisms and efficacy of dietary FODMAP restriction in IBS, Nat. Rev. Gastroenterol. Hepatol. 11 (2014) 256–266.

[82] M.B. Roberfroid, Inulin-type fructans: functional food ingredients, J. Nutr. 137 (2007) 2493S–2502S.

[83] J. van Loo, P. Coussement, L. de Leenheer, H. Hoebregs, G. Smits, On the presence of inulin and oligofructose as natural ingredients in the western diet, Crit. Rev. Food Sci. Nutr. 35 (1995) 525–552.

[84] J.J. Rumessen, Fructose and related food carbohydrates. Sources, intake, absorption, and clinical implications, Scand. J. Gastroenterol. 27 (1992) 819–828.

[85] S.J. Shepherd, M.C. Lomer, P.R. Gibson, Short-chain carbohydrates and functional gastrointestinal disorders, Am. J. Gastroenterol. 108 (2013) 707–717.

[86] U. Volta, F. Tovoli, R. Cicola, C. Parisi, A. Fabbri, M. Piscaglia, E. Fiorini, G. Caio, Serological tests in gluten sensitivity (nonceliac gluten intolerance), J. Clin. Gastroenterol. 46 (2012) 680–685.

[87] M. Uhde, M. Ajamian, G. Caio, R. De Giorgio, A. Indart, P.H. Green, E.C. Verna, U. Volta, A. Alaedini, Intestinal cell damage and systemic immune activation in individuals reporting sensitivity to wheat in the absence of coeliac disease, Gut. 65 (2016) 1930–1937.

[88] M. Uhde, G. Caio, R. De Giorgio, P.H. Green, U. Volta, A. Alaedini, Subclass Profile of IgG Antibody Response to Gluten Differentiates Nonceliac Gluten Sensitivity From Celiac Disease, Gastroenterology (2020).

[89] M.R. Barbaro, C. Cremon, A.M. Morselli-Labate, A. Di Sabatino, P. Giuffrida, G.R. Corazza, M. Di Stefano, G. Caio, G. Latella, C. Ciacci, D. Fuschi, M. Mastroroberto, L. Bellacosa, V. Stanghellini, U. Volta, G. Barbara, Serum zonulin and its diagnostic performance in non-coeliac gluten sensitivity, Gut. 69 (2020) 1966–1974.

[90] M. Vazquez-Roque, A.S. Oxentenko, Nonceliac Gluten Sensitivity', Mayo. Clin. Proc. 90 (2015) 1272–1277.

[91] N.I-SE Panel, J.A. Boyce, A. Assa'ad, A.W. Burks, S.M. Jones, H.A. Sampson, R.A. Wood, M. Plaut, S.F. Cooper, M.J. Fenton, S.H. Arshad, S.L. Bahna, L.A. Beck, C. Byrd-Bredbenner, C.A. Camargo, Jr., L. Eichenfield, G.T. Furuta, J.M. Hanifin, C. Jones, M. Kraft, B.D. Levy, P. Lieberman, S. Luccioli, K.M. McCall, L.C. Schneider, R.A. Simon, F.E. Simons, S.J. Teach, B.P. Yawn, J.M. Schwaninger, Guidelines for the diagnosis and management of food allergy in the United States: report of the NIAID-sponsored expert panel, J. Allergy Clin. Immunol. 126 (2010) S1–58.

[92] C.K. Yao, P.R. Gibson, S.J. Shepherd, Design of clinical trials evaluating dietary interventions in patients with functional gastrointestinal disorders, Am. J. Gastroenterol. 108 (2013) 748–758.

[93] J.R. Biesiekierski, E.D. Newnham, S.J. Shepherd, J.G. Muir, P.R. Gibson, Characterization of Adults With a Self-Diagnosis of Nonceliac Gluten Sensitivity, Nutr. Clin. Pract. 29 (2014) 504–509.

[94] H.M. Staudacher, K. Whelan, The low FODMAP diet: recent advances in understanding its mechanisms and efficacy in IBS, Gut. 66 (2017) 1517–1527.

[95] E.P. Halmos, When the low FODMAP diet does not work, J. Gastroenterol. Hepatol. 32 (1) (2017) 69–72 Suppl.

[96] E.P. Halmos, V.A. Power, S.J. Shepherd, P.R. Gibson, J.G. Muir, A diet low in FODMAPs reduces symptoms of irritable bowel syndrome, Gastroenterology 146 (2014) 67–75 e5.

[97] S.S. Rao, S. Yu, A. Fedewa, Systematic review: dietary fibre and FODMAP-restricted diet in the management of constipation and irritable bowel syndrome, Aliment. Pharmacol. Ther. 41 (2015) 1256–1270.

[98] P. Varju, N. Farkas, P. Hegyi, A. Garami, I. Szabo, A. Illes, M. Solymar, A. Vincze, M. Balasko, G. Par, J. Bajor, A. Szucs, O. Huszar, D. Pecsi, J. Czimmer, Low fermentable oligosaccharides, disaccharides, monosaccharides and polyols (FODMAP) diet improves symptoms in adults suffering from irritable bowel syndrome (IBS) compared to standard IBS diet: A meta-analysis of clinical studies, PLoS One 12 (2017) e0182942.

[99] M. O'Keeffe, C. Jansen, L. Martin, M. Williams, L. Seamark, H.M. Staudacher, P.M. Irving, K. Whelan, M.C. Lomer, Long-term impact of the low-FODMAP diet on gastrointestinal symptoms, dietary intake, patient acceptability, and healthcare utilization in irritable bowel syndrome, Neurogastroenterol. Motil. (2018) 30.

[100] M. O'Keeffe, M.C. Lomer, Who should deliver the low FODMAP diet and what educational methods are optimal: a review, J. Gastroenterol. Hepatol. 32 (Suppl 1) (2017) 23–26.

CHAPTER 10

Pediatric noncoeliac gluten sensitivity

Antonio Carroccio[a], Pasquale Mansueto[b], Aurelio Seidita[c]

[a]Unit of Internal Medicine, "V. Cervello" Hospital, Ospedali Riuniti "Villa Sofia-Cervello," Department of Health Promotion Sciences, Maternal and Infant Care, Internal Medicine and Medical Specialties (PROMISE), University of Palermo, Palermo, Italy

[b]Department of Health Promotion Sciences, Maternal and Infant Care, Internal Medicine and Medical Specialties (PROMISE), University of Palermo, Palermo, Italy

[c]Department for the Treatment and Study of Abdominal Diseases and Abdominal Transplantation, IRCCS-ISMETT (Istituto di Ricovero e Cura a Carattere Scientifico - Istituto Mediterraneo per i Trapianti e Terapie ad alta specializzazione), UPMC (University of Pittsburgh Medical Center) Italy, Palermo, Italy

10.1 Introduction

Functional gastrointestinal disorders remain a rather obscure and vague chapter of medicine, clouded with discordant, or even completely contradictory findings and opinions, depending on the different researchers and studies. The two main etiological hypotheses are an alteration of the brain-gut axis with visceral hypersensitivity, or a microscopic inflammation of the gastro-intestinal mucosa. However, what appears to link these and other hypotheses is the evidence that the symptoms reported by patients, both in adults and in the pediatric age, are very often presented in association with the intake of specific foods [1–4].

To further complicate this heterogeneous situation, several authors have reported cases of patients without the celiac disease (CeD) or wheat allergy (WA) criteria, who referred a marked improvement in gastro-intestinal functional symptoms on a gluten-free diet (GFD) [5,6]. This condition was initially defined as 'gluten sensitivity' (GS), only to be changed to 'non-celiac gluten sensitivity' (NCGS) [7], to better differentiate it from CeD, and finally redefined as 'non-celiac wheat sensitivity' (NCWS), as it is still not certain even today which of the components of wheat is the real culprit of this pathology [8–10]. To date, this is one of the most studied and controversial areas of research in the gastroenterological field, with data that, at least in part, seem to indicate that NCGS/NCWS in adults could represent a form of non-IgE-mediated allergic reaction that has its roots in some clinical reactions already in the pediatric age [11,12].

10.2 Definition and epidemiology of NCGS/NCWS in children

Although NCGS/NCWS is a condition studied primarily in adults, some of its features can be identified in children. According to the Salerno Experts' Criteria (2015),

Coeliac Disease and Gluten-Related Disorders.
DOI: https://doi.org/10.1016/B978-0-12-821571-5.00006-4

NCGS/NCWS is defined as a syndrome characterized by both intestinal and extra-intestinal symptoms triggered by gluten intake, which resolve once gluten is eliminated from the diet, after CeD and WA have been excluded [13].

The available data about the prevalence of NCGS/NCWS in the adult population are extremely variable and often flawed by an incorrect diagnostic methodology or based on self-reported gluten intolerance. In a survey conducted from 2009 to 2014 in the USA by researchers of the Mayo Clinic, 1.7% of the general population regularly followed gluten- or wheat-free diets in 2013/2014, a marked increase over previous years (0.5% in 2009/2010 and 1.0% in 2011/2012) [14]. Limiting analysis to patients complaining of gluten- or wheat-related symptoms, the worldwide prevalence of NCGS/NCWS ranges from 0.5 to 13% of the general population. All studies, however, agree on a higher prevalence in women [15–21].

In pediatric populations, the data seem to be highly comparable/virtually identical; in a New Zealand study it was found that out of a pediatric population of 916 children, 5% regularly excluded gluten from their diet, although only 1% had a diagnosis of CeD [22]. In a study on high school students in Sicily (average age 17 years) the NCGS/NCWS self-reported rate was 12.2%, with symptoms largely comparable to IBS [23]. Unlike in adults, NCGS/NCWS in children seems to be more frequent in males than in females [24,25]. Unfortunately, most of these data were collected on the basis of a perceived rather than diagnosed NCGS/NCWS. A different approach, based on well-defined diagnostic criteria, was used in 2018 by Francavilla et al. In this study the authors were able to define a prevalence of NCGS/NCWS ranging from 0.36% to 0.98% in a pediatric population of 1114 children enrolled at 5 pediatric gastroenterology centers in Italy [24,25].

However, a limitation of this study was that the initial work-up to identify a relationship between functional gastro-intestinal disorders (FGID) and NCGS/NCWS was based on parental self-reporting. Consequently, the inclusion criteria were based on a parent's hypothesis of food intolerance, thus probably not identifying other potential patients. An ideal study design should include a period of elimination diet in all subjects with FGID. In our experience, a true "food intolerance-hypersensitivity" in infants and children is generally not suspected and the real frequency of FGID linked to food "intolerance/hypersensitivity" is much higher than 1%.

10.3 Clinical features and diagnosis of NCGS/NCWS in children

10.3.1 Clinical features in adults and children

The main clinical features of NCGS/NCWS in adults include a series of intestinal and extra-intestinal symptoms that can be directly related to gluten intake (and therefore resolve with its elimination). What most differentiates NCGS/NCWS from other gastrointestinal diseases, such as IBS, is precisely the high degree of extraintestinal manifestations, which can represent the main, or sometimes the only, symptoms reported by patients [8,11,13,21].

Table 10.1 Summary of Francavilla R. et al. main evidence in pediatric noncoeliac gluten sensitivity.

	Francavilla et al., [24]	Francavilla et al., [26]
Years of the study	2012-2013	2013-2016
Number of Centers Involved	2	5
Assessment of GS	Open gluten challenge	Open gluten challenge followed by a Randomized Double-Blind Placebo Controlled food challenge with crossover
Median age of GS children	9.6 ± 3.9	11.5 ± 3.8
Sex of GS children: Male Female	10 (67%) 5 (33%)	4 (36%) 7 (64%)
Gastrointestinal symptoms: Abdominal pain Diarrhea Constipation Bloating Nausea/Vomiting	12 (80%) 11 (73%) 3 (20%) 4 (26%) 3 (20%)	10 (90%) 4 (36%) 0 (0%) 5 (45%) 1 (9%)
Extra-intestinal symptoms: Fatigue Muscle/Joint pain Headache Skin rash Failure to thrive	5 (33%) 3 (20%) 0 (0%) 0 (0%) 2 (13%)	11 (100%) 6 (54%) 9 (81%) 4 (36%) 0 (0%)
HLA DQ2/DQ8 positivity	7 (42%)	8 (72%)
Anti-gliadin IgG	10 (67%)	4 (36%)
Duodenal biopsy: Marsh 0 Marsh 1	11/15 (73%) 9 (82%) 2 (18%)	6/11 (54%) 4 (67%) 2 (33%)

GS: gluten sensitivity HLA: Human Leucocyte Antigen

In the pediatric field, the few studies carried out seem to indicate a greater prevalence of intestinal symptoms than extraintestinal ones. Among the former, the most frequent are abdominal pain (78-80%), chronic diarrhea (39-73%), bloating (26-35%), dyspepsia (20-39%) and mixed bowel habits (13-28%). Other, less frequently reported intestinal symptoms are nausea and vomiting (15-20%), constipation (10-20%) and regurgitation (1-3%). Extra-intestinal symptoms involve mainly the central and peripheral nervous systems, the skin and the musculoskeletal system. Among these are fatigue (33-85%), headache (20-71%), muscle and joint pain (20-57%) and skin rash (13-39%). Other potential manifestations, such as irritability, inappetence, mouth ulcers, and paresthesia have been reported in less than 10% of the children enrolled in the studies. Of relevance is the almost constant association of 2 or more symptoms (80%), among which the most common are abdominal pain and chronic diarrhea (53%) [24,26]; (see Table 10.1).

10.3.2 Diagnosis of NCGS/NCWS

What is most evident from a clinical point of view, despite the high number of studies to date, is that it is impossible to identify a serum or histologic marker able to allow diagnosis, which, therefore, can only be made by exclusion. In the pediatric age, as in adults, the starting point must be the use of serological and/or histological markers to exclude CeD (duodenal biopsy, IgA-endomysial and IgA-tissue transglutaminase antibodies, HLA-DQ2/DQ8 test) and WA (gluten- and wheat-specific IgE, skin prick, prick by prick and atopy patch tests to wheat) [11,13,26]. Therefore, the diagnostic criteria require the demonstration of both an improvement in the clinical picture on a gluten elimination diet, and the reappearance of symptoms triggered by the intake of gluten [8,13,26,27]. For this reason, a diagnosis of NCGS/NCWS can be considered only after performing a DBPC gluten challenge with crossover in both adult [8] and pediatric populations [26]. While following a strict GFD, patients are randomized in double blind to gluten or placebo intake for at least 2 consecutive weeks, at the end of which, and after at least one wash-out week, they have to take the other substance [26]. During these weeks, adult patients and children (or their parents) have to record all the symptoms, their appearance and severity using a questionnaire based on a combination of a Visual Analogic Scale (VAS, from 0 = no symptoms, to 10 = worst possible symptoms) [28] and the Faces Pain Scale (FPS, 6 faces, from a relaxed face to a face showing intense pain) [29]. According to the Salerno Experts' criteria, response to the challenge must be considered positive if there is a reduction of at least 30% in the global VAS between the administration of gluten and the administration of placebo [13]. Unfortunately, in daily clinical practice the DBPC challenge method is extremely difficult to apply, not only due to the purely technical problems related to the administration of gluten vs placebo (e.g. there are still no clearly defined criteria for gluten or placebo composition, packaging, administration modality, etc.) but, even more often, to the poor compliance of patients to this diagnostic methodology [30].

One clinical finding in adults, which still needs to be demonstrated in pediatric patients, is the association of NCGS/NCWS with hypersensitivity to other foods. Recently, we found a very high percentage (206 out of 276, 74.6%) of patients who experienced IBS-like symptoms both at DBPC gluten challenge and at DBPC cow's milk protein challenge. In addition, many of these seemed to present the same symptoms even at an open challenge with egg, chocolate, and other foods. This picture seems to suggest that several patients with NCWS/NCGS may suffer from multiple food hypersensitivity (FH) [8]. Multiple FH is, however, well known in pediatric populations [8–10]. For example, the elimination of wheat, eggs and other foods from the diet of children with chronic constipation due to CMP allergy, but whose symptoms did not resolve with a CMP-free diet alone, was shown to induce a clear improvement in symptoms, up to a complete resolution of the problems. This evidence, subsequently confirmed by DBPC food challenges, classifies these patients as suffering from multiple FH [31,32]; Giuseppe [33]. Similar results have been obtained in an adult population of patients suffering from chronic constipation unresponsive to laxatives, who were able to resolve their disorder after following both gluten- and CMP-free diets [34], However, multiple FH in children is not only associated with chronic constipation; cases of rectal bleeding, malabsorption and diarrhea, discomfort and abdominal pain, weight loss, growth retardation, anemia, and regurgitation have been reported [35,36].

It is of interest that patients with NCWS often have a history of allergic/atopic diseases. In a previous study, we demonstrated that 35.4% (73 of 206) of adult patients with NCWS and CMP hypersensitivity had a history of atopy, a much higher value compared to IBS controls. [8] Confirmatory data about the association between NCWS and atopy have come from a study by Massari et al. [37], which found that in a population of 262 adult patients with a known allergic/atopic history and gastrointestinal symptoms of undefined origin, 77 (29.4%) had duodenal lesions in the absence of positive serological markers for CeD. These patients showed a significant improvement in symptom control when undergoing a GF-diet.

These studies performed in adults recall the numerous data in children with non-IgE-mediated food allergy suffering from gastro-intestinal diseases, which resolve on a wheat-cow's milk free diet.

10.3.3 Serological features of NCGS/NCWS in children

As previously remarked, unlike for CeD or WA, to date there are no known serological markers suitable for NCGS/NCWS diagnosis. However, both in adults (50-60%) [8,38] and in children (40-66%) [24,26], the presence of IgG anti-gliadin antibodies (AGA) has been demonstrated, albeit with a lower titer than in CeD subjects (greater than 80%). Unfortunately, this not negligible finding has a low positive predictive value, since these antibodies can also be found in 10-14% of adult subjects [8] and in 13% of children [24] with IBS alone, as well as in subjects suffering from connective tissue diseases, autoimmune diseases or even in healthy controls [27]. The increased levels of AGA IgG seem to suggest that NCGS/NCWS could be included in the context of FH. In fact, numerous studies have shown that the presence of food-specific IgG can be a marker of food sensitivity and can thus guide the diagnosis and therapy of pediatric and adult patients suffering from food-related IBS [39–41], or gastroesophageal reflux [42,43]. However, it must be remembered that experience of the use of serum IgG in the diagnosis of food hypersensitivity is contradictory. Indeed, other studies have shown that the presence of food-specific IgG is of little assistance in the diagnosis of allergic disorders (whether IgE or non-IgE mediated) related to food intake and that it can also in fact be an indicator of tolerance [44,45]. For this reason, the main American [46,47] and European [48] allergy societies do not recommend the use of food-specific IgG as the only diagnostic marker for the diagnosis and consequent therapy of food allergies, in order to avoid any unnecessary elimination diets that can compromise the quality of life of a patient.

To return to the NCWS context, an increased serum anti-gliadin antibody level (mainly IgG) in a child with suspected wheat-related symptoms can be considered "suggestive". It should induce clinicians to attempt a wheat-free diet and to monitor the clinical course on elimination diet. If a clinical improvement is observed on a wheat-free diet, a subsequent ("open" or preferably DBPC) wheat challenge will be needed to confirm or exclude the NCWS diagnosis.

An additional finding from the Francavilla et al. studies, linked to the exclusion of CeD in the diagnostic flow-chart of these patients, is a significant, although not statistically significant, positivity for HLA-DQ2 (40-70%) [24,26]; however, given the nonsignificance of the data and the common finding of this antigen even in the general population, experts do not suggest routinely performing HLA-DQ2 testing in cases of suspected NCGS/NCWS. This assay, however, can be useful to exclude a CeD condition as its negative predictive value in CeD diagnosis is almost 99% [49,13,27].

Recently our group examined the possibility of using the in vitro basophil activation test (BAT) for the diagnosis of NCGS/NCWS. This test, unlike common allergy tests, evaluates the expression of some markers (e.g. CD63 or CD203c) in the basophil cytoplasmic membrane, related to their activation after adequate antigenic stimulation [50]. This test has proven to be sensitive and specific for the diagnosis of IgE-mediated allergic reactions against pollen, hymenoptera venom, latex and drugs, as well as being more specific than food-specific IgE and skin prick tests for foods [51–55]. In addition, BAT has been shown to be just as sensitive but more specific than routine tests for the diagnosis of CMP hypersensitivity [56]. In the context of NCGS/NCWS, this test has shown to have a far from negligible diagnostic ability, with 86% sensitivity, 88% specificity and 87% accuracy [57]. Unfortunately, these data have not been confirmed in subsequent studies due to the modification of the method, as it is no longer performed on separate leukocytes but on whole blood [11,12]. Although excessively influenced by the method (and therefore poorly reproducible), the evidence of basophil activation after a wheat in vitro challenge seems to strengthen the pathogenetic hypothesis of an "atypical" allergic response in NCGS/NCWS patients.

10.4 Histological features of NCGS/NCWS in children

In the few studies performed in children, histological findings on duodenal biopsies show a picture of normal (Marsh 0: 82% of biopsies) or slightly inflamed (Marsh 1: 18% of biopsies) mucosa [24,26]. Data on duodenal histology in adults have shown a much more frequent IEL infiltration. Sapone et al. [58] reported >30 IELs/100 enterocytes in 84.6% (22 of 26) of the biopsies of patients with NCGS, and Volta et al. Volta et al [38] identified 42% of patients (33 of 78) with a Marsh 1 lesion. In a study by our group, we found 89% of the patients diagnosed with NCGS/NCWS had more than 25 IELs/100 enterocytes [8]. All these studies, however, whether by Francavilla et al. in pediatric patients, or by Sapone et al. and Carroccio et al. in adults showed that the amount/degree of IEL infiltration in patients with NCGS/NCWS was significantly lower than in patients with CeD [8,24,58].

A more relevant meaning than the presence of IELs can probably be attributed to the finding of mucosal eosinophilic infiltration. In our experience, adult patients with NCGS/NCWS present an eosinophilic infiltrate in the colonic mucosa (60% in the lamina propria and 63% intraepithelial), which is not found in controls with IBS.

Patients with NCGS/NCWS were found to have significantly higher high-power field counts (40x) than CeD or IBS patients when eosinophil presence in the duodenal mucosa was assessed. In the same study, we also highlighted that patients with multiple FH including NCGS/NCWS had higher eosinophil values both in the colon and duodenum than patients with NCGS/NCWS alone (p = 0.0001) [8]. Unfortunately, no studies in pediatric patients have evaluated the infiltration of eosinophils in the intestinal mucosa of NCWS patients. However, it is noteworthy that a similar eosinophil infiltration in the intestinal mucosa has been detected in pediatric patients suffering from chronic constipation with an allergic pathogenesis [31–33] or from rectal bleeding due to allergic proctocolitis [59].

A possible role for eosinophils as a diagnostic marker, and a leading role in NCWS pathogenesis have recently been suggested by data on histological changes revealed by the use of confocal laser endomicroscopy (CLE). This method allows the study of the instantaneous changes occurring in the duodenal mucosa (e.g. IEL increase, extravasation of fluids, expression of proteins, etc.) after exposure to specific food antigens. Fritscher-Ravens et al. showed that in patients undergoing an elimination diet against the antigens identified by CLE there was a significant reduction in symptoms at 3 months and over a 1-year follow-up [60]. The same authors used CLE to study 108 IBS adults, of which 76 tested positive (CLE+), among these, over 60% tested positives when stimulated with wheat. They showed that the post-challenge degranulating eosinophil number was higher in CLE+ than in the other CLE- groups (p=0.023), with a significant difference between CLE+ and healthy controls, and that eosinophil cationic protein concentration was higher in the intestinal fluid of CLE+ than CLE- patients [12].

In this context, the putative role for eosinophils seems to be relevant because of their known role as antigen-presenting cells and for thetoxic granular proteins they release which are able to induce a dysfunction of the epithelial barrier, as already shown in other non-IgE mediated digestive allergic disease [61,62]. Unfortunately, CLE studies are difficult to perform and no studies have been performed in other centers or in pediatric patients.

10.5 Therapeutic approach to NCGS/NCWS in children

As in CeD, the only possible therapeutic option known for NCGS/NCWS is abstention from gluten/wheat by following a rigorous GFD, as no guidelines [27] or specific gluten tolerance thresholds (which seem extremely variable from subject to subject) have been defined [13].

In the Francavilla et al. studies, children undergoing a GFD showed a marked reduction in symptoms evaluated by multiple rating scales, with a reappearance of symptoms on gluten reintroduction [24,26]. Despite the clinical improvement, there is a negative aspect to this therapy, represented by the possible nutritional deficiencies and the increased intake of saturated fats that could be associated with a GFD. Specifically, low contents of vitamin A, niacin (vitamin B3), folate (vitamin B9), cobalamin (vitamin B12), vitamin E, iron, zinc, selenium, calcium, phosphorus, essential amino acids, and arachidonic acid have been reported [63–65]. In addition, an alteration of the intestinal microbial flora has been demonstrated, with a reduction in beneficial bacteria such as Bifidobacterium, Clostridium lituseburense, Faecalibacterium prausnitzii and Lactobacillus [66]. Such dietary abnormalities seem to expose patients to an increased risk of developing metabolic syndrome, hepatic steatosis [67], type 2 diabetes mellitus, obesity [68], osteopenia/osteoporosis [69], depression [70], and, more generally, to an increased risk of cardiovascular pathologies [71,72,68].

Therefore, in order to prevent major dietary deficiencies and potential future pathologies, all children suffering from NCGS/NCWS should follow a GFD only under medical prescription and supervised by expert nutritionists [25].

On the other hand, it should be recalled that in our experience, complete avoidance of gluten/wheat is not mandatory in all NCWS patients and that a different tolerance threshold could exist among different patients. Determining this threshold and individualizing dietary advice is a very important goal which also requires the support and professional competence of the dieticians.

10.6 Conclusion

NCGS/NCWS is a pathology still being defined both in adults and in children, and diagnosis is made even more difficult by the variability of the symptoms (intestinal and/or extraintestinal) and the absence of valid serological and/or histological markers, therefore elimination diets and DPBC gluten challenge are necessary. To date, the number of studies in pediatric patients is totally insufficient and most of the pathogenetic, clinical, and therapeutic data have been transposed from the studies performed in adults. From these, a complex framework seems to emerge, in which many patients suffer from multiple FH (including NCGS/NCWS), whose pathogenic bases seem, at least in part, to stem from an "atypical" non-IgE-mediated allergic reaction; in the case of other food antigens (e.g. CMP), this condition is well known as the cause of several intestinal disorders in the pediatric population. The therapeutic approach is based exclusively on a GFD, which can, however, lead to serious nutritional deficiencies, which can be especially detrimental in the developmental age; therefore, the dietary approach should only be carried out under medical prescription and under the guidance of expert nutritionists.

References

[1] L. Böhn, S. Störsrud, H. Törnblom, U. Bengtsson, M. Simrén, Self-reported food-related gastrointestinal symptoms in IBS are common and associated with more severe symptoms and reduced quality of life, Am. J. Gastroenterol. 108 (5) (2013) 634–641 https://doi.org/10.1038/ajg.2013.105.

[2] G. Friedman, Nutritional therapy of irritable bowel syndrome, Gastroenterol. Clin. North. Am. 18 (3) (1989) 513–524.

[3] A.A. Moukarzel, H. Lesicka, M.E. Ament, Irritable bowel syndrome and nonspecific diarrhea in infancy and childhood–relationship with juice carbohydrate malabsorption, Clin. Pediatrics. 41 (3) (2002) 145–150.

[4] W.G. Thompson, G.F. Longstreth, D.A. Drossman, K.W. Heaton, E.J. Irvine, S.A. Müller-Lissner, Functional bowel disorders and functional abdominal pain, Gut. 45 Suppl. (2) (1999) II43–II47.

[5] R. Troncone, B. Jabri, Coeliac disease and gluten sensitivity, J. Intern. Med. 269 (6) (2011) 582–590 https://doi.org/10.1111/j.1365-2796.2011.02385.x.

[6] E.F. Verdu, D. Armstrong, J.A. Murray, Between celiac disease and irritable bowel syndrome: the "no man's land" of gluten sensitivity, Am. J. Gastroenterol. 104 (6) (2009) 1587–1594 https://doi.org/10.1038/ajg.2009.188.

[7] J.F. Ludvigsson, D.A. Leffler, J.C. Bai, F. Biagi, A. Fasano, P.H.R. Green, M. Hadjivassiliou, K. Kaukinen, C.P. Kelly, J.N. Leonard, K.E.A. Lundin, J.A. Murray, D.S. Sanders, M.M. Walker, F. Zingone, C. Ciacci, The Oslo definitions for coeliac disease and related terms, Gut. 62 (1) (2013) 43–52 https://doi.org/10.1136/gutjnl-2011-301346.

[8] A. Carroccio, P. Mansueto, G. Iacono, M. Soresi, A. D'Alcamo, F. Cavataio, I. Brusca, A.M. Florena, G. Ambrosiano, A. Seidita, G. Pirrone, G.B. Rini, Non-celiac wheat sensitivity diagnosed by double-blind placebo-controlled challenge: exploring a new clinical entity, Am. J. Gastroenterol. 107 (12) (2012) 1898–1906 quiz 1907 https://doi.org/10.1038/ajg.2012.236.

[9] A. Carroccio, G. Rini, P. Mansueto, Non-celiac wheat sensitivity is a more appropriate label than non-celiac gluten sensitivity, Gastroenterology 146 (1) (2014) 320–321 https://doi.org/10.1053/j.gastro.2013.08.061.

[10] M.D.E. Potter, M.M. Walker, S. Keely, N.J Talley, What's in a name? "Non-coeliac gluten or wheat sensitivity": controversies and mechanisms related to wheat and gluten causing gastrointestinal symptoms or disease, Gut. 67 (12) (2018) 2073–2077 https://doi.org/10.1136/gutjnl-2018-316360.

[11] A. Carroccio, I. Brusca, P. Mansueto, A. D'alcamo, M. Barrale, M. Soresi, A. Seidita, S.M. La Chiusa, G. Iacono, D. Sprini, A comparison between two different in vitro basophil activation tests for gluten- and cow's milk protein sensitivity in irritable bowel syndrome (IBS)-like patients, Clin. Chem. Lab. Med. 51 (6) (2013) 1257–1263 https://doi.org/10.1515/cclm-2012-0609.

[12] A. Fritscher-Ravens, T. Pflaum, M. Mösinger, Z. Ruchay, C. Röcken, P.J. Milla, M. Das, M. Böttner, T. Wedel, D. Schuppan, Many Patients With Irritable Bowel Syndrome Have Atypical Food Allergies Not Associated With Immunoglobulin E, Gastroenterology 157 (1) (2019) 109–118 e5 https://doi.org/10.1053/j.gastro.2019.03.046 .

[13] C. Catassi, L. Elli, B. Bonaz, G. Bouma, A. Carroccio, G. Castillejo, C. Cellier, F. Cristofori, L. de Magistris, J. Dolinsek, W. Dieterich, R. Francavilla, M. Hadjivassiliou, W. Holtmeier, U. Körner, D.A. Leffler, K.E.A. Lundin, G. Mazzarella, C.J. Mulder, A. Fasano, Diagnosis of Non-Celiac Gluten Sensitivity (NCGS): The Salerno Experts' Criteria, Nutrients. 7 (6) (2015) 4966–4977 https://doi.org/10.3390/nu7064966.

[14] R.S. Choung, A. Unalp-Arida, C.E. Ruhl, T.L. Brantner, J.E. Everhart, J.A. Murray, Less Hidden Celiac Disease But Increased Gluten Avoidance Without a Diagnosis in the United States: Findings From the National Health and Nutrition Examination Surveys From 2009 to 2014, Mayo. Clin. Proc. (2016) https://doi.org/10.1016/j.mayocp.2016.10.012.

[15] I. Aziz, N.R. Lewis, M. Hadjivassiliou, S.N. Winfield, N. Rugg, A. Kelsall, L. Newrick, D.S. Sanders, A UK study assessing the population prevalence of self-reported gluten sensitivity and referral characteristics to secondary care, Eur. J. Gastroenterol. Hepatol. 26 (1) (2014) 33–39 https://doi.org/10.1097/01.meg.0000435546.87251.f7.

[16] F. Cabrera-Chávez, G.V.A. Dezar, A.P. Islas-Zamorano, J.G. Espinoza-Alderete, M.J. Vergara-Jiménez, D. Magaña-Ordorica, N. Ontiveros, Prevalence of Self-Reported Gluten Sensitivity and Adherence to a Gluten-Free Diet in Argentinian Adult Population, Nutrients. 9 (1) (2017) https://doi.org/10.3390/nu9010081.

[17] F. Cabrera-Chávez, D.M. Granda-Restrepo, J.G. Arámburo-Gálvez, A. Franco-Aguilar, D. Magaña-Ordorica, M. Vergara-Jiménez, J. de, N. Ontiveros, Self-Reported Prevalence of Gluten-Related Disorders and Adherence to Gluten-Free Diet in Colombian Adult Population, Gastroenterol. Res. Practice. 2016 (2016) 4704309 https://doi.org/10.1155/2016/4704309.

[18] D.V. DiGiacomo, C.A. Tennyson, P.H. Green, R.T. Demmer, Prevalence of gluten-free diet adherence among individuals without celiac disease in the USA: results from the Continuous National Health and Nutrition Examination Survey 2009-2010, Scand. J. Gastroenterol. 48 (8) (2013) 921–925 https://doi.org/10.3109/00365521.2013.809598.

[19] S. Golley, N. Corsini, D. Topping, M. Morell, P. Mohr, Motivations for avoiding wheat consumption in Australia: results from a population survey, Public. Health. Nutr. 18 (3) (2015) 490–499 https://doi.org/10.1017/S1368980014000652.

[20] T. van Gils, P. Nijeboer, C.E. IJssennagger, D.S. Sanders, C.J.J. Mulder, G. Bouma, Prevalence and Characterization of Self-Reported Gluten Sensitivity in The Netherlands, Nutrients. 8 (11) (2016).

[21] U. Volta, M.T. Bardella, A. Calabrò, R. Troncone, G.R. Corazza, An Italian prospective multicenter survey on patients suspected of having non-celiac gluten sensitivity, BMC Med. 12 (2014) 85 https://doi.org/10.1186/1741-7015-12-85.

[22] P. Tanpowpong, T.R. Ingham, P.K. Lampshire, F.F. Kirchberg, M.J. Epton, J. Crane, C.A. Camargo, Coeliac disease and gluten avoidance in New Zealand children, Arch. Dis. Childhood. 97 (1) (2012) 12–16 https://doi.org/10.1136/archdischild-2011-300248.

[23] A. Carroccio, O. Giambalvo, F.L. Blasca, R. Iacobucci, A. D'Alcamo, P. Mansueto, Self-Reported Non-Celiac Wheat Sensitivity in High School Students: Demographic and Clinical Characteristics, Nutrients 9 (7) (2017) https://doi.org/10.3390/nu9070771.

[24] R. Francavilla, F. Cristofori, S. Castellaneta, C. Polloni, V. Albano, S. Dellatte, F. Indrio, L. Cavallo, C. Catassi, Clinical, serologic, and histologic features of gluten sensitivity in children, J. Pediatrics. 164 (3) (2014) 463–467 e1 https://doi.org/10.1016/j.jpeds.2013.10.007 .

[25] A. Llanos-Chea, A. Fasano, Gluten and Functional Abdominal Pain Disorders in Children, Nutrients 10 (10) (2018) https://doi.org/10.3390/nu10101491.

[26] R. Francavilla, F. Cristofori, L. Verzillo, A. Gentile, S. Castellaneta, C. Polloni, V. Giorgio, E. Verduci, E. D¼Angelo, S. Dellatte, F. Indrio, Randomized Double-Blind Placebo-Controlled Crossover Trial for the Diagnosis of Non-Celiac Gluten Sensitivity in Children, Am. J. Gastroenterol. 113 (3) (2018) 421–430 https://doi.org/10.1038/ajg.2017.483.

[27] U. Volta, M.I. Pinto-Sanchez, E. Boschetti, G. Caio, R. De Giorgio, E.F. Verdu, Dietary Triggers in Irritable Bowel Syndrome: Is There a Role for Gluten? J. Neurogastroenterol. Motility. 22 (4) (2016) 547–557 https://doi.org/10.5056/jnm16069.

[28] P.A. McGrath, C.E. Seifert, K.N. Speechley, J.C. Booth, L. Stitt, M.C. Gibson, A new analogue scale for assessing children's pain: an initial validation study, Pain. 64 (3) (1996) 435–443.

[29] C.L. Hicks, C.L. von Baeyer, P.A. Spafford, I. van Korlaar, B. Goodenough, The Faces Pain Scale-Revised: toward a common metric in pediatric pain measurement, Pain. 93 (2) (2001) 173–183.

[30] M.M. Leonard, A. Sapone, C. Catassi, A. Fasano, Celiac Disease and Nonceliac Gluten Sensitivity: A Review, JAMA 318 (7) (2017) 647–656 https://doi.org/10.1001/jama.2017.9730.

[31] A. Carroccio, C. Scalici, E. Maresi, L. Di Prima, F. Cavataio, D. Noto, R. Porcasi, M.R. Averna, G. Iacono, Chronic constipation and food intolerance: a model of proctitis causing constipation, Scand. J. Gastroenterol. 40 (1) (2005) 33–42.

[32] G. Iacono, F. Cavataio, G. Montalto, A. Florena, M. Tumminello, M. Soresi, A. Notarbartolo, A. Carroccio, Intolerance of cow's milk and chronic constipation in children, New. Engl. J. Med. 339 (16) (1998) 1100–1104.

[33] G. Iacono, S. Bonventre, C. Scalici, E. Maresi, L. Di Prima, M. Soresi, G. Di Gesù, D. Noto, A. Carroccio, Food intolerance and chronic constipation: manometry and histology study, Eur. J. Gastroenterol. Hepatol. 18 (2) (2006) 143–150.

[34] A. Carroccio, L. Di Prima, G. Iacono, A.M. Florena, F. D'Arpa, C. Sciumè, A.B. Cefalù, D. Noto, M.R Averna, Multiple food hypersensitivity as a cause of refractory chronic constipation in adults, Scand. J. Gastroenterol. 41 (4) (2006) 498–504.

[35] A. Carroccio, G. Iacono, L. Di Prima, A. Ravelli, G. Pirrone, A.B. Cefalù, A.M. Florena, G.B. Rini, G. Di Fede, Food hypersensitivity as a cause of rectal bleeding in adults, Clin. Gastroenterol. Hepatol. Official Clin. Practice J. Am. Gastroenterol. Assoc. 7 (1) (2009) 120–122 https://doi.org/10.1016/j.cgh.2008.07.029.

[36] R.E. Story, Manifestations of food allergy in infants and children, Pediatr. Ann. 37 (8) (2008) 530–535 https://doi.org/10.3928/00904481-20080801-05.

[37] S. Massari, M. Liso, L. De Santis, F. Mazzei, A. Carlone, S. Mauro, F. Musca, M.P. Bozzetti, M. Minelli, Occurrence of nonceliac gluten sensitivity in patients with allergic disease, Int. Arch. Allergy. Immunol. 155 (4) (2011) 389–394 https://doi.org/10.1159/000321196.

[38] U. Volta, F. Tovoli, R. Cicola, C. Parisi, A. Fabbri, M. Piscaglia, E. Fiorini, G. Caio, Serological tests in gluten sensitivity (nonceliac gluten intolerance), J. Clin. Gastroenterol. 46 (8) (2012) 680–685 https://doi.org/10.1097/MCG.0b013e3182372541.

[39] W. Atkinson, T.A. Sheldon, N. Shaath, P.J. Whorwell, Food elimination based on IgG antibodies in irritable bowel syndrome: a randomised controlled trial, Gut 53 (10) (2004) 1459–1464.

[40] G. Iacono, A. Carroccio, F. Cavataio, G. Montalto, D. Lorello, I. Kazmierska, M. Soresi, IgG anti-betalactoglobulin (betalactotest): its usefulness in the diagnosis of cow's milk allergy, Ital. J. Gastroenterol. 27 (7) (1995) 355–360.

[41] S. Zar, L. Mincher, M.J. Benson, D. Kumar, Food-specific IgG4 antibody-guided exclusion diet improves symptoms and rectal compliance in irritable bowel syndrome, Scand. J. Gastroenterol. 40 (7) (2005) 800–807.

[42] F. Cavataio, G. Iacono, G. Montalto, M. Soresi, M. Tumminello, P. Campagna, A. Notarbartolo, A. Carroccio, Gastroesophageal reflux associated with cow's milk allergy in infants: which diagnostic examinations are useful? Am. J. Gastroenterol. 91 (6) (1996) 1215–1220.

[43] S. Salvatore, Y. Vandenplas, Gastroesophageal reflux and cow milk allergy: is there a link? Pediatrics. 110 (5) (2002) 972–984.

[44] L.P.C. Shek, L. Bardina, R. Castro, H.A. Sampson, K. Beyer, Humoral and cellular responses to cow milk proteins in patients with milk-induced IgE-mediated and non-IgE-mediated disorders, Allergy. 60 (7) (2005) 912–919.

[45] J.M. Skripak, S.D. Nash, H. Rowley, N.H. Brereton, S. Oh, R.G. Hamilton, E.C. Matsui, A.W. Burks, R.A. Wood, A randomized, double-blind, placebo-controlled study of milk oral immunotherapy for cow's milk allergy, J. Allergy Clin. Immunol. 122 (6) (2008) 1154–1160 https://doi.org/10.1016/j.jaci.2008.09.030.

[46] S.A. Bock, AAAAI support of the EAACI Position Paper on IgG4, J. Allergy. Clin. Immunol. 125 (6) (2010) 1410 https://doi.org/10.1016/j.jaci.2010.03.013.

[47] S. Carr, E. Chan, E. Lavine, W. Moote, CSACI Position statement on the testing of food-specific IgG, Allergy, Asthma, Clin. Immunol. : Official J. Can. Soc. Allergy Clin. Immunol. 8 (1) (2012) 12 https://doi.org/10.1186/1710-1492-8-12.

[48] S.O. Stapel, R. Asero, B.K. Ballmer-Weber, E.F. Knol, S. Strobel, S. Vieths, J. Kleine-Tebbe, Testing for IgG4 against foods is not recommended as a diagnostic tool: EAACI Task Force Report, Allergy. 63 (7) (2008) 793–796 https://doi.org/10.1111/j.1398-9995.2008.01705.x.

[49] C. Catassi Gluten Sensitivity. Annals of Nutrition & Metabolism, 67 (Suppl. 2), (2015) 16–26. https://doi.org/10.1159/000440990

[50] D.G. Ebo, C.H. Bridts, M.M. Hagendorens, N.E. Aerts, L.S. De Clerck, W.J. Stevens, Basophil activation test by flow cytometry: present and future applications in allergology, Cytomet. Part B, Clin. Cytomet. 74 (4) (2008) 201–210 https://doi.org/10.1002/cyto.b.20419.

[51] R. Bahri, A. Custovic, P. Korosec, M. Tsoumani, M. Barron, J. Wu, R. Sayers, A. Weimann, M. Ruiz-Garcia, N. Patel, A. Robb, M.H. Shamji, S. Fontanella, M. Silar, E.N.C. Mills, A. Simpson, P.J. Turner, S. Bulfone-Paus, Mast cell activation test in the diagnosis of allergic disease and anaphylaxis, J. Allergy. Clin. Immunol. 142 (2) (2018) 485–496 e16 https://doi.org/10.1016/j.jaci.2018.01.043 .

[52] M. Gawinowska, K. Specjalski, M. Chełmińska, J. Łata, M. Zieliński, Application of basophil activation test in diagnosing aspirin hypersensitivity, Pneumonol. Alergol. Pol. 83 (1) (2015) 66–73 https://doi.org/10.5603/PiAP.2015.0010.

[53] J. Leysen, V. Sabato, M.M. Verweij, K.J. De Knop, C.H. Bridts, L.S. De Clerck, D.G. Ebo, The basophil activation test in the diagnosis of immediate drug hypersensitivity, Expert. Rev. Clin. Immunol. 7 (3) (2011) 349–355 https://doi.org/10.1586/eci.11.14.

[54] A. Ocmant, S. Mulier, L. Hanssens, M. Goldman, G. Casimir, F. Mascart, L. Schandené, Basophil activation tests for the diagnosis of food allergy in children, Clin. Experiment. Allergy : J. Br. Soc. Allergy Clin. Immunol., 39 (8) (2009) 1234–1245 https://doi.org/10.1111/j.1365-2222.2009.03292.x.

[55] H. Ott, K. Tenbrock, J. Baron, H. Merk, S. Lehmann, Basophil activation test for the diagnosis of hymenoptera venom allergy in childhood: a pilot study, Klin. Padiatr. 223 (1) (2011) 27–32 https://doi.org/10.1055/s-0030-1268430.

[56] O. Ciepiela, J. Zwiazek, A. Zawadzka-Krajewska, I. Kotula, M. Kulus, U. Demkow, Basophil activation test based on the expression of CD203c in the diagnostics of cow milk allergy in children, Eur. J. Med. Res. 15 (2) (2010) 21–26 Suppl.

[57] A. Carroccio, I. Brusca, P. Mansueto, G. Pirrone, M. Barrale, L. Di Prima, G. Ambrosiano, G. Iacono, M.L. Lospalluti, S.M. La Chiusa, G. Di Fede, A cytologic assay for diagnosis of food hypersensitivity in patients with irritable bowel syndrome, Clin. Gastroenterol. Hepatol. Official Clin. Practice J. Am. Gastroenterol. Assoc., 8 (3) (2010) 254–260 https://doi.org/10.1016/j.cgh.2009.11.010.

[58] A. Sapone, K.M. Lammers, V. Casolaro, M. Cammarota, M.T. Giuliano, M. De Rosa, R. Stefanile, G. Mazzarella, C. Tolone, M.I. Russo, P. Esposito, F. Ferraraccio, M. Cartenì, G. Riegler, L. de Magistris, A. Fasano, Divergence of gut permeability and mucosal immune gene expression in two gluten-associated conditions: celiac disease and gluten sensitivity, BMC. Med. 9 (23) (2011) https://doi.org/10.1186/1741-7015-9-23.

[59] A. Ravelli, V. Villanacci, S. Chiappa, S. Bolognini, S. Manenti, M. Fuoti, Dietary protein-induced proctocolitis in childhood, Am. J. Gastroenterol. 103 (10) (2008) 2605–2612 https://doi.org/10.1111/j.1572-0241.2008.02035.x.

[60] A. Fritscher-Ravens, D. Schuppan, M. Ellrichmann, S. Schoch, C. Röcken, J. Brasch, J. Bethge, M. Böttner, J. Klose, P.J. Milla, Confocal endomicroscopy shows food-associated changes in the intestinal mucosa of patients with irritable bowel syndrome, Gastroenterology 147 (5) (2014) 1012–1020 e4 https://doi.org/10.1053/j.gastro.2014.07.046.

[61] N. Kalach, N. Kapel, A.-J. Waligora-Dupriet, M.-C. Castelain, M.O. Cousin, C. Sauvage, F. Ba, I. Nicolis, F. Campeotto, M.J. Butel, C. Dupont, Intestinal permeability and fecal eosinophil–derived neurotoxin are the best diagnosis tools for digestive non-IgE-mediated cow's milk allergy in toddlers, Clin. Chem. Lab. Med. 51 (2) (2013) 351–361.

[62] J. Travers, M.E. Rothenberg, Eosinophils in mucosal immune responses, Mucosal. Immunol. 8 (3) (2015) 464–475 https://doi.org/10.1038/mi.2015.2.

[63] M.T. Bardella, C. Fredella, L. Prampolini, N. Molteni, A.M. Giunta, P.A. Bianchi, Body composition and dietary intakes in adult celiac disease patients consuming a strict gluten-free diet, Am. J. Clin. Nutr. 72 (4) (2000) 937–939.

[64] A. Lerner, T. O'Bryan, T. Matthias, Navigating the Gluten-Free Boom: The Dark Side of Gluten Free Diet, Front. Pediatrics. 7 (2019) 414 https://doi.org/10.3389/fped.2019.00414.

[65] D. Wild, G.G. Robins, V.J. Burley, P.D. Howdle, Evidence of high sugar intake, and low fibre and mineral intake, in the gluten-free diet, Aliment. Pharmacol. & Ther. 32 (4) (2010) 573–581 https://doi.org/10.1111/j.1365-2036.2010.04386.x.

[66] G. De Palma, I. Nadal, M.C. Collado, Y. Sanz, Effects of a gluten-free diet on gut microbiota and immune function in healthy adult human subjects, Br. J. Nutr. 102 (8) (2009) 1154–1160 https://doi.org/10.1017/S0007114509371767.

[67] A. Ciccone, D. Gabrieli, R. Cardinale, M. Di Ruscio, F. Vernia, G. Stefanelli, S. Necozione, D. Melideo, A. Viscido, G. Frieri, G. Latella, Metabolic Alterations in Celiac Disease Occurring after Following a Gluten-Free Diet, Digestion 100 (4) (2019) 262–268 https://doi.org/10.1159/000495749.

[68] A. Taetzsch, S.K. Das, C. Brown, A. Krauss, R.E. Silver, S.B. Roberts, Are Gluten-Free Diets More Nutritious? An Evaluation of Self-Selected and Recommended Gluten-Free and Gluten-Containing Dietary Patterns, Nutrients 10 (12) (2018) https://doi.org/10.3390/nu10121881.

[69] L.M.S. Kotze, T. Skare, A. Vinholi, L. Jurkonis, R. Nisihara, Impact of a gluten-free diet on bone mineral density in celiac patients. Revista Espanola de Enfermedades Digestivas, Organo Oficial de La Sociedad Espanola de Patologia Digestiva 108 (2) (2016) 84–88.

[70] N.J.M. van Hees, E.J. Giltay, S.M.A.J. Tielemans, J.M. Geleijnse, T. Puvill, N. Janssen, W. van der Does, Essential amino acids in the gluten-free diet and serum in relation to depression in patients with celiac disease, PLoS. One. 10 (4) (2015) e0122619 https://doi.org/10.1371/journal.pone.0122619.

[71] C. Anania, L. Pacifico, F. Olivero, F.M. Perla, C. Chiesa, Cardiometabolic risk factors in children with celiac disease on a gluten-free diet, World J. Clin. Pediatrics 6 (3) (2017) 143–148 https://doi.org/10.5409/wjcp.v6.i3.143.

[72] T. Siriwardhane, K. Krishna, K. Devarajan, V. Ranganathan, V. Jayaraman, T. Wang, K. Bei, J.J. Rajasekaran, H. Krishnamurthy, Insights into cardiovascular risk and nutritional status in subjects with wheat-related disorders, Biomarkers : Biochem. Indicators Exposure Response, Suscept. Chem., 24 (3) (2019) 303–307 https://doi.org/10.1080/1354750X.2019.1578829.

CHAPTER 11

Neurological manifestations of gluten-related disorders

Marios Hadjivassiliou, Panagiotis Zis

Academic Department of Neurosciences, Sheffield Teaching hospitals NHS Trust and University of Sheffield, Royal Hallamshire Hospital, Sheffield, UK

11.1 Introduction

Gluten-related disorders (GRD) represent a spectrum of diverse clinical entities sharing a common trigger: ingestion of gluten. Most health-care workers and the public would be familiar with coeliac disease (CD) also known as gluten sensitive enteropathy. CD is the best recognized GRD, where a small bowel enteropathy occurs in genetically susceptible individuals after exposure to the protein gliadin (found in wheat, barley and rye) [1]. Classic CD is characterized by gastrointestinal symptoms such as diarrhea, abdominal bloating and pain, weight loss and evidence of malabsorption. The presence or not of enteropathy is irrelevant to the existence or the development of neurological manifestations which include cerebellar ataxia, sensorimotor axonal neuropathy, sensory ganglionopathy, encephalopathy and other less common neurological deficits. Such neurological manifestations are increasingly seen in clinical practice even in the absence of gastrointestinal symptoms. The absence of such symptoms, unfortunately, can be a barrier in alerting clinicians to the possibility of GRD with neurological manifestations. This is particularly true when neurological manifestations occur in isolation.

Serological evidence of gluten sensitivity without the presence of enteropathy has previously largely been dismissed by gastroenterologists. In the last few years, however, there has a been an improved understanding of gluten sensitivity in the absence of enteropathy, based primarily on the observation that such patients also benefit symptomatically from a gluten free diet (GFD). The terms non-coeliac gluten sensitivity (NCGS) or non-coeliac wheat sensitivity were introduced by gastroenterologists to describe patients primarily with GI symptoms related to the ingestion of gluten who do not have enteropathy but who symptomatically benefit from a gluten free diet [2]. However, when it comes to neurological manifestations, we consider as gluten sensitive (GS) only those patients with positive serology in the form of anti-liadin IgG and/or IgA (AGA) and/or transglutaminase 6 (TG6) IgG and/or IgA antibodies with or without

Coeliac Disease and Gluten-Related Disorders.
DOI: https://doi.org/10.1016/B978-0-12-821571-5.00008-8

positivity for transglutaminase 2 (TG2) or endomysial antibodies (EMA) [3]. Some but not all of these patients will also have enteropathy.

This chapter will focus on the commonest neurological manifestations providing clinical details on presentation and management. At the end of the chapter we will discuss the current knowledge on the pathophysiology of such manifestations.

11.2 Gluten ataxia

Gluten ataxia (GA) is defined as cerebellar ataxia with positive AGA [4–6] in the absence of alternative etiology for the ataxia and also evidence of improvement of the ataxia with GFD. In a series of 1500 patients with progressive ataxia evaluated over a period of more than 20 years at a specialist Ataxia Centre (Sheffield, UK), GA had a prevalence of 20% amongst all ataxias and more than 40% amongst idiopathic sporadic ataxias [7].

GA usually presents with pure cerebellar ataxia, occasionally in combination with a peripheral neuropathy [8] or, rarely, ataxia in combination with myoclonus [9-11]. GA is usually of insidious onset; however it can also be rapidly progressive mimicking paraneoplastic cerebellar degeneration. Gaze-evoked nystagmus is common. As there is predilection for vermian involvement, all patients have gait ataxia and the majority have lower limb ataxia. Less than 10% of patients with GA will have any GI symptoms, however about half of the patients will have enteropathy on biopsy.

Patients with GA usually have evidence of cerebellar atrophy on MR imaging with particular predilection for the cerebellar vermis. MR spectroscopy of the vermis is abnormal in all patients with GA (low N-acetyl aspartate/creatine area ratio). Even in patients with GA without cerebellar atrophy, MR spectroscopy is abnormal, suggesting that MR spectroscopy is a useful monitoring tool [12].

The current recommendation in the management of patients with GA is that such patients should be offered a specialist dietician review and to be advised to embark on a strict GFD, with regular follow up to ensure that the antibodies are eliminated, which usually takes 6 to 12 months.

The response to treatment with GFD depends on the duration of the ataxia prior to the diagnosis of sensitivity to gluten. In a systematic case-controlled study of 43 patients presenting with ataxia, with or without an enteropathy, it was shown that patients who adhere to a strict GFD (having serological evidence of elimination of the antigliadin antibodies whilst being on the diet) improved significantly in their performance in ataxia test scores and in the subjective global clinical impression scale compared to the control group, which comprised of patients opting not to embark on a GFD [13]. The two groups did not differ regarding age, gender and severity of ataxia and, interestingly, the improvement was apparent even after excluding patients with an enteropathy.

Apart from the clinical improvement, in a large study of 117 patients with GA, GFD was found to have a positive effect on MR spectroscopy of the cerebellum [12]. In particular, the N-acetylaspartate/creatine (NAA/Cr) area ratio from the cerebellar vermis increased in 62 out of 63 (98%) patients on strict GFD (as indicated by the elimination of AGA), in 9 of 35 (26%) patients on GFD but positive antibodies, and in only 1 of 19 (5%) patients not on GFD. The demonstration of increased NAA/Cr ratio on repeat scanning following strict GFD strengthens previous findings of clinical improvement of the ataxia in patients with GA. The presence of enteropathy is not a prerequisite for such improvement; indeed there were no differences in the response to GFD between those patients with or without enteropathy.

Most patients with GA respond well to GFD. Absence of such response should raise 3 possible concerns: by far the commonest would be ongoing exposure to gluten due to the GFD not being 100% strict. This is supported by the ongoing positive serology in such patients. Secondly, such patients may require additional treatment with immunosuppressive medication because the diet alone may not be able to suppress the immune diathesis that results in cerebellar damage. Such patients are often negative for the relevant antibodies and review by the dietitian suggests very strict adherence to the GFD. These patients benefit from the use of mycophenolate as an additional treatment of their ataxia. Finally, absence of response may be due to that fact that the diagnosis is not correct and that the patient has an alternative or additional cause for their ataxia (e.g. genetic).

Apart for mycophenolate, in a small uncontrolled case series of four patients with GA, Bürk et al. [14] reported the use of intravenous immunoglobulins (IVIG) as having a positive effect, however, there was no information about the diet and serological status in these patients at the time of treatment.

Myoclonic ataxia where the myoclonus is of cortical origin can be seen in refractory CD where GFD alone is not sufficient to suppress the immune overactivity. In the largest series of myoclonic ataxia secondary to refractory CD [11], five patients were treated with mycophenolate and one in addition with rituximab and IVIG. Whilst their ataxia and enteropathy improved, the myoclonus remained the most disabling feature of their illness. Two patients have been treated with cladribine resulting in significant neurological improvement.

11.3 Gluten neuropathy

Gluten neuropathy (GN) is defined as an otherwise idiopathic sporadic large-fiber neuropathy with serological evidence of sensitivity to gluten [15]. The commonest types are symmetrical sensorimotor axonal length dependent peripheral neuropathy (about 75% of cases), followed by sensory ganglionopathy, an asymmetric form of pure sensory neuropathy [16] where the pathology is within the dorsal root ganglia [17-19].

Involvement of small fibers (Aδ and C fibers) leads to small fiber neuropathy (SFN), which is characteristically painful. Similar to GN, the majority of patients with pure SFN related to gluten report a burning sensation in a length–dependent manner (mainly soles and palms), however a small proportion of patients present with a non–length dependent pattern of symptoms suggesting that the predominant pathology lies in the dorsal root ganglia [20]. Small fiber involvement may also cause autonomic dysfunction [21-23]. Though the exact prevalence of gluten-related pure SFN is yet to be confirmed, it is known that more than half of the patients with GN report pain [24], suggesting that small fibers are often involved.

Gluten neuropathy is a slowly progressive condition. About 25% of the patients will have evidence of enteropathy on biopsy (CD) but the presence or absence of an enteropathy does not influence the positive effect of a strict GFD. In a systematic, controlled study of the effect of GFD in a large series of patients with gluten neuropathy (of the sensorimotor axonal type), a clear clinical and neurophysiological improvement was demonstrated after 12 months in those patients adhering to a strict GFD with serological elimination of the AGA [17]. In particular, there was significant increase in the sural sensory nerve action potential, the pre–defined primary endpoint, as well as subjective improvement of the neuropathic symptoms. Subgroup analysis showed that the capacity for recovery is less when the neuropathy is severe. In a large series of patients with gluten sensitivity and sensory ganglionopathy, it has been shown that strict adherence to a GFD may result in stabilization of the neuropathy irrespective of the presence of enteropathy [19].

In the largest – to date – cohort of patients with GN, of both the axonal type and sensory ganglionopathy, it has been shown that GFD is protective as it is associated with a significant reduction in peripheral neuropathic pain associated with GN [24].

GN can cause a significant burden in the overall quality of life (QoL) of patients with GN. In a case-controlled study, it was shown that, compared to controls, patients had worse scores in the physical functioning, role limitations due to physical health, energy/fatigue, and general health subdomains of the SF-36 questionnaire. Interestingly after having adjusted for age, gender and disease severity being on a strict GFD correlates with better SF-36 scores in the pain domain and in the overall health change domain, suggesting that along with the clinical and the neurophysiological improvement, a strict GFD can lead to a better overall QoL [25].

A beneficial effect of IVIG as an adjuvant option to GFD for the treatment of neuropathic pain associated to gluten-related SFN (2 patients) and GN (1 patient) with concomitant GA has been suggested by Souayah et al [26]. Similarly, others have reported clinical improvement following IVIG and GFD implementation in an isolated case of a patient with a combination of GN and GA [27]. However, no reports or controlled trials of the use of IVIG in GN as a monotherapy have been reported, and therefore, the therapeutic potential of IVIG in GN in patients who are not adhering to GFD remains questionable.

11.4 Gluten encephalopathy

The term gluten encephalopathy was introduced in 2001 to describe a combination of frequent, often intractable headaches, cognitive complaints (sometimes patients describe these as "foggy brain") and excessive, for age, white matter abnormalities on brain MR imaging [28]. Despite the fact that the white matter abnormalities do not resolve following a GFD, such diet arrests progression of the MRI changes and both the associated cognitive difficulties and headaches usually resolve.

The headaches take the form of chronic migraines, often unresponsive to the usual medication. Headaches affect both adult and children patients with CD and GS approximately two times more frequent compared to controls [29-32]. Although there was no correlation between years on GFD and migraine severity, Lionetti et al., showed that a GFD has a beneficial effect with reduction of headache frequency in the majority of patients, with approximately 25% managing to become headache-free [29].

In a population-based retrospective cohort study that was conducted in Sweden, Lebwohl et al. [33] reported that among 28,638 patients with CD and 143,126 controls, headache-related visits occurred in 4.7% and 2.9% of each group, respectively, suggesting a hazard ratio of 1.7 (95% 1.6–1.8; $p < 0.0001$). However, in this study, there was no information provided regarding the criteria for headache diagnosis used and if diagnosed, its exact type.

The white matter changes, often seen in the context of gluten encephalopathy may prove to be responsible for the cognitive deficits that can be found in patients with CD [34]. A recent study confirmed cognitive deficits in patients with CD which appear to be established at diagnosis of CD after which they stabilize assuming patients adhere to a strict GFD. A possible association between progressive cognitive impairment and CD has been suggested in small case series [35-38] but analysis of Swedish epidemiological data showed that although patients with CD are not at increased risk for dementia overall, subgroup analysis revealed that they may be at increased risk of vascular dementia [39]. The white matter abnormalities encountered in the gluten sensitive patients are identical to what gets labeled as "vascular" disease which is really what defines vascular dementia in the context of cognitive deficits.

Therapeutic options for patients with advanced dementia are not available, and even a GFD at this advance stage of neural damage is unlikely to have a beneficial effect but may arrest progression. However, in newly diagnosed CD, cognitive performance improves with adherence to a GFD in parallel to mucosal healing [37]. Therefore, a GFD should be introduced as soon as possible in patients with GS or CD as, even in the absence of a mild cognitive impairment, it has a potentially protective effect [34].

An important issue with this topic is how severe a clinician may expect these issues to be in patients with "classic" CD, who are often treated purely by gastroenterology and without any referrals to neurology. A recent study used 3rd party population data from a national UK Biobank [40]. This ensured that participants with a CD diagnosis (who were otherwise healthy) were representative and also, as the data collection had been performed by another study team, there was no ascertainment or recruitment bias. The results confirmed the presence of a cognitive deficit (in reaction time) and white matter tract injury as evidenced via diffusion tensor imaging.

This confirms that neurological and cognitive deficits are present in classic CD. It should also be noted that the severity of cognitive deficits may not typically warrant a referral to neurology. This phenotype may be comparable to syndromes such as subjective cognitive impairment, which is studied as a predementia state where the patient is aware of a decline in cognition but conventional clinical dementia scales such as the MMSE or MoCA may not be sensitive enough to confirm their presence. Nonetheless, when cognitive assessment tools more commonly used in research are utilized they are able to detect such deficits when compared to controls [41]. Importantly, the psychological impact of such "subclinical" cognitive deficits on patient quality of life is well recognized [42]. This highlights the relevance of these outcomes to patient wellbeing, and the value of acknowledging these in a clinical setting.

11.5 Epilepsy

Epileptic seizures presenting in the context of gluten sensitivity include both patients with and without overt brain pathology who may or may not respond to antiepileptic drugs (AED). This spectrum includes a range of interesting pathological features including epilepsy and cerebral calcifications (CEC), hippocampal sclerosis and temporal lobe epilepsy (TLE) in the context of gluten sensitivity and those who apparently display no pathological clues to the specific cause of the epilepsy [43].

More than half of patients with epilepsy and GS or CD will respond positively to a GFD (defined as either decreased frequency of seizures with GFD, cessation of seizures with GFD, successful reduction of AED with initiation of GFD or cessation of AED following the introduction of GFD). The response to GFD could reflect resolution/reduction of a neurological insult caused by gluten ingestion or be the result of improved absorption of AED due to resolution of gastrointestinal disturbance.

When considering specifically patients with CEC, a syndrome which refers to patients with focal, medically refractory epilepsy and show parieto–occipitally brain calcifications on CT or MRI, response to AED alone appears to be poor with the majority (73%) being unresponsive to treatment. Response to GFD appears to be more effective, with 53% of patients on GFD demonstrating a good response. Patients with CEC are often children and have primarily been reported in patients from Italy, Argentina and Spain. Epilepsy onset is at a mean of 6 years with the focus being primarily occipital. Due to the unresponsiveness to AED the condition can evolve into epileptic encephalopathy. The most effective therapeutic intervention is the GFD. Interestingly, three prospective cohort studies appear to have demonstrated an inverse relationship between effectiveness of GFD and duration of epilepsy prior to GFD, perhaps due to increasing neurological damage [44-46].

11.6 Gluten myopathy

Myopathy cases in both adults and children have been described as isolated case reports over the last four decades [47,48]. In the majority, if not all the cases the myopathy was the first manifestation of CD and GFD was proven to be effective in resolving the patient's symptoms. It is speculated that such acquired myopathies are related to deficiency in fat-soluble vitamin D or E, as a result of malabsorption secondary to CD which is reversed following a GFD. However, there have also been cases of inflammatory myopathy (myositis with or without inclusion bodies) without enteropathy and, therefore, an immune mediated mechanism to account for the myopathy has also been suggested [49].

Hadjivassiliou et al. presented the largest to date series of patients who presented with a clinical picture of a myopathy and in whom diagnostic work-up led to the diagnosis of gluten sensitivity [50]. The mean age at onset of the myopathic symptoms was 54 years. Inflammatory myopathy was the most common finding on neuropathological examination. Not all patients had enteropathy, suggesting that GS without enteropathy can also be linked to myopathy. Although some patients were on immunosuppressive treatment (i.e. azathioprine, methotrexate, prednisolone) some showed clinical improvement of the myopathy just with GFD. The improvement was also associated with reduction or normalization of serum creatine kinase level. Myopathy, however is a rare manifestation of GS.

11.7 Pathophysiology

Post mortem data from patients with gluten ataxia demonstrate patchy loss of Purkinje cells throughout the cerebellar cortex, a rather end stage non-specific finding in many cerebellar disorders. However, findings supporting an immune mediated pathogenesis include diffuse infiltration mainly of T-lymphocytes within the cerebellar white matter as well as marked perivascular cuffing with inflammatory cells [51]. The peripheral nervous system also shows sparse lymphocytic infiltrates with perivascular cuffing being observed in sural nerve biopsy of patients with gluten neuropathy and in dorsal root ganglia in patients with sensory neuronopathy. Patients with CD produce an immune response to gluten involving both innate and adaptive immunity [52]. Antibodies to gliadin are part of this response, and their systemic levels appear to mirror the immune reaction triggered by gluten in the intestine. There is cross-reactivity of these antibodies with antigenic epitopes on Purkinje cells as demonstrated by the fact that serum from patients with GA and from patients with CD recognize Purkinje cells of both human and rat origin

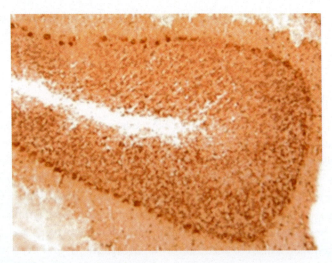

Figure 11.1 Evidence of Purkinje and granular layer staining using serum from a patient with gluten ataxia on rat cerebellum.

(Fig. 11.1) [53].This reactivity can also be seen using polyclonal AGA and can be eliminated by absorption with crude gliadin. However, when using sera from patients with GA there is evidence of additional antibodies targeting Purkinje cell epitopes since elimination of AGA alone is not sufficient to abolish such reactivity [53]. There is evidence that these additional antibodies are one or more transglutaminase isoenzymes (TG2, TG6) [54].

TG2 belongs to a family of enzymes that crosslink or modify proteins. Gluten proteins (from wheat, barley and rye), the immunological trigger of CD, are glutamine rich donor proteins amenable to deamidation. Deamidation of gluten peptides enhances binding with disease-relevant HLAs and thereby enhances presentation, leading to the development of gluten-specific Th1-like CD4$^+$ T cells [52]. Thus activation of TG2 and deamidation of gluten peptides appears to be central to disease development. In genetically predisposed individuals, this is at the center of the chronic inflammatory reaction manifesting with enteropathy in the context of CD. Apart from the gluten-specific T cell response, one of the hallmarks of GRD is an IgA autoantibody response to TG2 and other transglutaminase isozymes [55-57]. Events leading to the formation of autoantibodies against TG2 or TG6 are less clear. The recent characterization of an immature plasma cell response in the small intestine may explain the association of gluten related disorders with autoantibodies to transglutaminases [58]. Intestinal deposits of IgA antibodies targeting TG2 are present at all stages of CD, including early (when there is no overt enteropathy) and late stages including refractory CD. It is important to keep in mind that B cells have roles beyond antibody production including highly effective antigen presentation for T cell responses. Therefore, B cells may drive clonal expansion

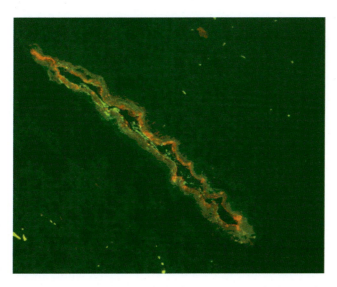

Figure 11.2 IgA deposits against transglutaminase (red/orange staining) on arterial vessel wall from the cerebellum of a patient with gluten ataxia.

of gluten–specific T cells which in turn may support development of B cells specific to transgutaminases. This potentially puts B cells in the center stage of GRD pathogenesis.

Questions also remain as to the contribution of these autoantibodies to organ-specific deficits. Anti–TG2 antibodies have been shown to be deposited in the small bowel mucosa of patients with GRD and may contribute to the development of the enteropathy. Such deposits have been found at extraintestinal tissues, such as muscle and liver [59] and more relevant around brain vessels in the context of GA (Fig. 11.2) [60]. This finding suggests that such autoantibodies could play a role in the pathogenesis of the whole spectrum of manifestations seen in GRD. Variations in the specificity of antibodies produced in individual patients could explain the wide spectrum of manifestations. Whilst TG2 has been shown to be the autoantigen in CD, the epidermal transglutaminase TG3 has been shown to be the autoantigen in DH [56]. Antibodies against TG6, a primarily brain expressed transglutaminase, have been shown to be present in patients with GA [6]. Similar to anti-TG2, the production of these anti-TG3 and anti-TG6 antibodies in DH and GA patients, respectively, is gluten–dependent which substantiates the link to a gluten-specific T cell population. In GA and DH, IgA deposits of TG6 and TG3 respectively seem to accumulate in the periphery of blood vessels at sites where in health the respective proteins are absent. It is likely that inflammation around vessels is at the heart of GA. Indeed perivascular cuffing with lymphocytes is a common finding in brain tissue from patients with GA but is also seen in peripheral nerve and muscle in patients with gluten neuropathy or myopathy [61]. In most sera reactive with more than one TG isoenzyme, distinct antibody populations are responsible for such reactivity rather than this being a result of cross-reactivity with different TG isozymes. This makes shared epitopes less likely to be the cause for B cell development to other transglutaminases and points to the possibility that TG isozymes other than TG2 can be the primary antigen in extraintestinal manifestations.

IgA deposition in brain vessels and the pathological finding of perivascular cuffing with inflammatory cells, may indicate that vasculature-centered inflammation may compromise the blood-brain barrier, allowing exposure of the CNS to pathogenic antibodies, and therefore be the trigger of nervous system involvement.

It is also possible that additional factors other than the autoantibodies themselves play a role. These may either affect vascular permeability, blood brain barrier integrity or antigen availability. An unrelated infection or other brain insult (e.g. stroke or head injury) that causes local inflammation may, in the presence of circulation-derived autoantibodies, bring about pathogenic immune complexes at the blood brain barrier [62]. This hypothesis is consistent with experimental evidence showing that antibody-mediated neuronal damage in mice harboring pathogenic antibodies does only occur upon compromise of the blood-brain barrier [63]. It appears therefore that regionally-specific vascular permeability may lead to localized neuronal damage (eg the cerebellum in gluten ataxia).

One may argue that the development and deposition of antibodies is an epiphenomenon rather than being pathogenic. One method to demonstrate the pathological effect of an antibody is the passive transfer of the disease through antibody injection into a naïve animal. A common problem in such studies is to be able to demonstrate whether it is these specific antibodies or other circulating autoantibodies that cause neuronal damage. Using a mouse model it has been shown that serum from GA patients as well as clonal monovalent anti-TG immunoglobulins derived using phage display, cause ataxia when injected intraventricularly in mice [54]. These data therefore provide evidence that anti-TG immunoglobulins (derived from patients) compromise neuronal function in selected areas of the brain once exposed to the CNS.

Further support for the role of TG6 in the pathogenesis of GA this comes from the identification of mutations in the gene encoding TG6 in families with autosomal dominant ataxia [64]. This form of spinocerebellar ataxia is now referred to as SCA35.

11.8 Conclusions

The neurological manifestations of CD and GS are diverse and very common. Those patients with classic CD (i.e. they have the classic gastrointestinal symptoms have the advantage of being diagnosed and treated early thus preventing future permanent neurological disability. Even at the time of diagnosis of CD, however, these patients have evidence of neurological deficits both clinical and on brain imaging. It is imperative to educate health professionals to suspect GS in every patient with neurological dysfunction, particularly those with ataxia or neuropathy.

References

[1] J.F. Ludvigsson, J.C. Bai, F. Biagi, et al., Diagnosis and management of adult coeliac disease: Guidelines from the British society of gastroenterology, Gut. 63 (8) (Aug, 2014) 1210–1228.

[2] A. Sapone, J.C. Bai, C. Ciacci, et al., Spectrum of gluten-related disorders: consensus on new nomenclature and classification, BMC Med. 10 (Feb 7, 2012) 13, doi:10.1186/1741-7015-10-13.

[3] M. Hadjivassiliou, A. Gibson, G.A.B. Davies-Jones, A.J. Lobo, T.J. Stephenson, A. Milford-Ward, Does cryptic gluten sensitivity play a part in neurological illness? Lancet. 347 (1996) 369–371.

[4] M. Hadjivassiliou, R.A. Grünewald, A.K. Chattopadhyay, et al., Clinical, radiological, neurophysiological and neuropathological characteristics of gluten ataxia, Lancet. 352 (1998) 1582–1585.

[5] M. Hadjivassiliou, R. Grünewald, B. Sharrack, et al., Gluten ataxia in perspective: epidemiology, genetic susceptibility and clinical characteristics, Brain. 126 (2003) 685–691.

[6] M. Hadjivassiliou, P. Aeschlimann, D.S. Sanders, et al., Transglutaminase 6 antibodies in the diagnosis of gluten ataxia, Neurology. 80 (19) (May 7, 2013) 1740–1745.

[7] M. Hadjivassiliou, J. Martindale, P. Shanmugarajah, et al., Causes of progressive cerebellar ataxia: prospective evaluation of 1500 patients, Neurol. Neurosurg. Psychiatry. 88 (4) (Apr, 2017) 301–309.

[8] P. Zis, P.G. Sarrigiannis, D.G. Rao, N. Hoggard, D.S. Sanders, M. Hadjivassiliou, Cerebellar ataxia with sensory ganglionopathy; does autoimmunity have a role to play? Cerebellum. Ataxias. 22. 4 (Dec, 2017) 20.

[9] C.S. Lu, P.D. Thompson, N.P. Quin, J.D. Parkes, CD. Marsden, Ramsay Hunt syndrome and coeliac disease: A new association, Mov. Disord. 1 (1986) 209–219.

[10] K.P. Bhatia, P. Brown, R. Gregory, et al., Progressive myoclonic ataxia associated with coeliac disease. The myoclonus is of cortical origin, but the pathology is in the cerebellum, Brain. 118 (Pt 5) (Oct, 1995) 1087–1093.

[11] P.G. Sarrigiannis, N. Hoggard, D. Aeschlimann, et al., Myoclonus ataxia and refractory coeliac disease, Cerebellum. Ataxias. 1 (1) (Sep, 2014) 11, doi:10.1186/2053-8871-1-11.

[12] M. Hadjivassiliou, R.A. Grunewald, D.S. Sanders, P. Shanmugarajah, N. Hoggard, Effect of gluten-free diet on MR spectroscopy in gluten ataxia, Neurology. 89 (2017) 1–5.

[13] M. Hadjivassiliou, G.A.B. Davies-Jones, D.S. Sanders, RAG. Grünewald, Dietary treatment of gluten ataxia, J. Neurol. Neurosurg. Psychiatry. 74 (9) (2003) 1221–1224.

[14] K. Bürk, A. Melms, J.B. Schulz, J. Dichgans, Effectiveness of intravenous immunoglobulin therapy in cerebellar ataxia associated with gluten sensitivity, Ann. Neurol. 50 (2001) 827–828.

[15] P. Zis, D.G. Rao, P.G. Sarrigiannis, et al., Transglutaminase 6 antibodies in gluten neuropathy, Dig. Liver. Dis. 49 (11) (Nov, 2017) 1196–1200.

[16] P. Zis, M. Hadjivassiliou, P.G. Sarrigiannis, A. Barker, DG. Rao, Rapid neurophysiological screening for sensory ganglionopathy: A novel approach, Brain. Behav. 7 (12) (Nov 24, 2017) e00880, doi:10.1002/brb3.880.

[17] M. Hadjivassiliou, R.H. Kandler, A.K. Chattopadhyay, et al., Dietary treatment of gluten neuropathy, Muscle Nerve. 34 (6) (Dec, 2006) 762–766.

[18] M. Hadjivassiliou, D.G. Rao, R.A. Grunewald, et al., Neurological dysfunction in Coeliac Disease and Non-Coeliac Gluten Sensitivity, Am. J. Gastroenterol. 111 (4) (Apr, 2016) 561–567.

[19] M. Hadjivassiliou, D.S. Rao, S.B. Wharton, D.S. Sanders, R.A. Grunewald, GAB. Davies-Jones, Sensory ganglionopathy due to gluten sensitivity, Neurology. 75 (2010) 1003–1008.

[20] P. Zis, P.G. Sarrigiannis, D.G. Rao, D.S. Sanders, M. Hadjivassiliou, Small fiber neuropathy in coeliac disease and gluten sensitivity, Postgraduate Med. 131 (7) (Oct 3, 2019) 496–500.

[21] M. Hadjivassiliou, A.K. Chattopadhyay, G.A.B. Davies-Jones, A. Gibson, R.A. Grunewald, AJ. Lobo, Neuromuscular disorder as a presenting feature of coeliac disease, J. Neurol. Neurosurg Psychiatry. 63 (1997) 770–775.

[22] T.H. Brannagan, A.P. Hays, S.S. Chin, H.W. Sander, R.L. Chin, M.P. etal, Small-fiber neuropathy/neuronopathy associated with celiac disease: skinbiopsy findings, Arch. Neurol. 62 (2005) 1574–1578.

[23] C.H. Gibbons, R. Freeman, Autonomic. neuropathy and celiac. disease., J. Neurol. Neurosurg Psychiatry 76 (2005) 579–581.

[24] P. Zis, P.G. Sarrigiannis, D.G. Rao, M. Hadjivassiliou, Gluten neuropathy: prevalence of neuropathic pain and the role of gluten-free diet, J. Neurol. (Jul 21, 2018), doi:10.1007/s00415-018-8978-5.

[25] P. Zis, P.G. Sarrigiannis, D.G. Rao, M. Hadjivassiliou, Quality of Life in Patients with Gluten Neuropathy: A Case-Controlled Study, Nutrients. 10 (6) (May 23, 2018) pii: E662, doi:10.3390/nu10060662.

[26] N. Souayah, R.L. Chin, T.H. Brannagan, et al., Effect of intravenous immunoglobulin on cerebellar ataxia and neuropathic pain associated with celiac disease, Eur. J. Neurol. 15 (12) (Dec, 2008) 1300–1303.

[27] D. Anandacoomaraswamy, J. Ullal, AI. Vinik, A 70-year-old male with peripheral neuropathy, ataxia and antigliadin antibodies shows improvement in neuropathy, but not ataxia, after intravenous immunoglobulin and gluten-free diet, J. Multidiscip. Healthc. 1 (1) (Oct, 2008) 9396.

[28] M. Hadjivassiliou, R.A. Grünewald, M. Lawden, et al., Headache and CNS white matter abnormalities associated with gluten sensitivity, Neurology. 56 (2001) 385–388.

[29] E. Lionetti, R. Francavilla, L. Maiuri, et al., Headache in paediatric patients with celiac disease and its prevalence as a diagnostic clue, J. Pediatr. Gastroenterol Nutr. 49 (2) (Aug, 2009) 202–207.

[30] A.K. Dimitrova, R.C. Ungaro, B. Lebwohl, et al., Prevalence of migraine in patients with celiac disease and inflammatory bowel disease, Headache. 53 (2) (Feb, 2013) 344–355.

[31] P. Zis, T. Julian, M. Hadjivassiliou, Headache associated with coeliac disease: A systematic review and meta-analysis, Nutrients. 10 (2018) 1445–1456.

[32] R. Nenna, L. Petrarca, P. Verdecchia, et al., Celiac disease in a large cohort of children and adolescents with recurrent headache: A retrospective study, Dig. Liver. Dis. 48 (5) (May, 2016) 495–498.

[33] B. Lebwohl, A. Roy, A. Alaedini, P.H.R. Green, JF. Ludvigsson, Risk of headache-related healthcare visits in patients with celiac disease: A population based observational study, Headache. 56 (2016) 849–858.

[34] I.D. Croall, C. Tooth, A. Venneri, et al., Cognitive impairment in coeliac disease with respect to disease duration and gluten-free diet adherence: a pilot study, Nutrients. 12 (2020) 2028, doi:10.3390/nu12072028.

[35] P. Collin, T. Pirttilä, T. Nurmikko, H. Somer, T. Erilä, O. Keyriläinen, Celiac disease, brain atrophy, and dementia, Neurology. 41 (3) (Mar, 1991) 372–375.

[36] W.T. Hu, J.A. Murray, M.C. Greenaway, J.E. Parisi, KA. Josephs, Cognitive impairment and celiac disease, Arch. Neurol. 63 (10) (Oct, 2006) 1440–1446.

[37] I.T. Lichtwark, E.D. Newnham, S.R. Robinson, et al., Cognitive impairment in coeliac disease improves on a gluten-free diet and correlates with histological and serological indices of disease severity, Aliment. Pharmacol. Ther. 40 (2) (Jul, 2014) 160–170.

[38] S. Makhlouf, M. Messelmani, J. Zaouali, R. Mrissa, Cognitive impairment in celiac disease and non-celiac gluten sensitivity: review of literature on the main cognitive impairments, the imaging and the effect of gluten free diet, Acta. Neurol. Belg. 118 (1) (Mar, 2018) 21–27.

[39] B. Lebwohl, J.A. Luchsinger, D.E. Freedberg, P.H. Green, JF. Ludvigsson, Risk of dementia in patients with celiac disease: a population-based cohort study, J. Alzheimers. Dis. 49 (1) (2016) 179–185.

[40] I.D. Croall, D.S. Sanders, M. Hadjivassiliou, N. Hoggard, Cognitive Deficit and White Matter Changes in Persons With Celiac Disease: A Population-Based Study, Gastroenterology. 158 (8) (2020) 2112–2122.

[41] L.A. Rabin, C.M. Smart, P.K. Crane, et al., Subjective Cognitive Decline in Older Adults: An Overview of Self-Report Measures Used Across 19 International Research Studies, J. Alzheimer's. Dis: JAD. 48 (Suppl 1(01)) (2015) S63–S86.

[42] N.L. Hill, C. McDermott, J. Mogle, et al., Subjective cognitive impairment and quality of life: a systematic review, Int. Psychogeriatr. 29 (12) (2017) 1965–1977.

[43] T. Julian, M. Hadjivassiliou, P. Zis Gluten sensitivity and epilepsy: a systematic review. J Neurol. doi.org: 10.1007/s00415-018-9025-2.

[44] L. Licchetta, F. Bisulli, L. Di Vito, et al., Epilepsy in coeliac disease: not just a matter of calcifications, Neurol. Sci. 32 (6) (Dec, 2011) 1069–1074.

[45] H.A. Arroyo, S. De Rosa, V. Ruggieri, M.T. de Dávila, N. Fejerman, Argentinean Epilepsy and Celiac Disease Group. Epilepsy, occipital calcifications, and oligosymptomatic celiac disease in childhood, J. Child. Neurol. 17 (11) (Nov, 2002) 800–806.

[46] G. Gobbi, F. Bouquet, L. Greco, et al., Coeliac disease, epilepsy, and cerebral calcifications. The Italian Working Group on Coeliac Disease and Epilepsy, Lancet. 340 (8817) (Aug 22, 1992) 439–443.

[47] D. Hardoff, B. Sharf, A. Berger, Myopathy as a presentation of coeliac disease, Dev Med Child Neurol. 22 (6) (Dec, 1980) 781–783.

[48] R. Sandyk, MJ. Brennan, Isolated ocular myopathy and celiac disease in childhood, Neurology. 33 (6) (Jun, 1983) 792.

[49] K.A. Kleopa, K. Kyriacou, E. Zamba-Papanicolaou, T. Kyriakides, Reversible inflammatory and vacuolar myopathy with vitamin E deficiency in celiac disease, Muscle Nerve. 31 (2) (Feb, 2005) 260–265.

[50] M. Hadjivassiliou, A.K. Chattopadhyay, R.A. Grünewald, et al., Myopathy associated with gluten sensitivity, Muscle Nerve. 35 (4) (Apr, 2007) 443–450.

[51] M.D. Rouvroye, P. Zis, A. Van-Dam, A.J.M. Rozemuller, G. Bouma, M. Hadjivassiliou, The Neuropathology of Gluten-Related Neurological disorders: a systematic review, Nutrients. 12 (2020) 822.

[52] B. Jabri, LM. Sollid, Tissue-mediated control of immunopathology in coeliac disease, Nat. Rev. Immunol. 9 (12) (2009) 858–870 Dec.

[53] M. Hadjivassiliou, S. Boscolo, G.A.B. Davies-Jones, et al., The humoral response in the pathogenesis of gluten ataxia, Neurology. 58 (2002) 1221–1226.

[54] S. Boscolo, A. Lorenzon, D. Sblattero, et al., Anti-Transglutaminase antibodies cause ataxia in mice, PLoS One. 5 (3) (2010) e9698, doi:10.1371/journal.pone.0009698.

[55] W. Dietrich, T. Ehnis, M. Bauer, et al., Identification of tissue transglutaminase as the autoantigen of celiac disease, Nature Med. 3 (1997) 797–801.

[56] M. Sárdy, S. Kárpáti, B. Merkl, et al., Epidermal transglutaminase (TGase3) is the autoantigen of Dermatitis Herpetiformis, J. Exp. Med. 195 (2002) 747–757.

[57] M. Hadjivassiliou, P. Aeschlimann, A. Strigun, et al., Autoantibodies in gluten ataxia recognise a novel neuronal transglutaminase, Ann. Neurol. 64 (2008) 332–343.

[58] R. Di Niro, L. Mesin, N.Y. Zheng, et al., High abundance of plasma cells secreting transglutaminase 2-specific IgA autoantibodies with limited somatic hypermutation in celiac disease intestinal lesions, Nat. Med. 18 (3) (2012) 441–445.

[59] I.R. Korponay-Szabó, T. Halttunen, Z. Szalai, et al., In vivo targeting of intestinal and extraintestinal transglutaminase 2 by coeliac autoantibodies, Gut. 53 (2004) 641–648.

[60] M. Hadjivassiliou, M. Maki, D.S. Sanders, et al., Autoantibody targeting of brain and intestinal transglutaminase in gluten ataxia, Neurology. 66 (2006) 373–377.

[61] M. Hadjivassiliou, D.S. Sanders, R.A. Grunewald, et al., Gluten sensitivity: from gut to brain, Lancet. Neurol. 9 (2010) 318–330.

[62] J.F. Ludvigsson, M. Hadjivassiliou, Can head trauma trigger coeliac disease? Nation-wide case-control study, BMC neurology. 13 (2013) 105.

[63] C. Kowal, L.A. Degiorgio, J.Y. Lee, et al., Human lupus autoantibodies against NMDA receptors mediate cognitive impairment, Proc. Natl. Acad. Sci. U. S. A. 103 (52) (2006) 19854–19859.

[64] J.L. Wang, X. Yang, K. Xia, et al., TGM6 identified as a novel causative gene of spinocerenbellar ataxias using exome sequencing, Brain. 133 (2010) 3510–3518.

CHAPTER 12

The role of gluten in multiple sclerosis, psoriasis, autoimmune thyroid diseases and type 1 diabetes

Moschoula Passali[a,b,c], **Julie Antvorskov**[c], **Jette Frederiksen**[a,b], **Knud Josefsen**[c]

[a]University of Copenhagen, Department of Clinical Medicine, Faculty of Health Sciences, Copenhagen, Denmark
[b]Rigshospitalet-Glostrup, Multiple Sclerosis Clinic, Department of Neurology, Glostrup, Denmark
[c]Rigshospitalet, The Bartholin Institute, Copenhagen, Denmark

12.1 Introduction

Wheat is one of the most important staple foods and a major source of dietary carbohydrate in the western world [1]. The unique viscoelastic properties of gluten proteins confer wheat flour with high baking quality allowing for the production of a wide variety of products [2, 3]. Although gluten has long been of great interest to the food industry, the market for gluten-free products has seen an enormous growth during the last years [4]. Abstaining from gluten or wheat is the current treatment for celiac disease (CD) and wheat allergy respectively [5]. However, it has been hypothesized that gluten may also contribute to deteriorating the course of other immune-mediated disorders [6,7]. In this chapter, we present and discuss the currently available data addressing a potential involvement of gluten in multiple sclerosis (MS), psoriasis, autoimmune thyroid diseases (ATD) and type 1 diabetes (T1D).

12.2 Gluten

In the field of food chemistry, gluten is defined as the protein network that remains after wheat dough has been washed with water [3,8]. Gluten consists primarily of the water insoluble but ethanol soluble prolamin together with the water- and ethanol-insoluble glutelin fractions of the storage proteins found in the starchy endosperm of wheat [3]. According to the Codex Alimentarius [9] gluten is defined as "a protein fraction from wheat, rye, barley, oats or their crossbred varieties and derivatives thereof, to which some persons are intolerant and that is insoluble in water and 0.5M NaCl." The word gluten is today considered to be a common term for the prolamins and glutelins of wheat, rye, barley and in some cases also oats (Table 12.1).

Coeliac Disease and Gluten-Related Disorders.
DOI: https://doi.org/10.1016/B978-0-12-821571-5.00003-9

Table 12.1 Nomenclature of the prolamins and glutenins found in wheat, rye, barley, and oats.

	Wheat	Rye	Barley	Oats
Prolamins	gliadins	secalins	hordeins	avenins
Glutelins	glutenins	secalinins	hordenins	oat glutelins

The primary structure of gluten proteins is characterized by repetitive sequence sections with high content of the amino acids proline and glutamine [10,11]. Proline rich sections of gluten peptides are partly resistant to proteolytic degradation by the enzymes of the human gastrointestinal tract [3], why relatively long gluten peptides reach the small intestine. In most people this is believed to be unproblematic, however, in individuals with CD gluten peptides trigger an immune mediated small intestinal enteropathy [12]. Proteins from oats have very low immunogenic activity, and although being a topic of debate, it is believed that only a minimal number of patients with CD (if any) will develop an immunological reaction following the ingestion of oats that have not been contaminated with gluten from wheat, rye or barley [13,14].

Hypotheses regarding a role for gluten in autoimmune disorders are to a certain degree based on in vitro studies illustrating that gliadins, the prolamin fraction from wheat, can stimulate the secretion of pro-inflammatory cytokines from cell lines [15–17]. Molecular mimicry due to cross-reactivity between gluten-antibodies and various autoantigens has also been proposed as a potential mechanism [18–20]. In addition, ex vivo studies suggest that gliadin may affect the permeability of the human intestinal epithelium [21,22] through a mechanism which involves release of the protein zonulin [23]. Furthermore, gluten immunogenic peptides have been detected in the urine of healthy individuals [24] supporting that gliadin peptides can at least to some degree enter the bloodstream of non-celiac individuals. The exact mechanism by which gliadin elicits autoimmune diseases, is however not known, and, except for celiac disease, the role of gliadin has not been established in their pathogenesis.

12.3 Multiple sclerosis

MS is a chronic, autoimmune, demyelinating disease affecting the central nervous system [25]. Patients with MS have a great interest in dietary modifications as a potential way of ameliorating the course of their disease [26]. Although no evidence-based dietary recommendations have been developed yet, a systematic review of web pages providing dietary advice for patients with MS found that 31.3% (10 out of 32) of the included web pages recommended patients to remove gluten from their diet [27]. In addition, dietary surveys have reported that a gluten-free diet (GFD) was adopted by 5.6% [28] and 16.4% [29] of the surveyed American and Australian patients respectively.

12.3.1 Celiac disease and multiple sclerosis

12.3.1.1 Studies of comorbidity

Two French publications have studied the presence of comorbidities among patients with CD finding the prevalence of MS to be 0.11% (1/ 924) [30] and 0.14% (1/ 741) [31]. The crude prevalence of MS in the french population has similarly been estimated to be 0.15% [32]. An Italian study investigating the prevalence of neurological symptoms among 160 patients with CD identified one patient with MS (0.6%) [33]. Additionally, a Swedish case-control study including 14,371 patients with CD and 70,096 matched reference individuals found no significant association between CD and subsequent diagnosis of MS (hazard ratio (HR) = 0.9; 95% confidence interval (CI) = (0.3–2.3)) [34]. The latter results are in agreement with a publication using the National Patient Registry of Denmark, which identified 6 cases of comorbidity between CD and MS resulting in an odds ratio (OR) of 1.0 for the two diseases [35]. Similarly, a more recent Danish population-based study with data from 1997-2016 did not find increased prevalence of MS among patients with CD [36].

12.3.1.2 Gluten-related antibodies in multiple sclerosis

Class A immunoglobulins against tissue transglutaminase (IgA-tTG) are highly specific for CD [37,38], why measurements of these antibodies among patients with MS can contribute to identifying cases of comorbidity. Twelve studies [39–50] measuring IgA-tTG among patients with MS found no evidence to support a relationship between CD and MS. One study compared mean values of serum concentration of IgA-tTG and IgG-tTG between 30 patients with MS and 25 healthy controls (HC) and found significantly elevated mean values among patients with MS for both classes of tTG antibodies [51]. However, it is not stated if any of the MS patients included in the study were later diagnosed with CD [51]. The only study to find an increased prevalence of CD among patients with MS is that by Rodrigo et al. [52] finding 8 (11.1%) cases of CD among 72 patients with MS. However, the latter study used a lower threshold [53] for the seropositivity of antibodies (measured by ELISA), which could possibly explain the higher prevalence found [52].

   Data from five case-control studies [40,42,45,46,50] support that there is no significant difference in the number of seropositive individuals for IgA-AGA between patients with MS and HC. Using the above studies we have calculated the prevalence of IgA-AGA to be 3.8% (23/610) among patients with MS and 2.4% (14/584) among HC (not significant (NS)). Similarly, only one [50] out of four [42,45,46] studies found increased prevalence of IgG-AGA among patients with MS compared to HC. We calculated the prevalence of IgG-AGA to be 3.6% (19/525) among patients with MS (n = 544) and 2.2% (12/536) among HC (NS). Furthermore, one cohort study found no AGA positive individuals among 60 patients with MS [39], whereas in another publication screening for AGA antibodies revealed their prevalence to be 10% [54] among 100 patients with MS and 12.5% among 1200 HC [55]. Four studies [42,47,49,50] have compared mean values of

AGA antibodies between patients with MS and HC, but the results are contradicting. We can therefore not exclude that patients with MS may have higher concentrations of IgA-AGA or IgG-AGA than healthy individuals, however, still remaining within normal reference values.

12.3.2 Can patients with multiple sclerosis benefit from a gluten-free diet?

To date, only one clinical trial has investigated whether patients with MS can benefit from a GFD [56]. The study compared a group of 36 patients who had followed a GFD with a group of 36 patients who had followed a regular diet for a median of 4.5 years (mean 5.3±1.6). The GFD group had a significantly lower expanded disability status scale (EDSS) score at follow-up (GFD vs regular diet: 1.5±1.4 vs 2.1±1.5, p = 0.001) which translates to better neurological function. Interestingly, and in contrast to the regular diet group, the EDSS score of the GFD group did not deteriorate over the course of the study time (mean baseline EDSS score was 1.7 for both groups). In addition, only 28% of individuals in the GFD group showed MRI activity compared to 67% of individuals in the regular diet group (p = 0.001, MRI activity was defined as the presence of new contrast-enhanced lesions or an increase in the number of T2-weighted lesions). No significant effect on annual relapse rate was observed. Although the above results are highly promising, the fact that the control group consisted of patients, who did not manage to adhere to the GFD during the first six months of the study, is a major limitation [56]. Equally important, we find the title of the publication describing the study as a "randomized clinical trial" highly misleading and this together with other minor issues has reduced our capacity to trust this publication.

A few studies have evaluated the effects of the The Wahls Protocol, a lifestyle intervention including adherence to a gluten-free, modified paleolithic diet among patients with relapsing remitting MS [57] or progressive MS [58–60]. The studies report that a gluten-free, modified paleolithic diet can significantly improve endpoints such as mood, fatigue, lipid profiles and quality of life. However, it is not possible to evaluate the extent to which the elimination of gluten has contributed to these results. In addition, as more disease-specific endpoints such as EDSS score and MRI data were not included, it is hard to evaluate whether the observed results can translate to reduced disease activity. Despite their limitations, the studies on The Wahls Protocol highlight the importance of diet as a tool that can contribute to improving patients' quality of life. This is especially relevant for patients with primary progressive or secondary progressive MS, as treatments for these patient groups are still scarce [61].

12.4 Psoriasis

Psoriasis is a chronic, inflammatory disease of the skin characterized by the development of erythematous scaly lesions. Several different types of psoriasis have been described with psoriasis vulgaris (plaque psoriasis) being the most common [62]. According to a U.S. national survey published in 2017, a GFD was the most common special diet to be used by patients with psoriasis [63]. More specifically, 38% of the patients who responded to the survey (n = 1206) reported avoiding gluten with 53.4% (247/459) of them claiming to have experienced clearance or improvement of their psoriasis as a response to the diet [63].

12.4.1 Celiac disease and psoriasis

12.4.1.1 Studies of comorbidity

A recent systematic review and meta-analysis [64] has identified 18 studies examining the association between CD and psoriasis. Five [65–69] out of nine [70–73] studies investigating the prevalence of CD among patients with psoriasis or psoriatic arthritis reported a significant association with the meta-analysis presenting an OR of 2.16 (95%CI = (1.74–2.69)) [64]. Two studies have so far reported the incidence of CD among patients with psoriasis with only one of them finding significant results (HR = 1.9, 95%CI = (1.6–2.2) [74] & HR = 1.20, 95%CI = (0.91–1.59) [68]). Additionally, four [36,75–77] out of eight [78–81] identified studies found an increased prevalence of psoriasis among patients with CD and the metaanalysis calculated an overall OR of 1.8 (95%CI = (1.36–2.38)) [64]. Both studies reporting the risk of psoriasis (incidence) among patients with CD found significant results with HR = 1.72, 95%CI = (1.54–1.92) [75] and HR = 1.9, 95%CI = (1.5–2.3) [74].

12.4.1.2 Gluten-related antibodies in psoriasis

Ten case-control studies [67,70,71,82–88] have compared the prevalence of IgA-tTG between patients with psoriasis and HC. Most studies fail to reveal a significant difference, however we have calculated the mean prevalence of IgA-tTG to be 4.3% (38/884) among patients with psoriasis and 1.2% (14/1153) among HC (p<0.001). In addition, five cross-sectional cohort studies [89–93] have estimated the prevalence of IgA-tTG seropositivity among patients with psoriasis. From these we calculated the prevalence of IgA-tTG in psoriasis to be 3.7% (26/708). Finally, a study comparing mean IgA-tTG antibody titers found significantly higher values among 67 patients with psoriasis compared to 85 HC (0.94±1.13 vs. 0.85±0.58, p<0.05) [84].

Studies reporting the prevalence of IgA-AGA seropositivity among patients with psoriasis have previously been summarized in a meta-analysis published in 2014 [94]. The metaanalysis supports both that patients with psoriasis have higher mean values of IgA-AGA compared to HC, and that seropositivity for IgA-AGA is more prevalent among patients with psoriasis compared to HC [94]. We found similar results using an updated list of case-control studies [82–84,86,95–99] and calculated an increased prevalence of IgA- and IgG-AGA in patients with psoriasis compared to HC (IgA-AGA: 10.3% (n = 737) vs 3.7% (n = 561), p<0.001 and IgG-AGA: 12.0% (n = 452) vs 6.0% (n = 369), p<0.001). Furthermore, cross-sectional cohort studies have estimated the prevalence of AGA seropositivity among patients with psoriasis (0% (n = 80) [100] & 3% (n = 30) [91]) and palmoplantar pustulosis (18% (n = 123) [89] & 0% (n = 32) [91]). Lastly, four [82,84,101,102] out of five [98] studies comparing mean concentrations of IgA-AGA found significantly higher levels among patients with psoriasis compared to HC. However, for IgG-AGA this was the case in only one [84] out of four studies [98,101,102].

12.4.2 Gluten-related antibodies as markers of disease activity in psoriasis

In a study of 130 patients with psoriasis, 16.2% were positive for at least one of the following antibodies: IgG-AGA, IgA-AGA, IgA-tTG [92]. Systemic therapy was currently given or had previously been given to more seropositive patients than seronegative patients (48% (n = 21) vs 22% (n = 109), p = 0.04). Likewise, PUVA phototherapy was administered to more seropositive patients than seronegative patients (57% (n = 21) vs 30% (n = 109), p = 0.03). No significant differences were identified for UVB phototherapy, the presence of arthralgia or arthritis [92]. A similar study with 120 patients with psoriasis (8 seropositive for IgA-AGA, 5 seropositive for IgG-AGA) found no associations between seropositivity for AGA and high disease severity at the reported time or past treatment for high disease severity [98]. A smaller study (n = 41) reports a significant relationship between seropositivity for IgA-AGA and disease duration (p<0.001), however, seropositivity was not related to psoriasis area and severity index (PASI) scores [82].

A study measured the proliferative response of peripheral blood mononuclear cells from patients with psoriasis (n = 37) and HC (n = 37) upon stimulation with selected wheat peptides [103]. No significant differences were found between the two groups, however the five highest responses against peptide p62–75 were observed among patients with psoriasis [103]. In another publication using an array with 75 autoantigens, IgG4 antigliadin antibodies were the only to be elevated in the serum of 12 patients with severe psoriasis (PASI>30) [104]. IgG4 antigliadin antibodies were not detected in sera from 12 HC, and later validation using a cohort with 73 psoriasis patients and 75 HC supported that IgG4 antigliadin antibodies may be of use as biomarkers of psoriasis. Additionally, for patients with the highest levels of IgG4 antigliadin antibodies, antibody levels were correlated with PASI score (r = 0.65, p<0.001) [104].

12.4.3 Can gluten intake affect the risk of psoriasis?

A publication based on the Nurses' Health Study II investigated whether intake of gluten, calculated through food frequency questionnaires, was associated with the incidence of psoriasis, psoriatic arthritis or atopic dermatitis [105]. The multivariate HR were 1.15 (95%CI = (0.98–1.36)), 1.12 (95%CI = (0.78–1.62)) and 0.91 (95%CI = (0.66–1.25)) for psoriasis, psoriatic arthritis and atopic dermatitis, respectively, when comparing the highest and lowest gluten intake quintiles. No dose-response association was observed, however, potential effects of a strict GFD were not investigated in this study. Based on this publication [105] reducing the intake of gluten will not reduce the risk of psoriasis, psoriatic arthritis and atopic dermatitis among adult women.

12.4.4 Can patients with psoriasis benefit from a gluten-free diet?

In 2018 the Medical Board of the National Psoriasis Foundation conducted a systematic review with the aim of developing nutritional recommendations for patients with psoriasis or psoriatic arthritis [106]. On the basis of three publications [67,107,108] describing two independent studies, the board weakly recommends a GFD for patients with positive levels of gluten-related antibodies [106].

The first study to investigate whether patients with psoriasis can benefit from a GFD reported a clinical improvement in 73% (22/30) of patients who adhered to a GFD for three months [108]. The included patients were positive for IgA-AGA or IgG-AGA, and the PASI score was reduced from 5,5±4,5 to 3,6±3,0 (p = 0,001). No effect was seen among six patients with no AGA antibodies who also followed a GFD. The study was planned as a crossover trial, however, the second part of the study consisting of three months on a gluten-containing diet could not be finalized as 18 out of 30 of the AGA positive patients needed intensified medical treatment [108]. Immunohistochemical staining of skin biopsies from 19 of the above AGA positive patients before and after the GFD showed a significant reduction in Ki67 positive cells in the involved dermis following the GFD [107]. In addition, expression of tTG was higher in involved compared to uninvolved dermis (5.06±3.80% vs. 0.67±0.54%, n = 13, p = 0.0002) and tTG expression decreased by 50% after the GFD [107]. The same research group later described 16 patients with palmoplantar pustulosis that adhered to a GFD [89]. For AGA-seropositive patients who strictly adhered to the GFD (n = 9), the palmoplantar pustulosis greatly improved or even cleared. However, improvements were only seen among two out of four patients who were not strict about the diet and none of the seronegative patients (n = 3) [89].

An Italian, primary care, multicenter study identified nine psoriasis patients with concomitant CD, who were put on a GFD [67]. One patient was lost to follow-up, but after three months on a GFD one patient had no skin lesions, two experienced an improvement in their PASI score by at least 50%, and five experienced an improvement in their PASI score by at least 75%. Only one of the above patients experienced worsening at follow-up after six months on the GFD, whereas two further improved and the remaining five patients maintained their clinical improvement. A more recent study by Kolchak et al. [95] investigated the effects of 12 months on a GFD on PASI score among 13 patients positive for IgA against deamidated gliadin peptides. A 56% improvement in the PASI score was observed in the group of five patients with the highest (>30 U/ml) levels of antibodies. The group with the remaining eight patients with antibody levels between 11,5–30.0 U/ml experienced a 36% improvement in the PASI score. Unfortunately, the participants of this study were not tested for the presence of antibodies against tTG, why the presence of cases with CD cannot be excluded. Overall, evidence suggests that psoriasis patients with gluten-related antibodies may benefit from a GFD, however, larger trials are still lacking.

12.5 Autoimmune thyroid diseases

Several different types of ATD exist with Hashimoto's thyroiditis (HT) and Graves' disease being examples of common thyroid disorders leading to hypothyroidism and hyperthyroidism respectively [109].

12.5.1 Celiac disease and autoimmune thyroid diseases

12.5.1.1 Studies of comorbidity

A systematic review and meta-analysis of 27 studies [110] calculated the median prevalence of CD in ATD to be 3.2%, however, this number decreased to 1.6% (CI 1.3–1.9) for biopsy-verified CD, since this obviously represents a more strict diagnostic criterion. The meta-analysis also reported a higher prevalence of biopsy-verified CD among patients with hyperthyroidism (2.6%, 95% CI = 0.7–4.4) than in patients with hypothyroidism (1.4%, 95% CI = 1.0–1.9), however, the CI of the two prevalence estimates are overlapping [110]. Likewise, another study found that the prevalence of CD was higher among patients with Grave's disease (1.1%) compared to HC (0.3%) (OR = 3.81, 95% CI = 1.17–12.41) [111].

In another meta-analysis [112], patients with CD had a significantly higher prevalence of thyroid disease, euthyroid ATD and hypothyroidism, but not hyperthyroidism, than controls (OR = 3.08, 95% CI = (2.76–3.56); OR = 4.35, 95% CI = (2.88-6.56); OR = 3.38, 95% CI = (2.73–4.19) & OR = 1.28, 95% CI = (0.37–4.46) respectively). We hypothesize that the late age of disease onset for hyperthyroidism could possibly explain why the prevalence of CD has been found to be higher among patients with hyperthyroidism [111], whereas patients with CD do not seem to have increased prevalence of hyperthyroism [112].

The prevalence of ATD has also been reported to be high among patients with non-celiac gluten/ wheat sensitivity [113,114] and dermatitis herpetiformis [115]. Furthermore, first-degree relatives of patients with CD may also have increased risk of ATD [116]. Interestingly, a publication has reported a higher prevalence of autoimmune thyroiditis among seronegative (26.9%) compared to seropositive (9.7%, tTG and endomysium antibodies) patients with CD (p = 0.002) [117]. Lastly, it has been debated whether late diagnosis and thus late treatment of CD can increase the risk of developing other autoimmune diseases [30,79,80,118], but a meta-analysis supports that this is not the case for thyroid disease (OR $= 1.08, 95\%$CI $= (0.61–1.92)$) [112].

12.5.1.2 Gluten-related antibodies in autoimmune thyroid diseases

Most case-control studies find a higher prevalence of IgA-tTG among patients with ATD compared to HC and we calculated that 5.4% of patients with ATD (n $= 2041$) had such antibodies compared to 1.7% of HC (n $= 1908$) (p < 0.001) [119–126]. Similarly, from 14 identified cohort studies we have calculated the mean prevalence of IgA-tTG autoantibodies in patients with ATD (n $= 2667$) to be 4.4% [127–140].

Based on five [119,127,131,138,141] and three [127,138,141] cohort-studies we have calculated the prevalence of IgA-AGA and IgG-AGA in ATD to be 4.1% and 6.0% respectively. Only one [128] out of two [123] identified case-control studies found a significantly higher prevalence of IgA-AGA in patients with ATD compared to HC. Similarly, out of four case-control studies measuring IgG-AGA, one study found significantly higher prevalence in ATD [123], two found a trend toward a slightly higher prevalence in ATD [122,126] and one found a trend toward a slightly higher prevalence in HC [123]. Lastly, a study that measured IgG antibodies against 125 foods found that patients with HT more commonly had antibodies against barley compared to HC (93.2% vs 71.0%, p<0.0001), but no difference was found in antibodies against wheat or gliadin [142].

12.5.2 Gluten-related antibodies as markers of disease activity in autoimmune thyroid diseases

A mechanistic study from 2008 illustrated that IgA-tTG from patients with CD not only correlate with titers of TPO antibodies, but they also have the ability to bind to thyroid tissue [143]. Furthermore, in another study IgG-tTG and IgA-AGA could predict levels of TPO and thyroglobulin antibodies [123]. Interestingly, a publication has shown that in patients with T1D, the presence of AGA antibodies was associated with chronic thyroiditis (38% vs. 2.7%, p $= 0.005$) and thyroid peroxidase (TPO) antibodies (69% vs. 27%, p $= 0.01$) [144]. The association between CD, ATD and T1D has already been acknowledged [125,145] and patients with a dual diagnosis of both CD and T1D have been found to have increased risk of developing ATD [146]. This is further supported by a study illustrating that both tTG antibodies (p $= 0.023$) and glutamic acid decarboxylase (GAD) antibodies (p < 0.00001) correlated with the titer of TPO antibodies [121]. Last but not least, patients treated with high dosage of levothyroxine have been found to have significantly higher levels of IgA-AGA compared to patients receiving low levels of levothyroxine (medians: 19.69 vs 13.00, p $= 0.033$) [128]. This topic is further discussed below.

12.5.3 Can patients with autoimmune thyroid diseases benefit from a gluten-free diet?

A few publications have studied the effects of a GFD among patients with concomitant ATD and CD, but none in patients with ATD alone.

A trial from 2019 investigated the effect of a GFD for six months among 34 untreated women with HT that were positive for IgA–tTG, but had no CD-related symptoms [147]. Sixteen women followed a GFD and were compared to 18 women who maintained a normal diet [147]. The authors observed that the levels of TPO- and thyroglobulin antibodies decreased, 25-hydroxyvitamin D increased, and the structure parameter inference approach (SPINA)–GT index improved. Thyrotropin and free triiodothyronine levels did not change [147].

In an Italian multicenter study, thyroid disease was reversed in some patients with newly diagnosed CD after one year on a GFD (n = 128) [148]. Similarly, in a publication from 1999 it was noted that symptoms related to hypothyroidism and thyroxine dosage improved among three patients with concomitant CD and ATD who followed a GFD for six months [149]. Thyroglobulin and TPO antibody status only changed for one patient who had an additional follow-up at 18 months [149]. Another study observed a progressive reduction in TPO antibodies in CD patients with ATD after 6, 12 and 24 months on a GFD (76.9% (10/13), 46.1% (6/13) and 15.3% (2/13), respectively) [150], whereas in smaller studies of shorter duration, a GFD did not have any effect on thyroid antibodies [129,151] or thyroid volume [151].

It has been reported that patients with concomitant CD and HT need a 50% higher dose of levothyroxine to reach target TSH values when compared to patients with HT alone [152]. This could possibly be explained by a higher disease activity illustrated by a higher TSH (5.7 vs 7.26, p = 0.0099) and lower free T4 (1.12 vs 0.01, p < 0.0001) when compared to patients with HT alone [152]. However, apart from possibly exaggerating the course of ATD, CD might also reduce the absorption of levothyroxine in these patients as the need for extra levothyroxine was prevented by introduction of a GFD (n = 21) [152]. Likewise, one [153] out of two [134] identified studies on the topic found that patients requiring high doses of levothyroxine to maintain an euthyroid state were more likely to have CD [153].

Altogether, patients with concomitant CD and ATD are advised to adhere to a GFD. This may improve absorption of levothyroxine and possibly also contribute to reducing levels of TPO and thyroglobulin antibodies. There are no data to evaluate the effects of a GFD in patients with ATD without CD.

12.6 Type 1 diabetes

T1D is a common chronic disease often diagnosed in childhood leading to a lifelong need for exogenous insulin. The disease is caused by a T-cell mediated selective destruction of the insulin-producing beta cells in the pancreas. The importance of gluten in the pathogenesis of T1D has been consistently shown in animal models and could potentially be caused by changes in the intestinal microbiome, intestinal permeability or by a direct effect on different immune cell populations [154].

12.6.1 Celiac disease and type 1 diabetes

12.6.1.1 Studies of comorbidity

The association between CD and T1D is well documented and can possibly be explained by shared genetic risk factors such as the haplotypes HLA-DR3-DQ2 and HLA-DR4-DQ8. A Swedish population-based study has shown that patients with CD have increased risk of developing T1D before the age of 20 (HR = 2.4, 95%CI = (1.9–3.0)) [155]. Meta-analysis of a systematic review has calculated the prevalence of biopsy-confirmed CD (6.0%, 95%CI = (5.0–6.9%)) and the weighted prevalence of CD (4.7%, 95%CI = (4.0–5.5)) in T1D [156]. Interestingly, the prevalence of CD was found to be higher among children compared to adults with T1D (children vs adults: 6.2%, 95%CI = (6.1– 6.3%) vs 2.7%, 95%CI = (2.1–3.3%)) [156].

12.6.1.2 Gluten-related antibodies in type 1 diabetes

Reports on the prevalence of gluten-related antibodies among patients with T1D vary greatly. In a meta-analysis from 2019 the weighted prevalence of any gluten-related antibody was calculated to be 10.2% (95%CI = (8.4–12.7), whereas the prevalence of IgA-tTG and endomysium antibodies where reported to be 9.8% (95%CI = (8.2–11.6) and 5.3% (95%CI = (4.3–6.4)) respectively [157]. In addition, the meta-analysis calculated the prevalence of IgA-AGA and IgG-AGA to be 9.7% (95%CI = (5.1–15.5) and 12.7% (95%CI = (6.1–21.0)) respectively. Last-mentioned results on AGA are in agreement with results from our calculations.

Biologic as well as technical factors may contribute to the variation across publications estimating the prevalence of gluten-related antibodies among patients with T1D. Technical differences with regards to the analytic assays used reduce the comparability among studies. In addition, genetics and environmental factors may influence the prevalence of gluten-related antibodies among healthy populations, stressing the importance of including a group of HC in such studies. Furthermore, some publications suggest that age as well as the duration of T1D may affect the prevalence of gluten-related antibodies [157,158], whereas others report that antibody status can change with time [159]. Lastly, Tiberti *et al.* observed that the diagnosis of T1D among IgA-tTG seropositive individuals was associated with lower titers of IgG-tTG and deamidated gliadin antibodies compared to CD patients without T1D [160].

12.6.2 Can gluten intake affect the risk of Type 1 Diabetes?

1n human studies there seems to be a connection between early life gluten exposure and the subsequent risk for developing T1D. A recent Danish cohort-study found that high gluten intake by mothers during pregnancy doubled the risk of T1D development in their offspring compared to women with the lowest gluten intake during pregnancy [161]. A similar Norwegian mother-child cohort could not confirm these findings, but found that a higher intake of gluten by the child itself, at an early age, was associated with a higher risk of T1D [162]. Two earlier studies investigated the effect of prenatal gluten exposure on the development of islet autoantibodies in genetically predisposed individuals and found no association [163,164]. Prenatal exposure to gluten could be important because the autoimmune process against the islets may start in fetal life. Thus, seroconvertion is seen as early as 9–12 months of age in children later developing T1D [165–167]. Many other human studies have addressed the connection between infant feeding patterns and subsequent development of islet autoantibodies/ T1D and there seems to be a link between the dose, timing and context of gluten introduction (i.e. breastfeeding when gluten is introduced) [154].

12.6.3 Can patients with type 1 diabetes benefit from a gluten-free diet?

Several trials have studied the effects of a GFD on T1D progression among individuals without concomitant CD. In 2003 a study evaluated the effects of six months on a GFD followed by six months on a gluten-containing diet among 17 individuals with high risk of T1D (first-degree relatives of T1D patients with at least two β-cell autoantibodies) [168]. The diet did not have a significant effect on autoantibody titers but patients´ insulin sensitivity measured as HOMA-IR (homeostasis model of insulin resistance) increased at the end of the GFD and subsequently decreased (p<0.005) after 6 months of normal gluten-containing diet. Similarly, insulin response measured by glucose tolerance test increased (p = 0.04) in 12 out of 14 patients after the GFD and non-significantly decreased (p = 0.07) in 10 out of 13 patients after reintroduction of the normal gluten-containing diet [168].

Another study from 2004 found no effect of one year on a GFD on autoantibody titers among seven children (first-degree relatives of T1D patients) with β-cell autoantibodies [169]. In addition, these seven children had the same risk of progressing to T1D at five years follow-up when compared to a similar control group. Of note, the seven children adhered to the GFD only for 12 months [169]. In the above studies, the effects of a GFD were investigated in subjects that had already developed autoantibodies, however, it is also relevant that the role of gluten (and the GFD) is studied prior to the development of autoantibodies in the very early events that initiate the autoimmune process.

According to a case study from 2012 a GFD may have prolonged remission in a 5-year-old non-celiac boy with T1D [170]. A GFD was introduced 2-3 weeks after diagnosis when his HbA1c was 7.8%. HbA1c stabilized at 5.8%-6.0% and fasting blood glucose stabilized at 4.0-5.0 mmol/l without insulin therapy. Twenty months after diagnosis he was still without need for daily insulin therapy [170]. Based on this finding a Danish intervention study was performed to evaluate the effects of 12 months on a GFD among children with newly diagnosed T1D [171]. Fifteen children were recruited and after 12 months on a GFD significantly more children were on partial remission (insulin dose-adjusted A1c (IDAA1c) $\leq$ 9) compared to two previous cohorts (p<0.05). Additionally, HbA1c was 21% lower compared to the previous cohorts, but there was no difference in stimulated C peptide [171]. Although the results of this study are promising, the lack of an internal control group is a major limitation.

A Czech, non-randomized controlled intervention trial was conducted to test whether a GFD could contribute to decelerating the decline in beta-cell capacity among children with newly diagnosed T1D without CD [172]. Twenty out of the 26 children, who began adhering to a GFD within a median of 38 days after disease onset, managed to complete the 12 months long intervention with satisfactory compliance. When compared to 19 children remaining on a standard diet, the GFD was associated with a prolonged partial remission period. Furthermore, at 12 months follow-up the intervention group had a lower mean HbA1c (by 0.7%; p = 0.02), a lower IDAA1c (by 1.37; p = 0.01) as well as a tendency toward a lower insulin dose (by 0.15U/kg/day; p = 0.07) compared to the control group [172]. Overall, it seems likely that a GFD can benefit individuals with newly diagnosed or at risk of developing T1D by contributing to preserve the function of the remaining beta-cells.

Besides above publications addressing the effects of a GFD among patients with T1D or predisposition to the disease in the absence of CD, several studies have investigated whether diagnosis of CD and as a result treatment with a GFD improves diabetic control among patients with concomitant CD and T1D [173–191]. Results from these studies are contradicting, but many of the publications are limited by low sample sizes, poor or no information regarding compliance to the GFD and lack of a relevant control group. In addition, these publications also describe the clinical profile of patients with both diseases and in this case the effects of a GFD could also be explained by resolution of inflammation due to treatment of CD.

12.7 Concluding remarks

There is strong evidence supporting that psoriasis, ATD and T1D are all associated with CD, but there seems to be no association between CD and MS. So far, only one clinical trial has studied the effects of a GFD among patients with MS and while the results seem to be highly promising, our capacity to trust this publication is low. Although larger trials are still missing, it is possible that the presence of gluten-related antibodies may be an effect modifying factor that could potentially help identify a subgroup of patients with psoriasis that may benefit from a GFD. The effects of a GFD among non-celiac patients with ATD have not been studied yet, however, a GFD may contribute to reducing titers of thyroid-related antibodies among patients with confirmed CD or tTG seropositivity.

Furthermore, a GFD may also contribute to improving absorption of levothyroxine among patients with CD. With regards to T1D a GFD may improve metabolic control, whereas an effect on diabetes-related autoantibodies seems to be less likely. Altogether, the currently available evidence is not sufficient to allow for the development of gluten-related recommendations for patients with MS, psoriasis, ATD or T1D.

References

[1] P.R. Shewry, SJ. Hey, The contribution of wheat to human diet and health, Food Energy Secur. 4 (2015) 178–202.

[2] SPR. Wheat, J. Exp. Bot. 60 (2009) 1537–1553 https://doi.org/10.1093/jxb/erp058.

[3] P. Koehler, H. Wieser, K. Konitzer, Gluten—The Precipitating Factor, Celiac. Disease and Gluten. (2014) 97–148 https://doi.org/10.1016/b978-0-12-420220-7.00002-x.

[4] B. Niland, BD. Cash, Health Benefits and Adverse Effects of a Gluten-Free Diet in Non-Celiac Disease Patients, Gastroenterol. Hepatol. 14 (2018) 82–91.

[5] B. Lebwohl, D.S. Sanders, PHR. Green, Coeliac disease, Lancet. North. Am. Ed. 391 (2018) 70–81 https://doi.org/10.1016/s0140-6736(17)31796-8.

[6] A. Mokarizadeh, P. Esmaeili, H. Soraya, K. Hassanzadeh, A. Jalili, M. Abdi, et al., Antibody against α-gliadin 33-mer peptide: Is the key initiating factor for development of multiple sclerosis during gluten sensitivity? Journal of Medical Hypotheses and Ideas. 9 (2015) 38–44 https://doi.org/10.1016/j.jmhi.2015.02.002.

[7] A. Lerner, Y. Shoenfeld, T. Matthias, Adverse effects of gluten ingestion and advantages of gluten withdrawal in nonceliac autoimmune disease, Nutr. Rev. 75 (2017) 1046–1058.

[8] H.-D. Belitz, W. Grosch, P. Schieberle, Cereals and Cereal Products., in: H-D Belitz, W Grosch, P Schieberle (Eds.), Food Chemistry., Fourth edition, 2009, pp. 670–745.

[9] Codex Alimentarius International Food Standards, Standard for foods for special dietary use for persons intolerant to gluten, Codex stan, 2008, pp. 1–5. 118-1979.

[10] H. Wieser, Chemistry of gluten. proteins., Food Microbiol. 24 (2007) 115–119 https://doi.org/10.1016/j.fm.2006.07.004.

[11] K.A. Scherf, P. Koehler, H. Wieser, Gluten and wheat sensitivities – An overview, J. Cereal Sci. 67 (2016) 2–11 https://doi.org/10.1016/j.jcs.2015.07.008.

[12] J.F. Ludvigsson, D.A. Leffler, J.C. Bai, F. Biagi, A. Fasano, P.H.R. Green, et al., The Oslo definitions for coeliac disease and related terms, Gut. 62 (2013) 43–52.

[13] S.E.J. Cooper, N.P. Kennedy, B.M. Mohamed, M. Abuzakouk, J. Dunne, G. Byrne, et al., Immunological indicators of coeliac disease activity are not altered by long-term oats challenge, Clinical & Experimental Immunology. 171 (2013) 313–318 https://doi.org/10.1111/cei.12014.

[14] R.D. Fritz, Y. Chen, Oat safety for celiac disease patients: theoretical analysis correlates adverse symptoms in clinical studies to contaminated study oats, Nutr. Res. 60 (2018) 54–67.

[15] L. Jelínková, L. Tučková, J. Cinová, Z. Flegelová, H. Tlaskalová-Hogenová, Gliadin stimulates human monocytes to production of IL-8 and TNF-α through a mechanism involving NF-κB, FEBS Lett. 571 (2004) 81–85 https://doi.org/10.1016/j.febslet.2004.06.057.

[16] L. Tučková, Z. Flegelová, H. Tlaskalová-Hogenová, Z. Zídek, Activation of macrophages by food antigens: enhancing effect of gluten on nitric oxide and cytokine production, J. Leukoc. Biol. 67 (2000) 312–318 https://doi.org/10.1002/jlb.67.3.312.

[17] M. Nikulina, C. Habich, S.B. Flohé, F.W. Scott, H. Kolb, Wheat gluten causes dendritic cell maturation and chemokine secretion, J. Immunol. 173 (2004) 1925–1933.

[18] A. Vojdani, D. Kharrazian, PS. Mukherjee, The prevalence of antibodies against wheat and milk proteins in blood donors and their contribution to neuroimmune reactivities, Nutrients. 6 (2013) 15–36.

[19] A. Vojdani, I. Tarash, Cross-Reaction between Gliadin and Different Food and Tissue Antigens, Food and Nutrition Sciences. 04 (2013) 20–32 https://doi.org/10.4236/fns.2013.41005.

[20] A. Vojdani, Reaction of food-specific antibodies with different tissue antigens, International Journal of Food Science & Technology. 55 (2020) 1800–1815 https://doi.org/10.1111/ijfs.14467.

[21] J. Hollon, E.L. Puppa, B. Greenwald, E. Goldberg, A. Guerrerio, A. Fasano, Effect of gliadin on permeability of intestinal biopsy explants from celiac disease patients and patients with non-celiac gluten sensitivity, Nutrients. 7 (2015) 1565–1576.

[22] S. Drago, R. El Asmar, M. Di Pierro, M. Grazia Clemente, A. Tripathi, A. Sapone, et al., Gliadin, zonulin and gut permeability: Effects on celiac and non-celiac intestinal mucosa and intestinal cell lines, Scand. J. Gastroenterol. 41 (2006) 408–419.

[23] K.M. Lammers, R. Lu, J. Brownley, B. Lu, C. Gerard, K. Thomas, et al., Gliadin induces an increase in intestinal permeability and zonulin release by binding to the chemokine receptor CXCR3, Gastroenterology. 135 (2008) 194–204 e3.

[24] L. Moreno M de, Á. Cebolla, A. Muñoz-Suano, C. Carrillo-Carrion, I. Comino, Á. Pizarro, et al., Detection of gluten immunogenic peptides in the urine of patients with coeliac disease reveals transgressions in the gluten-free diet and incomplete mucosal healing, Gut. 66 (2017) 250–257.

[25] A.L. Hernandez, K.C. O'Connor, D.A. Hafler. Multiple Sclerosis. The Autoimmune Diseases (2014) 735–756. https://doi.org/10.1016/b978-0-12-384929-8.00052-6.

[26] J.N. Brenton, MD. Goldman, A study of dietary modification: Perceptions and attitudes of patients with multiple sclerosis, Mult. Scler. Relat. Disord. 8 (2016) 54–57.

[27] J.M. Beckett, M.-L. Bird, J.K. Pittaway, KD. Ahuja, Diet and Multiple Sclerosis: Scoping Review of Web-Based Recommendations, Interact. J. Med. Res. 8 (2019) e10050.

[28] K.C. Fitzgerald, T. Tyry, A. Salter, S.S. Cofield, G. Cutter, R.J. Fox, et al., A survey of dietary characteristics in a large population of people with multiple sclerosis, Multiple Sclerosis and Related Disorders. 22 (2018) 12–18 https://doi.org/10.1016/j.msard.2018.02.019.

[29] E.M. Leong, S.J. Semple, M. Angley, W. Siebert, J. Petkov, RA. McKinnon, Complementary and alternative medicines and dietary interventions in multiple sclerosis: what is being used in South Australia and why? Complement. Ther. Med. 17 (2009) 216–223.

[30] J. Cosnes, C. Cellier, S. Viola, J.-F. Colombel, L. Michaud, J. Sarles, et al., Incidence of autoimmune diseases in celiac disease: protective effect of the gluten-free diet, Clin. Gastroenterol. Hepatol. 6 (2008) 753–758.

[31] J.-B. Escudié, B. Rance, G. Malamut, S. Khater, A. Burgun, C. Cellier, et al., A novel data-driven workflow combining literature and electronic health records to estimate comorbidities burden for a specific disease: a case study on autoimmune comorbidities in patients with celiac disease, BMC Med. Inform. Decis. Mak. 17 (2017) 140.

[32] S. Foulon, G. Maura, M. Dalichampt, F. Alla, M. Debouverie, T. Moreau, et al., Prevalence and mortality of patients with multiple sclerosis in France in 2012: a study based on French health insurance data, J. Neurol. 264 (2017) 1185–1192.

[33] U. Volta, R. De Giorgio, N. Petrolini, V. Stangbellini, G. Barbara, A. Granito, et al., Clinical findings and anti-neuronal antibodies in coeliac disease with neurological disorders, Scand. J. Gastroenterol. 37 (2002) 1276–1281.

[34] J.F. Ludvigsson, T. Olsson, A. Ekbom, SM. Montgomery, A population-based study of coeliac disease, neurodegenerative and neuroinflammatory diseases, Aliment. Pharmacol. Ther. 25 (2007) 1317–1327.

[35] W.W. Eaton, N.R. Rose, A. Kalaydjian, M.G. Pedersen, PB. Mortensen, Epidemiology of autoimmune diseases in Denmark, J. Autoimmun. 29 (2007) 1–9 https://doi.org/10.1016/j.jaut.2007.05.002.

[36] L. Grode, B.H. Bech, T.M. Jensen, P. Humaidan, I.E. Agerholm, O. Plana-Ripoll, et al., Prevalence, incidence, and autoimmune comorbidities of celiac disease: a nation-wide, population-based study in Denmark from 1977 to 2016, Eur. J. Gastroenterol. Hepatol. 30 (2018) 83–91.

[37] K. Giersiepen, M. Lelgemann, N. Stuhldreher, L. Ronfani, S. Husby, S. Koletzko, et al., Accuracy of diagnostic antibody tests for coeliac disease in children: summary of an evidence report, J. Pediatr. Gastroenterol. Nutr. 54 (2012) 229–241.

[38] ID. Hill, What are the sensitivity and specificity of serologic tests for celiac disease? Do sensitivity and specificity vary in different populations? Gastroenterology. 128 (2005) S25–S32.

[39] F.B. Ahmadabadi, B. Shahbazkhani, A. Tafakhori, A. Khosravi, M.M. Mashhadi, A genetic study of celiac disease in patients with multiple sclerosis in comparison with celiac patients and healthy controls, Govaresh. 22 (2018) 256–260.

[40] M. Banati, P. Csecsei, E. Koszegi, H.H. Nielsen, G. Suto, L. Bors, et al., Antibody response against gastrointestinal antigens in demyelinating diseases of the central nervous system, Eur. J. Neurol. 20 (2013) 1492–1495.

[41] E.B. Roth, E. Bodil Roth, E. Theander, E. Londos, M. Sandberg-Wollheim, L. Å, et al., Pathogenesis of Autoimmune Diseases: Antibodies Against Transglutaminase, Peptidylarginine Deiminase and Protein-bound Citrulline in Primary Sjögren's Syndrome, Multiple Sclerosis and Alzheimer's Disease, Scand. J. Immunol. 67 (2008) 626–631 https://doi.org/10.1111/j.1365-3083.2008.02115.x.

[42] A. Borhani Haghighi, N. Ansari, M. Mokhtari, B. Geramizadeh, KB. Lankarani, Multiple sclerosis and gluten sensitivity, Clin. Neurol. Neurosurg. 109 (2007) 651–653.

[43] P. de Oliveira, D.R. de Carvalho, I.V. Brandi, R. Pratesi, Serological prevalence of celiac disease in Brazilian population of multiple sclerosis, neuromyelitis optica and myelitis, Mult. Scler. Relat. Disord. 9 (2016) 125–128.

[44] M. Khoshbaten, M. Farhoudi, M. Nikanfar, H. Ayromlou, S. Shaafi, S.A. Sadreddini, et al., Celiac disease and multiple sclerosis in the northwest of Iran, Bratisl. Lek. Listy. 113 (2012) 495–497.

[45] A. Nicoletti, F. Patti, S. Lo Fermo, A. Sciacca, P. Laisa, A. Liberto, et al., Frequency of celiac disease is not increased among multiple sclerosis patients, Multiple Sclerosis Journal. 14 (2008) 698–700 https://doi.org/10.1177/1352458507087268.

[46] C. Pengiran Tengah, R.J. Lock, D.J. Unsworth, A.J. Wills, Multiple sclerosis and occult gluten sensitivity, Neurology. 62 (2004) 2326–2327.

[47] K.-L. Reichelt, D. Jensen, IgA antibodies against gliadin and gluten in multiple sclerosis, Acta. Neurol. Scand. 110 (2004) 239–241.

[48] S. Salvatore, S. Finazzi, A. Ghezzi, A. Tosi, A. Barassi, C. Luini, et al., Multiple sclerosis and celiac disease: is there an increased risk? Mult. Scler. 10 (2004) 711–712.

[49] D. Sánchez, L. Tucková, P. Sebo, M. Michalak, A. Whelan, I. Sterzl, et al., Occurrence of IgA and IgG autoantibodies to calreticulin in coeliac disease and various autoimmune diseases, J. Autoimmun. 15 (2000) 441–449.

[50] D.B.-A. Shor, O. Barzilai, M. Ram, D. Izhaky, B.S. Porat-Katz, J. Chapman, et al., Gluten Sensitivity in Multiple Sclerosis, Ann. N. Y. Acad. Sci. 1173 (2009) 343–349 https://doi.org/10.1111/j.1749-6632.2009.04620.x.

[51] R.S. Aboud, M.K. Ismael, HJ. Mohammed, Detection of Human Leukocyte Antigen and Celiac Disease Auto Antibodies in serum of Patients with Multiple Sclerosis, Iraqi. Journal of Science. 55 (2014) 1477–1483.

[52] L. Rodrigo, C. Hernández-Lahoz, D. Fuentes, N. Alvarez, A. López-Vázquez, S. González, Prevalence of celiac disease in multiple sclerosis, BMC Neurology. 11 (2011) https://doi.org/10.1186/1471-2377-11-31.

[53] E. Fernández, Comparison of six human anti-transglutaminase ELISA-tests in the diagnosis of celiac disease in the Saharawi population, World J. Gastroenterol. 11 (2005) 3762 https://doi.org/10.3748/wjg.v11.i24.3762.

[54] M. Hadjivassiliou, D.S. Sanders, RA. Grünewald, Multiple sclerosis and occult gluten sensitivity, Neurology. 64 (2005) 933–934 author reply 933–934.

[55] D.S. Sanders, D. Patel, T.J. Stephenson, A.M. Ward, E.V. McCloskey, M. Hadjivassiliou, et al., A primary care cross-sectional study of undiagnosed adult coeliac disease, Eur. J. Gastroenterol. Hepatol. 15 (2003) 407–413.

[56] R. L, L. Rodrigo, Randomised Clinical Trial Comparing the Efficacy of A Gluten-Free Diet Versus A Regular Diet in A Series of Relapsing-Remitting Multiple Sclerosis Patients, International Journal of Neurology and Neurotherapy. 1 (2014) https://doi.org/10.23937/2378-3001/1/1/1012.

[57] A.K. Irish, C.M. Erickson, T.L. Wahls, L.G. Snetselaar, WG. Darling, Randomized control trial evaluation of a modified Paleolithic dietary intervention in the treatment of relapsing-remitting multiple sclerosis: a pilot study, Degener. Neurol. Neuromuscul Dis. 7 (2017) 1–18.

[58] J.E. Lee, B. Bisht, M.J. Hall, L.M. Rubenstein, R. Louison, D.T. Klein, et al., A Multimodal, Nonpharmacologic Intervention Improves Mood and Cognitive Function in People with Multiple Sclerosis, J. Am. Coll. Nutr. 36 (2017) 150–168.

[59] K.F. Maxwell, T. Wahls, R.W. Browne, L. Rubenstein, B. Bisht, C.A. Chenard, et al., Lipid profile is associated with decreased fatigue in individuals with progressive multiple sclerosis following a diet-based intervention: Results from a pilot study, PLoS One. 14 (2019) e0218075 https://doi.org/10.1371/journal.pone.0218075.

[60] B. Bisht, W.G. Darling, E.T. Shivapour, S.K. Lutgendorf, L.G. Snetselaar, C.A. Chenard, et al., Multimodal intervention improves fatigue and quality of life in subjects with progressive multiple sclerosis: a pilot study, Degener. Neurol. Neuromuscul Dis. 5 (2015) 19–35.

[61] A.J. Thompson, S.E. Baranzini, J. Geurts, B. Hemmer, O. Ciccarelli, Multiple sclerosis, Lancet. 391 (2018) 1622–1636.

[62] S.A. Birlea, M. Serota, DA. Norris, Nonbullous Skin Diseases: Alopecia Areata, Vitiligo, Psoriasis, and Urticaria, The Autoimmune Diseases., 2020, pp. 1211–1234 https://doi.org/10.1016/b978-0-12-812102-3.00061-0.

[63] L. Afifi, M.J. Danesh, K.M. Lee, K. Beroukhim, B. Farahnik, R.S. Ahn, et al., Dietary Behaviors in Psoriasis: Patient-Reported Outcomes from a U.S. National Survey, Dermatol. Ther. 7 (2017) 227–242.

[64] P. Acharya, M. Mathur, Association between psoriasis and celiac disease: A systematic review and meta-analysis, J. Am. Acad. Dermatol. (2019) https://doi.org/10.1016/j.jaad.2019.11.039.

[65] S. Birkenfeld, J. Dreiher, D. Weitzman, AD. Cohen, Coeliac disease associated with psoriasis, Br. J. Dermatol. 161 (2009) 1331–1334 https://doi.org/10.1111/j.1365-2133.2009.09398.x.

[66] J.J. Wu, T.U. Nguyen, K.-Y.T. Poon, LJ. Herrinton, The association of psoriasis with autoimmune diseases, J. Am. Acad. Dermatol. 67 (2012) 924–930.

[67] R. De Bastiani, M. Gabrielli, L. Lora, L. Napoli, C. Tosetti, E. Pirrotta, et al., Association between coeliac disease and psoriasis: Italian primary care multicentre study, Dermatology. 230 (2015) 156–160.

[68] A. Egeberg, C.E.M. Griffiths, L. Mallbris, G.H. Gislason, L. Skov, The association between psoriasis and coeliac disease, Br. J. Dermatol. 177 (2017) e329–e330.

[69] J.F. Merola, V. Herrera, JB. Palmer, Direct healthcare costs and comorbidity burden among patients with psoriatic arthritis in the USA, Clin. Rheumatol. 37 (2018) 2751–2761.

[70] V. Ojetti, J. Aguilar Sanchez, C. Guerriero, B. Fossati, R. Capizzi, C. De Simone, et al., High Prevalence of Celiac Disease in Psoriasis, Am. J. Gastroenterol. 98 (2003) 2574–2575.

[71] M.A. Montesu, C. Dessì-Fulgheri, C. Pattaro, V. Ventura, R. Satta, F. Cottoni, Association between psoriasis and coeliac disease? A case-control study, Acta. Derm. Venereol. 91 (2011) 92–93.

[72] C. Blegvad, A. Egeberg, T.E. Tind Nielsen, G.H. Gislason, C. Zachariae, A.-M. Nybo Andersen, et al., Autoimmune Disease in Children and Adolescents with Psoriasis: A Cross-sectional Study in Denmark, Acta. Derm. Venereol. 97 (2017) 1225–1229.

[73] A. Zohar, A.D. Cohen, H. Bitterman, I. Feldhamer, S. Greenberg-Dotan, I. Lavi, et al., Gastrointestinal comorbidities in patients with psoriatic arthritis, Clin. Rheumatol. 35 (2016) 2679–2684.

[74] D. Aletaha, A.J. Epstein, M. Skup, P. Zueger, V. Garg, R. Panaccione, Risk of Developing Additional Immune-Mediated Manifestations: A Retrospective Matched Cohort Study, Adv. Ther. 36 (2019) 1672–1683.

[75] J.F. Ludvigsson, B. Lindelöf, F. Zingone, C. Ciacci, Psoriasis in a Nationwide Cohort Study of Patients with Celiac Disease, J. Invest. Dermatol. 131 (2011) 2010–2016 https://doi.org/10.1038/jid.2011.162.

[76] T. Iqbal, M.A. Zaidi, G.A. Wells, J. Karsh, Celiac disease arthropathy and autoimmunity study, J. Gastroenterol. Hepatol. 28 (2013) 99–105.

[77] A. Assa, Y. Frenkel-Nir, D. Tzur, L.H. Katz, R. Shamir, Large population study shows that adolescents with celiac disease have an increased risk of multiple autoimmune and nonautoimmune comorbidities, Acta. Paediatr. 106 (2017) 967–972.

[78] P. Collin, T. Reunala, E. Pukkala, P. Laippala, O. Keyrilainen, A. Pasternack, Coeliac disease–associated disorders and survival, Gut. 35 (1994) 1215–1218 https://doi.org/10.1136/gut.35.9.1215.

[79] A. Ventura, G. Magazzù, L. Greco, Duration of exposure to gluten and risk for autoimmune disorders in patients with celiac disease. SIGEP Study Group for Autoimmune Disorders in Celiac Disease, Gastroenterology. 117 (1999) 297–303.

[80] C. Sategna Guidetti, E. Solerio, N. Scaglione, G. Aimo, G. Mengozzi, Duration of gluten exposure in adult coeliac disease does not correlate with the risk for autoimmune disorders, Gut. 49 (2001) 502–505 https://doi.org/10.1136/gut.49.4.502.

[81] S. Bibbò, G.M. Pes, P. Usai-Satta, R. Salis, S. Soro, B.M. Quarta Colosso, et al., Chronic autoimmune disorders are increased in coeliac disease: A case-control study, Medicine (Baltimore). 96 (2017) e8562.

[82] N. Nagui, E. El Nabarawy, D. Mahgoub, H.M. Mashaly, N.E. Saad, D.F. El-Deeb, Estimation of (IgA) anti-gliadin, anti-endomysium and tissue transglutaminase in the serum of patients with psoriasis, Clin. Exp. Dermatol. 36 (2011) 302–304.

[83] S. Akbulut, G. Gür, F. Topal, E. Senel, F.E. Topal, N. Alli, et al., Coeliac disease-associated antibodies in psoriasis, Ann. Dermatol. 25 (2013) 298–303.

[84] S. Singh, G.K. Sonkar, S.S. Usha, Celiac disease-associated antibodies in patients with psoriasis and correlation with HLA Cw6, J. Clin. Lab. Anal. 24 (2010) 269–272.

[85] C.M. Hull, M. Liddle, N. Hansen, L.J. Meyer, L. Schmidt, T. Taylor, et al., Elevation of IgA anti-epidermal transglutaminase antibodies in dermatitis herpetiformis, Br. J. Dermatol. 159 (2008) 120–124.

[86] C. Cardinali, D. Degl'innocenti, M. Caproni, P. Fabbri, Is the search for serum antibodies to gliadin, endomysium and tissue transglutaminase meaningful in psoriatic patients? Relationship between the pathogenesis of psoriasis and coeliac disease, Br. J. Dermatol. 147 (2002) 180–195 https://doi.org/10.1046/j.1365-2133.2002.47947.x.

[87] K. Juzlova, J. Votrubova, M. Dzambova, D. Gopfertova, J. Hercogova, Z. Smerhovsky, Gastrointestinal comorbidities in patients with psoriasis in the Czech Republic: The results of 189 patients with psoriasis and 378 controls, Biomed Pap. Med. Fac. Univ. Palacky. Olomouc. Czech. Repub. 160 (2016) 100–105.

[88] L. Riente, D. Chimenti, F. Pratesi, A. Delle Sedie, S. Tommasi, C. Tommasi, et al., Antibodies to tissue transglutaminase and Saccharomyces cerevisiae in ankylosing spondylitis and psoriatic arthritis, J. Rheumatol. 31 (2004) 920–924.

[89] G. Michaëlsson, G. Kristjánsson, I. Pihl Lundin, E. Hagforsen, Palmoplantar pustulosis and gluten sensitivity: a study of serum antibodies against gliadin and tissue transglutaminase, the duodenal mucosa and effects of gluten-free diet, Br. J. Dermatol. 156 (2007) 659–666.

[90] J. Teichmann, M.J. Voglau, U. Lange, Antibodies to human tissue transglutaminase and alterations of vitamin D metabolism in ankylosing spondylitis and psoriatic arthritis, Rheumatol. Int. 30 (2010) 1559–1563.

[91] P. Weisenseel, A.V. Kuznetsov, T. Ruzicka, JC. Prinz, Palmoplantar pustulosis is not inevitably associated with antigliadin antibodies, Br. J. Dermatol. 156 (2007) 1399–1400 https://doi.org/10.1111/j.1365-2133.2007.07908.x.

[92] W.K. Woo, S.A. McMillan, R.G.P. Watson, W.G. McCluggage, J.M. Sloan, JC. McMillan, Coeliac disease-associated antibodies correlate with psoriasis activity, Br. J. Dermatol. 151 (2004) 891–894.

[93] F. Zamani, S. Alizadeh, A. Amiri, R. Shakeri, M. Robati, S.M. Alimohamadi, et al., Psoriasis and coeliac disease; is there any relationship? Acta. Derm. Venereol. 90 (2010) 295–296.

[94] B.K. Bhatia, J.W. Millsop, M. Debbaneh, J. Koo, E. Linos, W. Liao, Diet and psoriasis, part II: celiac disease and role of a gluten-free diet, J. Am. Acad. Dermatol. 71 (2014) 350–358.

[95] N.A. Kolchak, M.K. Tetarnikova, M.S. Theodoropoulou, A.P. Michalopoulou, DS. Theodoropoulos, Prevalence of antigliadin IgA antibodies in psoriasis vulgaris and response of seropositive patients to a gluten-free diet, J. Multidiscip Healthc. 11 (2018) 13–19.

[96] K.F. Kia, R.P. Nair, R.W. Ike, R. Hiremagalore, J.T. Elder, CN. Ellis, Prevalence of Antigliadin Antibodies in Patients with Psoriasis is Not Elevated Compared with Controls, Am. J. Clin. Dermatol. 8 (2007) 301–305 https://doi.org/10.2165/00128071-200708050-00005.

[97] A. Lesiak, I. Bednarski, M. Pałczyńska, E. Kumiszcza, M. Kraska-Gacka, A. Woźniacka, et al., Are interleukin-15 and -22 a new pathogenic factor in pustular palmoplantar psoriasis? Postepy Dermatol Alergol. 33 (2016) 336–339.

[98] S.J. Sultan, Q.M. Ahmad, ST. Sultan, Antigliadin antibodies in psoriasis, Australas. J. Dermatol. 51 (2010) 238–242 https://doi.org/10.1111/j.1440-0960.2010.00648.x.

[99] A. Kalayciyan, A. Kotogyan, Psoriasis, enteropathy and antigliadin antibodies, Br. J. Dermatol. 154 (2006) 778–779.

[100] J.M. Khan, S.U. Rao, M.U. Rathore, W. Janjua, Association between psoriasis and coeliac disease related antibodies, J. Ayub Med. Coll. Abbottabad 26 (2014) 203–206.

[101] G. Michaelsson, B. Gerden, M. Ottosson, A. Parra, O. Sjoberg, G. Hjelmquist, et al., Patients with psoriasis often have increased serum levels of IgA antibodies to gliadin, Br. J. Dermatol. 129 (1993) 667–673 https://doi.org/10.1111/j.1365-2133.1993.tb03329.x.

[102] A. Damasiewicz-Bodzek, T. Wielkoszyński, Serologic markers of celiac disease in psoriatic patients, J. Eur. Acad. Dermatol. Venereol. 22 (2008) 1055–1061.

[103] J. Skavland, P.R. Shewry, J. Marsh, B. Geisner, JA. Marcusson, In vitro screening for putative psoriasis-specific antigens among wheat proteins and peptides, Br. J. Dermatol. 166 (2012) 67–73.

[104] J. Qiu, Y. Yuan, Y. Li, C. Haley, U.N. Mui, R. Swali, et al., Discovery of IgG4 Anti-Gliadin Autoantibody as a Potential Biomarker of Psoriasis Using an Autoantigen Array, Proteomics Clin. Appl. 14 (2020) e1800114.

[105] A.M. Drucker, A.A. Qureshi, J.M. Thompson, T. Li, E. Cho, Gluten intake and risk of psoriasis, psoriatic arthritis, and atopic dermatitis among United States women, J. Am. Acad. Dermatol. 82 (2020) 661–665.

[106] A.R. Ford, M. Siegel, J. Bagel, K.M. Cordoro, A. Garg, A. Gottlieb, et al., Dietary Recommendations for Adults With Psoriasis or Psoriatic Arthritis From the Medical Board of the National Psoriasis Foundation: A Systematic Review, JAMA Dermatol. 154 (2018) 934–950.

[107] G. Michaëlsson, S. Ahs, I. Hammarström, I.P. Lundin, E. Hagforsen, Gluten-free diet in psoriasis patients with antibodies to gliadin results in decreased expression of tissue transglutaminase and fewer Ki67+ cells in the dermis, Acta. Derm. Venereol. 83 (2003) 425–429.

[108] G. Michaëlsson, B. Gerdén, E. Hagforsen, B. Nilsson, I. Pihl-Lundin, W. Kraaz, et al., Psoriasis patients with antibodies to gliadin can be improved by a gluten-free diet, Br. J. Dermatol. 142 (2000) 44–51.

[109] A.P. Weetman, Thyroid Disease, in: NR Rose, IR Mackay (Eds.), The Autoimmune Diseases., Elsevier, 2014, pp. 557–574.

[110] A. Roy, M. Laszkowska, J. Sundström, B. Lebwohl, P.H.R. Green, O. Kämpe, et al., Prevalence of Celiac Disease in Patients with Autoimmune Thyroid Disease: A Meta-Analysis, Thyroid. 26 (2016) 880–890.

[111] S.M. Ferrari, P. Fallahi, I. Ruffilli, G. Elia, F. Ragusa, S. Benvenga, et al., The association of other autoimmune diseases in patients with Graves' disease (with or without ophthalmopathy): Review of the literature and report of a large series, Autoimmun. Rev. 18 (2019) 287–292 https://doi.org/10.1016/j.autrev.2018.10.001.

[112] X. Sun, L. Lu, R. Yang, Y. Li, L. Shan, Y. Wang, Increased Incidence of Thyroid Disease in Patients with Celiac Disease: A Systematic Review and Meta-Analysis, PLoS One. 11 (2016) e0168708.

[113] A. Carroccio, A. D'Alcamo, F. Cavataio, M. Soresi, A. Seidita, C. Sciumè, et al., High Proportions of People With Nonceliac Wheat Sensitivity Have Autoimmune Disease or Antinuclear Antibodies, Gastroenterology. 149 (2015) 596–603 e1 https://doi.org/10.1053/j.gastro.2015.05.040.

[114] G. Losurdo, M. Principi, A. Iannone, A. Giangaspero, D. Piscitelli, E. Ierardi, et al., Predictivity of Autoimmune Stigmata for Gluten Sensitivity in Subjects with Microscopic Enteritis: A Retrospective Study, Nutrients. 10 (2018) https://doi.org/10.3390/nu10122001.

[115] Z. Zettinig, F. Weissel, V. Dudczak, Dermatitis herpetiformis is associated with atrophic but not with goitrous variant of Hashimoto's thyroiditis, Eur. J. Clin. Invest. 30 (2000) 53–57 https://doi.org/10.1046/j.1365-2362.2000.00590.x.

[116] S. Soni, A. Agarwal, A. Singh, V. Gupta, R. Khadgawat, P.K. Chaturvedi, et al., Prevalence of thyroid autoimmunity in first-degree relatives of patients with celiac disease, Indian. J. Gastroenterol. 38 (2019) 450–455.

[117] R. Ríos León, L. Crespo Pérez, E. Rodríguez de Santiago, G. Roy Ariño, A. De Andrés Martín, C. García Hoz Jiménez, et al., Genetic and flow cytometry analysis of seronegative celiac disease: a cohort study, Scand. J. Gastroenterol. 54 (2019) 563–570.

[118] M. Viljamaa, K. Kaukinen, H. Huhtala, S. Kyrönpalo, P. Rasmussen, P. Collin, Coeliac Disease, autoimmune diseases and gluten exposure, Scand. J. Gastroenterol. 40 (2005) 437–443 https://doi.org/10.1080/00365520510012181.

[119] C.L. Ch'ng, M. Biswas, A. Benton, M.K. Jones, JGC. Kingham, Prospective screening for coeliac disease in patients with Graves' hyperthyroidism using anti-gliadin and tissue transglutaminase antibodies, Clin. Endocrinol. 62 (2005) 303–306.

[120] S. Guliter, F. Yakaryilmaz, Z. Ozkurt, R. Ersoy, D. Ucardag, O. Caglayan, et al., Prevalence of coeliac disease in patients with autoimmune thyroiditis in a Turkish population, World J. Gastroenterol. 13 (2007) 1599 https://doi.org/10.3748/wjg.v13.i10.1599.

[121] R.K. Marwaha, M.K. Garg, N. Tandon, R. Kanwar, A. Narang, A. Sastry, et al., Glutamic acid decarboxylase (anti-GAD) & tissue transglutaminase (anti-TTG) antibodies in patients with thyroid autoimmunity, Indian J. Med. Res. 137 (2013) 82–86.

[122] G. Ravaglia, P. Forti, F. Maioli, U. Volta, G. Arnone, G. Pantieri, et al., Increased prevalence of coeliac disease in autoimmune thyroiditis is restricted to aged patients, Exp. Gerontol. 38 (2003) 589–595.

[123] S. Hadizadeh Riseh, M. Abbasalizad Farhang, M. Mobasseri, M. Asghari Jafarabadi, The Relationship between Thyroid Hormones, Antithyroid Antibodies, Anti-Tissue Transglutaminase and Anti-Gliadin Antibodies in Patients with Hashimoto's Thyroiditis, Acta. Endocrinologica (Bucharest). 13 (2017) 174–179 https://doi.org/10.4183/aeb.2017.174.

[124] S. Sari, E. Yesilkaya, O. Egritas, A. Bideci, B. Dalgic, Prevalence of Celiac Disease in Turkish Children with Autoimmune Thyroiditis, Dig. Dis. Sci. 54 (2009) 830–832 https://doi.org/10.1007/s10620-008-0437-1.

[125] Z. Zhao, J. Zou, L. Zhao, Y. Cheng, H. Cai, M. Li, et al., Celiac Disease Autoimmunity in Patients with Autoimmune Diabetes and Thyroid Disease among Chinese Population, PLoS One. 11 (2016) e0157510.

[126] U. Volta, G. Ravaglia, A. Granito, P. Forti, F. Maioli, N. Petrolini, et al., Coeliac Disease in Patients with Autoimmune Thyroiditis, Digestion. 64 (2001) 61–65 https://doi.org/10.1159/000048840.

[127] M. Hadithi, Coeliac disease in Dutch patients with Hashimoto's thyroiditis and vice versa, World J. Gastroenterol. 13 (2007) 1715 https://doi.org/10.3748/wjg.v13.i11.1715.

[128] J. Jiskra, Z. Límanová, Z. Vaníčková, P. Kocna, IgA and IgG antigliadin, IgA anti-tissue transglutaminase and antiendomysial antibodies in patients with autoimmune thyroid diseases and their relationship to thyroidal replacement therapy, Physiol. Res. 52 (2003) 79–88.

[129] E. Mainardi, A. Montanelli, M. Dotti, R. Nano, G. Moscato, Thyroid-related autoantibodies and celiac disease: a role for a gluten-free diet? J. Clin. Gastroenterol. 35 (2002) 245–248.

[130] A. Mankaï, M. Chadli-Chaieb, F. Saad, L. Ghedira-Besbes, M. Ouertani, H. Sfar, et al., Screening for celiac disease in Tunisian patients with Graves' disease using anti-endomysium and anti-tissue transglutaminase antibodies, Gastroentérologie Clinique et Biologique. 30 (2006) 961–964 https://doi.org/10.1016/s0399-8320(06)73357-7.

[131] M. Mehrdad, F. Mansour-Ghanaei, F. Mohammadi, F. Joukar, S. Dodangeh, R. Mansour-Ghanaei, Frequency of Celiac Disease in Patients with Hypothyroidism, Journal of Thyroid Research. 2012 (2012) 1–6 https://doi.org/10.1155/2012/201538.

[132] Y. Sahin, O. Evliyaoglu, T. Erkan, F.C. Cokugras, O. Ercan, T. Kutlu, The frequency of celiac disease in children with autoimmune thyroiditis, Acta. Gastroenterol. Belg. 81 (2018) 5–8.

[133] N. Sattar, F. Lazare, M. Kacer, L. Aguayo-Figueroa, V. Desikan, M. Garcia, et al., Celiac Disease in Children, Adolescents, and Young Adults with Autoimmune Thyroid Disease, J. Pediatr. 158 (2011) 272–275 e1 https://doi.org/10.1016/j.jpeds.2010.08.050.

[134] B.R. Sharma, A.S. Joshi, P.K. Varthakavi, M.D. Chadha, N.M. Bhagwat, PS. Pawal, Celiac autoimmunity in autoimmune thyroid disease is highly prevalent with a questionable impact, Indian J Endocrinol Metab. 20 (2016) 97–100.

[135] A.C. Spadaccino, D. Basso, S. Chiarelli, M.P. Albergoni, A. D'Odorico, M. Plebani, et al., Celiac disease in North Italian patients with autoimmune thyroid diseases, Autoimmunity. 41 (2008) 116–121.

[136] H. Tuhan, S. Işık, A. Abacı, E. Şimşek, A. Anık, Ö. Anal, et al., Celiac disease in children and adolescents with Hashimoto Thyroiditis, Turk. Pediatri. Ars. 51 (2016) 100–105.

[137] O. Twito, Y. Shapiro, A. Golan-Cohen, Y. Dickstein, R. Ness-Abramof, M. Shapiro, Anti-thyroid antibodies, parietal cell antibodies and tissue transglutaminase antibodies in patients with autoimmune thyroid disease, Arch. Med. Sci. 14 (2018) 516–520.

[138] R. Valentino, S. Savastano, M. Maglio, F. Paparo, F. Ferrara, M. Dorato, et al., Markers of potential coeliac disease in patients with Hashimoto's thyroiditis, Eur. J. Endocrinol. (2002) 479–483 https://doi.org/10.1530/eje.0.1460479.

[139] A. Ventura, M.F. Ronsoni, M.B.C. Shiozawa, E.B. Dantas-Corrêa, S. Canalli MHB da, L. Schiavon L de, et al., Prevalence and clinical features of celiac disease in patients with autoimmune thyroiditis: cross-sectional study, Sao. Paulo. Med. J. 132 (2014) 364–371.

[140] R. Zubarik, E. Ganguly, M. Nathan, J. Vecchio, Celiac disease detection in hypothyroid patients requiring elevated thyroid supplementation: A prospective cohort study, Eur. J. Intern. Med. 26 (2015) 825–829.

[141] G.F. Meloni, P.A. Tomasi, A. Bertoncelli, G. Fanciulli, G. Delitala, T. Meloni, Prevalence of silent celiac disease in patients with autoimmune thyroiditis from Northern Sardinia, J. Endocrinol. Invest. 24 (2001) 298–302.

[142] D. Kaličanin, L. Brčić, A. Barić, S. Zlodre, M. Barbalić, V.T. Lovrić, et al., Evaluation of Correlations Between Food-Specific Antibodies and Clinical Aspects of Hashimoto's Thyroiditis, J. Am. Coll. Nutr. 38 (2019) 259–266 https://doi.org/10.1080/07315724.2018.1503103.

[143] A.J. Naiyer, J. Shah, L. Hernandez, S.-Y. Kim, E.J. Ciaccio, J. Cheng, et al., Tissue transglutaminase antibodies in individuals with celiac disease bind to thyroid follicles and extracellular matrix and may contribute to thyroid dysfunction, Thyroid. 18 (2008) 1171–1178.

[144] K. Sjöberg, R. Wassmuth, S. Reichstetter, K.F. Eriksson, U.B. Ericsson, S. Eriksson, Gliadin antibodies in adult insulin-dependent diabetes–autoimmune and immunogenetic correlates, Autoimmunity. 32 (2000) 217–228.

[145] J. Chong, J.S. Li Voon, K.S. Leong, M. Wallymahmed, R. Sturgess, I.A. MacFarlane, Is coeliac disease more prevalent in young adults with coexisting Type 1 diabetes mellitus and autoimmune thyroid disease compared with those with Type 1 diabetes mellitus alone? Diabet. Med. 19 (2002) 334–337 https://doi.org/10.1046/j.1464-5491.2002.00671.x.

[146] M. Kurien, K. Mollazadegan, D.S. Sanders, JF. Ludvigsson, Celiac Disease Increases Risk of Thyroid Disease in Patients With Type 1 Diabetes: A Nationwide Cohort Study, Diabetes Care. 39 (2016) 371–375.

[147] R. Krysiak, W. Szkróbka, B. Okopień, The Effect of Gluten-Free Diet on Thyroid Autoimmunity in Drug-Naïve Women with Hashimoto's Thyroiditis: A Pilot Study, Exp. Clin. Endocrinol. Diabetes 127 (2019) 417–422 https://doi.org/10.1055/a-0653-7108.

[148] C. Sategna-Guidetti, U. Volta, C. Ciacci, P. Usai, A. Carlino, L. De Franceschi, et al., Prevalence of thyroid disorders in untreated adult celiac disease patients and effect of gluten withdrawal: an Italian multicenter study, Am. J. Gastroenterol. 96 (2001) 751–757.

[149] R. Valentino, S. Savastano, A.P. Tommaselli, M. Dorato, M.T. Scarpitta, M. Gigante, et al., Prevalence of Coeliac Disease in Patients with Thyroid Autoimmunity, Hormone Research in Paediatrics. 51 (1999) 124–127 https://doi.org/10.1159/000023344.

[150] A. Ventura, E. Neri, C. Ughi, A. Leopaldi, A. Città, T. Not, Gluten-dependent diabetes-related and thyroid-related autoantibodies in patients with celiac disease, J. Pediatr. 137 (2000) 263–265.

[151] S. Metso, H. Hyytiä-Ilmonen, K. Kaukinen, K. Huhtala, P. Jaatinen, J. Salmi, et al., Gluten-free diet and autoimmune thyroiditis in patients with celiac disease. A prospective controlled study, Scand. J. Gastroenterol. 47 (2012) 43–48.

[152] C. Virili, G. Bassotti, M.G. Santaguida, R. Iuorio, S.C. Del Duca, V. Mercuri, et al., Atypical celiac disease as cause of increased need for thyroxine: a systematic study, J. Clin. Endocrinol. Metab. 97 (2012) E419–E422.

[153] R. Zubarik, M. Nathan, H. Vahora, E.K. Ganguly, J. Vecchio, Su1435 Hypothyroid Patients Requiring Elevated Doses of Levothyroxine to Maintain a Euthyroid State Should Be Tested for Celiac Disease (CD), Gastroenterology. 146 (2014) S −468 https://doi.org/10.1016/s0016-5085(14)61678-4.

[154] J.C. Antvorskov, K. Josefsen, K. Engkilde, D.P. Funda, K. Buschard, Dietary gluten and the development of type 1 diabetes, Diabetologia. 57 (2014) 1770–1780 https://doi.org/10.1007/s00125-014-3265-1.

[155] J.F. Ludvigsson, J. Ludvigsson, A. Ekbom, SM. Montgomery, Celiac Disease and Risk of Subsequent Type 1 Diabetes: A general population cohort study of children and adolescents, Diabetes Care. 29 (2006) 2483–2488 https://doi.org/10.2337/dc06-0794.

[156] P. Elfström, J. Sundström, JF. Ludvigsson, Systematic review with meta-analysis: associations between coeliac disease and type 1 diabetes, Aliment. Pharmacol. Ther. 40 (2014) 1123–1132.

[157] C. Nederstigt, B.S. Uitbeijerse, L.G.M. Janssen, E.P.M. Corssmit, E.J.P. de Koning, OM. Dekkers, Associated auto-immune disease in type 1 diabetes patients: a systematic review and meta-analysis, Eur. J. Endocrinol. 180 (2019) 135–144.

[158] C. Tiberti, F. Panimolle, M. Bonamico, T. Filardi, L. Pallotta, R. Nenna, et al., Long-standing type 1 diabetes: patients with adult-onset develop celiac-specific immunoreactivity more frequently than patients with childhood-onset diabetes, in a disease duration-dependent manner, Acta Diabetol. 51 (2014) 675–678.

[159] C.M. Smith, C.F. Clarke, L.E. Porteous, H. Elsori, DJS. Cameron, Prevalence of coeliac disease and longitudinal follow-up of antigliadin antibody status in children and adolescents with type 1 diabetes mellitus, Pediatr. Diabetes 1 (2000) 199–203 https://doi.org/10.1046/j.1399543x.2000.010405.x.

[160] C. Tiberti, F. Panimolle, M. Bonamico, B. Shashaj, T. Filardi, F. Lucantoni, et al., IgA anti-transglutaminase autoantibodies at type 1 diabetes onset are less frequent in adult patients and are associated with a general celiac-specific lower immune response in comparison with nondiabetic celiac patients at diagnosis, Diabetes Care. 35 (2012) 2083–2085.

[161] J.C. Antvorskov, T.I. Halldorsson, K. Josefsen, J. Svensson, C. Granström, B.O. Roep, et al., Association between maternal gluten intake and type 1 diabetes in offspring: national prospective cohort study in Denmark, BMJ. 362 (2018) k3547.

[162] N.A. Lund-Blix, G. Tapia, K. Mårild, A.L. Brantsaeter, P.R. Njølstad, G. Joner, et al., Maternal and child gluten intake and association with type 1 diabetes: The Norwegian Mother and Child Cohort Study, PLoS Med. 17 (2020) e1003032.

[163] M.M. Lamb, M.A. Myers, K. Barriga, P.Z. Zimmet, M. Rewers, JM. Norris, Maternal diet during pregnancy and islet autoimmunity in offspring, Pediatr. Diabetes. 9 (2008) 135–141.

[164] S.M. Virtanen, L. Uusitalo, M.G. Kenward, J. Nevalainen, U. Uusitalo, C. Kronberg-Kippilä, et al., Maternal food consumption during pregnancy and risk of advanced β-cell autoimmunity in the offspring, Pediatr. Diabetes. 12 (2011) 95–99.

[165] A.-G. Ziegler, E. Bonifacio, B.-B.S. Group, Age-related islet autoantibody incidence in offspring of patients with type 1 diabetes, Diabetologia. 55 (2012) 1937–1943.

[166] V. Parikka, K. Näntö-Salonen, M. Saarinen, T. Simell, J. Ilonen, H. Hyöty, et al., Early seroconversion and rapidly increasing autoantibody concentrations predict prepubertal manifestation of type 1 diabetes in children at genetic risk, Diabetologia. 55 (2012) 1926–1936.

[167] A.G. Ziegler, M. Rewers, O. Simell, T. Simell, J. Lempainen, A. Steck, et al., Seroconversion to multiple islet autoantibodies and risk of progression to diabetes in children, JAMA. 309 (2013) 2473–2479.

[168] M.-R. Pastore, E. Bazzigaluppi, C. Belloni, C. Arcovio, E. Bonifacio, E. Bosi, Six Months of Gluten-Free Diet Do Not Influence Autoantibody Titers, but Improve Insulin Secretion in Subjects at High Risk for Type 1 Diabetes, The Journal of Clinical Endocrinology & Metabolism. 88 (2003) 162–165 https://doi.org/10.1210/jc.2002-021177.

[169] M. Füchtenbusch, A.-G. Ziegler, M. Hummel, Elimination of dietary gluten and development of type 1 diabetes in high risk subjects, Rev. Diabet. Stud. 1 (2004) 39–41.

[170] S.M. Sildorf, S. Fredheim, J. Svensson, K. Buschard, Remission without insulin therapy on gluten-free diet in a 6-year old boy with type 1 diabetes mellitus, Case Reports. 2012 (2012) bcr0220125878–bcr0220125878 https://doi.org/10.1136/bcr.02.2012.5878.

[171] J. Svensson, S.M. Sildorf, C.B. Pipper, J.N. Kyvsgaard, J. Bøjstrup, F.M. Pociot, et al., Potential beneficial effects of a gluten-free diet in newly diagnosed children with type 1 diabetes: a pilot study, Springerplus. 5 (2016) 994.

[172] V. Neuman, S. Pruhova, M. Kulich, S. Kolouskova, J. Vosahlo, M. Romanova, et al., Gluten-free diet in children with recent-onset type 1 diabetes: A 12-month intervention trial, Diabetes Obes. Metab. 22 (2020) 866–872.

[173] N. Abid, O. McGlone, C. Cardwell, W. McCallion, D. Carson, Clinical and metabolic effects of gluten free diet in children with type 1 diabetes and coeliac disease, Pediatr. Diabetes. 12 (2011) 322–325.

[174] C.L. Acerini, M.L. Ahmed, K.M. Ross, P.B. Sullivan, G. Bird, DB. Dunger, Coeliac disease in children and adolescents with IDDM: clinical characteristics and response to gluten-free diet, Diabet. Med. 15 (1998) 38–44.

[175] R. Amin, N. Murphy, J. Edge, M.L. Ahmed, C.L. Acerini, DB. Dunger, A longitudinal study of the effects of a gluten-free diet on glycemic control and weight gain in subjects with type 1 diabetes and celiac disease, Diabetes Care. 25 (2002) 1117–1122.

[176] S.F. Bakker, M.E. Tushuizen, M.E. von Blomberg, C.J. Mulder, S. Simsek, Type 1 diabetes and celiac disease in adults: glycemic control and diabetic complications, Acta Diabetol. 50 (2013) 319–324.

[177] V.L. Goh, D.E. Estrada, T. Lerer, F. Balarezo, FA. Sylvester, Effect of gluten-free diet on growth and glycemic control in children with type 1 diabetes and asymptomatic celiac disease, J. Pediatr. Endocrinol. Metab. 23 (2010) 1169–1173.

[178] D. Hansen, B. Brock-Jacobsen, E. Lund, C. Bjørn, L.P. Hansen, C. Nielsen, et al., Clinical benefit of a gluten-free diet in type 1 diabetic children with screening-detected celiac disease: a population-based screening study with 2 years' follow-up, Diabetes Care. 29 (2006) 2452–2456.

[179] P. Kaur, A. Agarwala, G. Makharia, S. Bhatnagar, N. Tandon, Effect of gluten-free diet on metabolic control and anthropometric parameters in type 1 diabetes with subclinical celiac disease: A randomized controlled trial, Endocr. Pract. 26 (2020) 660–667.

[180] K. Kaukinen, J. Salmi, J. Lahtela, U. Siljamäki-Ojansuu, A.M. Koivisto, H. Oksa, et al., No effect of gluten-free diet on the metabolic control of type 1 diabetes in patients with diabetes and celiac disease. Retrospective and controlled prospective survey, Diabetes Care. 22 (1999) 1747–1748.

[181] J.S. Leeds, A.D. Hopper, M. Hadjivassiliou, S. Tesfaye, DS. Sanders, High prevalence of microvascular complications in adults with type 1 diabetes and newly diagnosed celiac disease, Diabetes Care. 34 (2011) 2158–2163.

[182] A. Mohn, M. Cerruto, D. Iafusco, F. Prisco, S. Tumini, O. Stoppoloni, et al., Celiac disease in children and adolescents with type I diabetes: importance of hypoglycemia, J. Pediatr. Gastroenterol. Nutr. 32 (2001) 37–40.

[183] P. Narula, L. Porter, J. Langton, V. Rao, P. Davies, C. Cummins, et al., Gastrointestinal symptoms in children with type 1 diabetes screened for celiac disease, Pediatrics. 124 (2009) e489–e495.

[184] A. Pham-Short, K. C Donaghue, G. Ambler, A. K Chan, S. Hing, J. Cusumano, et al., Early elevation of albumin excretion rate is associated with poor gluten-free diet adherence in young people with coeliac disease and diabetes, Diabet. Med. 31 (2014) 208–212.

[185] C. Poulain, C. Johanet, C. Delcroix, C. Lévy-Marchal, N. Tubiana-Rufi, Prevalence and clinical features of celiac disease in 950 children with type 1 diabetes in France, Diabetes Metab. 33 (2007) 453–458.

[186] I. Sanchez-Albisua, J. Wolf, A. Neu, H. Geiger, I. Wäscher, M. Stern, Coeliac disease in children with Type 1 diabetes mellitus: the effect of the gluten-free diet, Diabet. Med. 22 (2005) 1079–1082.

[187] T. Saukkonen, S. Väisänen, H.K. Åkerblom, E Savilahti and the Childhood Diabetes in Finland Study Group. Coeliac disease in children and adolescents with type 1 diabetes: a study of growth, glycaemic control, and experiences of families, Acta Paediatr. 91 (2007) 297–302 https://doi.org/10.1111/j.1651-2227.2002.tb01718.x.

[188] S. Sun, R. Puttha, S. Ghezaiel, M. Skae, C. Cooper, R. Amin, et al., The effect of biopsy-positive silent coeliac disease and treatment with a gluten-free diet on growth and glycaemic control in children with Type 1 diabetes, Diabet. Med. 26 (2009) 1250–1254.

[189] O.I. Saadah, M. Zacharin, A. O'Callaghan, M.R. Oliver, AG. Catto-Smith, Effect of gluten-free diet and adherence on growth and diabetic control in diabetics with coeliac disease, Arch. Dis. Child. 89 (2004) 871–876.

[190] E. Valletta, D. Ulmi, I. Mabboni, F. Tomasselli, L. Pinelli, Early diagnosis and treatment of celiac disease in type 1 diabetes. A longitudinal, case-control study, Pediatr. Med. Chir. 29 (2007) 99–104.

[191] E. Westman, G.R. Ambler, M. Royle, J. Peat, A. Chan, Children with coeliac disease and insulin dependent diabetes mellitus—xgrowth, diabetes control and dietary intake, J. Pediatr. Endocrinol. Metab. 12 (1999) 433–442.

CHAPTER 13

Irritable bowel syndrome

Anupam Rej, David Sanders
Academic Unit of Gastroenterology, Royal Hallamshire Hospital, Sheffield Teaching Hospital NHS Foundation Trust, Sheffield S10 2JF, United Kingdom

13.1 Introduction

The growth of gluten has been exponential over the last decade, with up to a quarter of individuals in the United States wanting gluten free food, with the market worth some 6.6 billion dollars [1]. Whilst the mainstay of management for celiac disease (CD) is a gluten free diet (GFD) [2], there has been a rise in interest in the GFD outside CD, such as in nonceliac gluten sensitivity (NCGS), irritable bowel syndrome (IBS) and healthy individuals.

IBS is common, with a global prevalence of approximately 10% [3]. The pathophysiology of IBS remains poorly understood, but factors thought to be involved include alterations in the gut microbiome, visceral hypersensitivity, gut immunity, enteric sensory and motor disturbances, as well as alterations in pain processing [4].

Diet appears to be a significant trigger for symptoms in individuals with IBS, being reported in up to 84 percent of individuals [5-7]. In addition to this, patients wish to learn about food items to avoid, with approximately 60 percent of individuals being keen on this [8]. As a result, there has been a large interest in the role of dietary therapies in IBS. The aim of this chapter is to provide an overview of a GFD for the management for IBS, as well as exploring the role of other dietary therapies.

13.2 Gluten free diet

The role of gluten outside the diagnosis of CD, has been noted for over 40 years [9], with there being a recent interest in the role of a GFD in IBS, particularly over the last decade [10]. The reporting of symptoms, both intestinal and extra-intestinal, attributed to the ingestion of gluten-based products outside the diagnosis of CD appears to be a global phenomenon, reported at approximately 10% [11]. This has been in view of the growing evidence for the role of a GFD in IBS. There have been several randomized controlled trials to date, assessing the role of a GFD in patients presenting with IBS, seen in Table 13.1 [12-17].

There appears to be a promising role of a GFD from the studies performed to date. For example, an open label prospective study in 41 patients with diarrhea predominant

Coeliac Disease and Gluten-Related Disorders.
DOI: https://doi.org/10.1016/B978-0-12-821571-5.00010-6

Table 13.1 Key randomized controlled trials assessing the gluten free diet in irritable bowel syndrome.

Lead Author for Study	Location	Study Duration	Study number	Findings
Biesiekierski [12]	Australia	6 weeks	34	Significantly worsening of overall symptoms (p=0.047), pain (p=0.016), bloating (p=0.031), satisfaction and stool consistency (p=0.024) with gluten within 1 week
Carroccio [19]	Italy	5 weeks	920	30% patients (n=276) identified as wheat sensitive
Vazquez-Roque [13]	USA	4 weeks	45	Greater bowel movements per day on GCD (p=0.04), greater in HLA-DQ2/8 positive than negative patients (p=0.019). GCD associated with higher small bowel permeability
Biesiekierski [14]	Australia	5 weeks	37	No dose dependent effects of gluten noted
Shahbazkhani [15]	Iran	5 weeks	72	Worsening of overall symptoms, including satisfaction with stool consistency, tiredness, nausea and bloating with gluten than placebo (p=0.001)
Zanwar [16]	India	4 weeks	60	Gluten led to worse abdominal pain, bloating and tiredness compared to placebo (p<0.05)
Barone [17]	Italy	9 weeks	26	Worsening of symptoms with an increase in VAS score of >30% compared to placebo in 46% of individuals following gluten challenge

GCD, gluten containing diet; HLA, human leukocyte antigen; VAS, visual analogue scale.

IBS (IBS-D), demonstrated a both clinical and statistically significant improvement in IBS symptoms severity score (IBS-SSS) after 6 weeks of a dietitian led GFD, with an improvement in the IBS-SSS from 286 to 131 (p<0.001) [18].

Studies have been performed to assess the mechanistic action in which gluten may trigger symptoms. A 4-week randomized controlled trial (RCT), in 45 patients with IBS-D, demonstrated that individuals on a gluten containing diet (GCD) had significantly more bowel movements per day (p=0.04), with an association with higher small bowel permeability, in comparison to individuals on a GFD [13]. Alterations in tight junction proteins following gluten in IBS has been demonstrated, which may potentially explain changes in intestinal permeability seen following the administration of gluten in individuals with IBS [20]. However, further studies are required to explore the mechanism of action of gluten in symptom generation in IBS.

Whilst there appears to be utility of using a GFD to manage patients with IBS, there remains a number of areas of uncertainty, which are discussed below.

13.2.1 Nutritional adequacy

The effects of a GFD on nutritional adequacy in individuals with IBS is unclear, with limited data available. However, effects on nutritional adequacy of a GFD in CD have been explored, highlighting potential concerns for implementation of this diet in individuals with IBS. Deficiencies of folate, calcium, iron, magnesium, zinc and fiber intake have been demonstrated using a GFD [21-23]. In addition to this, there have been concerns of a GFD with regards to the accumulation of heavy metals, with higher levels of total arsenic, mercury, lead and cadmium on individuals on a GFD [24]. Whilst there are potential concerns of a GFD in IBS, it must be noted that any diet, including a habitual diet, may have macronutrient or micronutrient deficiencies. For example, 95% of the general population are nonadherent to fiber recommendations [25], with dietary inadequacies likely related to habitual poor food choices in addition to potential inadequacies of the GFD itself [23]. Well-designed studies, comparing nutritional intake of a GFD to a habitual diet in patients with IBS, are required before any conclusions can be drawn.

13.2.2 Microbiota

There is uncertainty in the literature with regards to the effects of a GFD in IBS on alterations in gut microbiota composition. Studies have been performed in healthy individuals, as well as individuals with CD, assessing the impact of a GFD on gut microbiota composition [26-29]. It has been demonstrated that a GFD may lead to reductions in potentially beneficial gut bacterial populations, such as Bifidobacterium and Faecalibacterium prausnitzii [29,30]. Whilst the etiology for these changes is unclear, it has been hypothesized that as gluten exerts a prebiotic action, with its exclusion resulting in changes to gut microbiota composition, leading to an interest in the role of pre- or probiotics to supplement a GFD [31]. It is important to note however, that findings of the effect of a GFD on the gut microbiota must be interpreted with caution. The gut microbiota composition and metabolomic activity is highly individualized, as well as there being variability in gut microbiota composition over time [32]. Therefore, results from studies cannot be extrapolated from one population to others [32].

13.2.3 Wheat components

Whilst gluten has been postulated as a trigger for symptoms in individuals with IBS, it has also been suggested that other components of wheat other than gluten may trigger symptoms in IBS. A large proportion of individuals with IBS may be sensitive to wheat, as demonstrated by a large study in 920 patients [19]. During this study, 276 patients (30%) were identified to have been suffering from wheat sensitivity, highlighting the utility of a wheat free diet in IBS [19].

Whilst gluten is the main storage protein in wheat [33], several other components are also present in wheat which may trigger symptoms, and include alpha-amylase trypsin inhibitors (ATIs) [34], wheat germ agglutinins (WGAs) [35], and fructans [36]. Symptom generation maybe due to a nocebo response [37].

ATIs may lead to intestinal inflammation through the activation of toll-like receptor 4, leading to the release of pro-inflammatory cytokines [34]. WGAs are capable of crossing the intestinal barrier, have been found to affect enterocyte permeability and induce inflammatory responses by immune cells. However, human data demonstrating the effect of WGAs on inflammatory markers are lacking [35].

Fructans, an oligo-saccharide, are a type of fermentable oligo-, di-, mono- saccharide and polyol (FODMAP). It has been suggested that fructans, rather than gluten, may trigger symptoms in individuals [36]. Individuals with IBS may have an anticipatory nocebo response with any dietary therapy, and it is therefore essential that studies are well designed to mitigate this [37].

13.2.4 Long term outcomes

Currently, there is a lack of data assessing long term outcomes for the GFD in IBS. Whilst there is a lack of data on long term outcomes, it appears that patients are keen to continue to use the GFD to manage symptoms of IBS. In individuals who took part in an open-label study assessing the GFD in IBS-D, it was noted that the majority (72%) of individuals were still on a GFD after completion of the study at 18 months, with similar anthropometric and biochemical features in comparison to baseline [18]. However, long term studies are required assessing the efficacy of a GFD in IBS.

13.2.5 Adherence

Adherence to a GFD in IBS has been noted to be around 60% [38], with a similar reported adherence rate to individuals with CD [39]. In addition to this, patients appear to have a preference for this diet over more complex diets such as the low FODMAP diet [40], with significantly more individuals having tried the GFD rather than the low

FODMAP diet [41], highlighting that the GFD may be easy to implement. However, it is worth noting that the ease of implementation of these diets have not been assessed head to head in the literature.

13.3 Low FODMAP diet

Over the last decade, there has been a surge in interest in the role of a low FODMAP diet (LFD) in managing patients with IBS, with this diet being recommended as one of the key dietary therapies to manage IBS [42]. FODMAPs are short chain carbohydrates, which are poorly absorbed, increasing small bowel water content and intestinal transit, and fermented in the large bowel leading to intestinal gas production and distention [43,44]. Whilst healthy individuals and individuals with IBS develop luminal distention following FODMAPs, as demonstrated by MRI imaging [45], it is thought that colonic hypersensitivity is the likely pathological mechanism of symptom generation. In view of this, a LFD is thought to reduce this luminal distention and improve patient symptoms.

There have been several RCTs evaluating the LFD in IBS, demonstrating its benefit [46-52]. The majority of studies have evaluated the short-term effects of the low FODMAP diet, assessing the low FODMAP diet in the elimination phase, when there is a strict reduction of all high FODMAP-containing foods. The response rate to the effects of the LFD in the short-term has been reported at between 50 to 76 percent [10]. Whilst this has been promising, an important aspect of the LFD is the reintroduction phase, where FODMAPs are gradually reintroduced to tolerance [53]. In view of that, there have been an emerging number of studies evaluating the efficacy of the LFD in the long term, demonstrating sustained symptom relief in the long term [54-58]. It also appears that the majority of individuals are following an adapted LFD rather than a strict LFD in the long term, which is recommended in the literature due to the restrictive nature of the strict LFD [54,56,57]. Table 13.2 outlines the key studies assessing the LFD in IBS.

Similar to the GFD, there remains uncertainty to certain aspects of the LFD which are discussed below.

13.3.1 Nutritional adequacy

Total energy intake has been shown to be reduced in individuals on the LFD, compared to traditional dietary advice [48,49,52], as well as a reduction in calcium intake in the short term strict LFD phase [46]. The effects of inadequate nutrient intake of the LFD in the strict reduction phase may be expected, as the LFD involves the substitution of selected food across a large number of food groups [59]. As a result of this, the personalization phase is required, which maybe more nutritionally adequate. A study, assessing the LFD at long term follow up demonstrated this [54], although further studies required to validate this.

Table 13.2 Key Trials Assessing the low FODMAP diet in Irritable Bowel Syndrome.

Study	Location	Study Duration	Study Number	Findings
Short term studies				
Staudacher [46]	United Kingdom	4 weeks	41	Individuals on LFD reported a greater adequate control of symptoms in comparison to habitual diet (68% vs 23%, p=0.005)
Halmos [64]	Australia	21 days	38	Individuals with IBS had lower gastrointestinal symptom scores whilst on a LFD compared to an Australian diet (p<0.001)
Bohn [48]	Sweden	6 weeks	75	Severity of IBS symptoms reduced following the LFD and traditional IBS diet (p<0.0001), with no difference between groups (p=0.62)
Eswaran [49]	United States	4 weeks	92	Adequate relief of IBS-D symptoms demonstrated in 52% of LFD, with no difference between modified NICE diet (p=0.13)
McIntosh [50]	Canada	3 weeks	40	IBS-SSS improved on LFD (289 vs 208, p<0.001) but not on high FODMAP diet (271 vs 290, p=0.57)
Hustoft [51]	Norway	6 weeks	20	Improvement of all symptoms following 3 weeks of LFD (p<0.001)
Staudacher [47]	United Kingdom	4 weeks	104	Higher proportion of individuals on LFD had adequate symptom relief compared to sham diet (61% vs 38%, p=0.042)
Zahedi [52]	Iran	6 weeks	110	Greater improvement in IBS-SSS on low FODMAP diet versus traditional dietary advice (IBS-SSS at 6 weeks 108 vs 150, p<0.001)
Long term studies				
de Roest [58]	New Zealand	16 months	90	Improvement in abdominal pain, bloating, flatulence and diarrhea (p<0.001) following LFD
Peters [56]	Australia	6 months	74	Improvement in overall symptoms on LFD, with no difference between LFD and hypnotherapy/combination (p=0.67)
Maagaard [57]	Denmark	16 months	180	Significant reduction in chronic continuous disease course in IBS group following LFD (p<0.001)
Harvie [55]	New Zealand	6 months	50	Reduction in IBS-SSS on LFD vs normal diet at 3 months (p<0.0002), sustained at 6 months.
O'Keeffe [54]	United Kingdom	6–18 months	103	Satisfactory relief of symptoms following LFD reported at 61% at short term follow up and 57% at long term follow up

LFD, low fermentable oligo-, di-, mono– saccharides and polyols diet; IBS, irritable bowel syndrome; IBS-D, diarrhea–predominant irritable bowel syndrome; NICE, national institute for health and care excellence; IBS-SSS, irritable bowel syndrome symptom severity score; FODMAP, fermentable oligo-, di, mono-saccharides, and polyols.

13.3.2 Microbiota

The LFD is thought to potentially result in a reduction of natural prebiotics, leading to a change in the gut microbiota, which lack their natural substrate for their metabolomic activity [60]. A reduction in total bacterial abundance [61], as well as a reduction in beneficial bacteria such as Bifidobacterium has been demonstrated in the short term on the LFD [46]. In view of this probiotics have been suggested as a potential solution to be used in conjunction with the LFD, with a placebo-controlled study demonstrating this [47]. Recent research has suggested that responsiveness to a LFD may be predicted by bacterial profiles and their metabolomic activity [62,63].

Like the GFD, interpretations of gut microbiota analysis must be taken with caution. A number of factors may limit interpretation, including different sample collections, storage and analysis. Also, the FODMAP content is likely to be heterogenous between different studies [60].

13.3.3 Adherence

Adherence to LFD has been reported to be around 75% [58,65]. Whilst reported adherence is high, the LFD can be challenging to correctly implement, in view of the complexity of the diet. The LFD should be ideally implemented by a dietitian [10]. However, it appears that most gastroenterologists do not refer patients to a dietitian for LFD advice, and in fact provide an educational handout [41], which goes against what is recommended in the literature [66]. Indeed, information provided by general practitioners and gastroenterologists can be challenging and hard to implement for patients, highlighting that patients may not fully understand or implement this diet correctly [67]. Also, this diet may be expensive, with the cost of the diet greater for the majority of individuals following the LFD [68].

13.3.4 Efficacy versus other dietary and nondietary therapies

A number of studies have assessed the LFD to other dietary therapies, with differing outcomes [48,49,52]. A multi-center, parallel, single-blind study in Sweden in 75 patients with IBS demonstrated an improvement in IBS symptom severity following both the LFD and traditional dietary advice (p<0.0001), but no differences between interventions (p=0.62) [48]. Similar findings were also noted in a study in the United States (US), where there was no difference in the proportion of responders to both the LFD and modified National Institute for Health Care and Excellence (mNICE) diet [49]. In contrast, a study in Iran demonstrated that individuals on a LFD had greater benefits in IBS improvement in comparison to general dietary advice [52]. The difference in results seen may in part be due to the heterogeneity of the content of the dietary therapies delivered, with some studies including the reduction of gas performing foods as part of traditional dietary advice [69].

In addition to dietary therapies, it is unclear whether the LFD is more efficacious in comparison to nondietary therapies. Studies have been performed demonstrating equal efficacy of gut-directed hypnotherapy and yoga, in comparison to the LFD [56,70].

13.4 Traditional dietary advice

Whilst there has been a huge interest in the role of the GFD and LFD to manage patients with IBS, traditional dietary advice remains commonplace to manage symptoms of IBS, and is the first line management in some national guidelines [71,72]. Advice for this diet includes regular meals, alteration of fiber intake, adequate fluid intake, assessment of alcohol and caffeine intake, reduction of fat intake, as well as assessment of spice intake [71,72]. The basis for this guidance comes from clinical experience and small studies, with a lack of RCTs assessing the efficacy of each component of this diet [71]. Despite this, there appears to be merit of using this diet, with some studies demonstrating efficacy of this diet on a similar level to the LFD [48,49]. However, there is conflicting data on this [52], with further head-to-head studies against traditional dietary advice required, before any conclusions may be drawn.

13.5 Challenges of dietary studies

Whilst there have been several studies demonstrating the efficacy of dietary therapies for IBS, in particular the GFD and LFD, the evidence to recommend these dietary therapies is still graded as very low or insufficient [73]. This may in part be due to several different study designs used, differing population groups being assessed and different end points being used. It has becoming increasingly challenging to perform blinded studies, with a high awareness of the GFD and LFD, as well as high placebo and nocebo responses [37,74]. An example of the importance of study design is highlighted by the studies performed by Biesiekierski and colleagues [12,14]. An initial study by this group demonstrated the worsening of symptoms in individuals with IBS after having received gluten in comparison to placebo [12]. However, subsequent to this, another study by the same group demonstrated that there was no effect of gluten in individuals who had been placed on a LFD [14]. These findings could potentially be explained by the study design; participants received high gluten, low gluten or placebo challenges, which may have led to an anticipatory nocebo response. Also, individuals had a high visual analogue scale (VAS) at baseline, which may not be truly representative of this patient group [14].

Whilst it has been suggested that less stringent criteria maybe used to assess dietary trials in comparison to drug trials, the design of these studies is crucial [37]. In view of this, criteria have been developed to aid researchers in the development of robust dietary trials [74]. It is important to note that dietary therapies are not without their potential risks, and therefore robust trials are required to validate the use of these therapies.

Figure 13.1 Dietary options for Individuals with irritable bowel syndrome.

13.6 Choice of dietary therapy

There appears to be a growing evidence for the use of several dietary therapies in IBS, with traditional dietary advice, LFD and GFD being commonly used therapies in view of their evidence base (Fig. 13.1). In addition to this, the wheat free diet may be beneficial in IBS, as highlighted by a large study in individuals with IBS. However, it is still unclear which dietary therapy is the optimal therapy for patients with IBS. It is likely that individuals with IBS are a varied population group, with dietary therapies being required to be tailored to the individual.

In terms of ease of use, traditional dietary advice and a GFD are likely to be easier to implement than a restrictive LFD, particularly in its initial phase. Also, it appears that patients seem to prefer a GFD over a LFD [40]. However, it has been suggested that fructan reduction maybe the key to symptom relief in patients with IBS, rather than gluten [36]. This has led to consideration of a more personalized 'bottom up' approach, which may be a less restrictive approach to the LFD [75]. One way to achieve this maybe through the GFD, which would lead to a fructan reduction, as wheat is the largest source of fructans, with bread being the major dietary source [76]. It appears that this is occurring already, with a long-term study assessing the LFD demonstrating fructans to be a key component being reduced in the long term, with gluten free products being one of the components used to achieve this [54].

However, it is important to note that one dietary therapy cannot be advocated over another currently. There is conflicting data in the literature comparing traditional dietary advice to the LFD [48,49,52], with no head to head trials comparing traditional dietary advice, LFD and GFD. In view of this patient choice is important in determining dietary therapies, as well as identifying triggers. Whilst patient choice is important, dietetic input is essential to deliver these therapies. Without their input, obsessive behaviors such as orthorexia nervosa may develop, which can be prevalent in this cohort, with the need to identify patients at risk of eating disorders, by using tools such as the SCOFF questionnaire [42,77].

13.7 Conclusion

There has been a huge interest in the role of a GFD, with increasing evidence of its efficacy in individuals with IBS. Likewise, over the last decade, there has been growing evidence for the use of the LFD. This has given patients several options to help manage their IBS symptoms using dietary therapies. Whilst these dietary therapies have developed a growing evidence base, several questions remain, including long-term effects of these diets, as well as effects on nutritional adequacy and gut microbiota composition. Patient choice is essential in the delivery of these therapies, in conjunction with expert dietetic support. There is emerging interest in implementing less restrictive dietary therapies, with the GFD at the forefront to potentially deliver this. However, there is currently no evidence to confidently advocate one dietary therapy over another, with further research required.

References

[1] I. Aziz, K. Dwivedi, DS. Sanders, From coeliac disease to noncoeliac gluten sensitivity; should everyone be gluten free? Curr. Opin. Gastroenterol. 32 (2) (2016) 120–127.

[2] B. Lebwohl, D.S. Sanders, PHR. Green, Coeliac disease, Lancet 391 (10115) (2018) 70–81.

[3] R.M. Lovell, AC. Ford, Global prevalence of and risk factors for irritable bowel syndrome: a meta-analysis, Clin. Gastroenterol. Hepatol. 10 (7) (2012) 712–721.e714.

[4] DA. Drossman, Functional Gastrointestinal Disorders: History, Pathophysiology, Clinical Features and Rome IV, Gastroenterology (2016).

[5] M. Simrén, A. Månsson, A.M. Langkilde, et al., Food-related gastrointestinal symptoms in the irritable bowel syndrome, Digestion 63 (2) (2001) 108–115.

[6] K.W. Monsbakken, P.O. Vandvik, PG. Farup, Perceived food intolerance in subjects with irritable bowel syndrome– etiology, prevalence and consequences, Eur. J. Clin. Nutr. 60 (5) (2006) 667–672.

[7] L. Böhn, S. Störsrud, H. Törnblom, U. Bengtsson, M. Simrén, Self-reported food-related gastrointestinal symptoms in IBS are common and associated with more severe symptoms and reduced quality of life, Am. J. Gastroenterol. 108 (5) (2013) 634–641.

[8] A. Halpert, C.B. Dalton, O. Palsson, et al., What patients know about irritable bowel syndrome (IBS) and what they would like to know. National Survey on Patient Educational Needs in IBS and development and validation of the Patient Educational Needs Questionnaire (PEQ), Am. J. Gastroenterol. 102 (9) (2007) 1972–1982.

[9] A. Ellis, BD. Linaker, Non-coeliac gluten sensitivity? Lancet 1 (8078) (1978) 1358–1359.

[10] A. Rej, A. Avery, A.C. Ford, et al., Clinical application of dietary therapies in irritable bowel syndrome, J. Gastrointestin. Liver Dis. 27 (3) (2018) 307–316.

[11] I. Aziz, The Global Phenomenon of Self-Reported Wheat Sensitivity, Am. J. Gastroenterol. 113 (7) (2018) 945–948.

[12] J.R. Biesiekierski, E.D. Newnham, P.M. Irving, et al., Gluten causes gastrointestinal symptoms in subjects without celiac disease: a double-blind randomized placebo-controlled trial, Am. J. Gastroenterol. 106 (3) (2011) 508–514, quiz 515.

[13] M.I. Vazquez-Roque, M. Camilleri, T. Smyrk, et al., A controlled trial of gluten-free diet in patients with irritable bowel syndrome-diarrhea: effects on bowel frequency and intestinal function, Gastroenterology. 144 (5) (2013) 903–911.e903.

[14] J.R. Biesiekierski, S.L. Peters, E.D. Newnham, O. Rosella, J.G. Muir, P.R. Gibson, No effects of gluten in patients with self-reported non-celiac gluten sensitivity after dietary reduction of fermentable, poorly absorbed, short-chain carbohydrates, Gastroenterology. 145 (2) (2013) 320–328.e321-323.

[15] B. Shahbazkhani, A. Sadeghi, R. Malekzadeh, et al., Non-Celiac Gluten Sensitivity Has Narrowed the Spectrum of Irritable Bowel Syndrome: A Double-Blind Randomized Placebo-Controlled Trial, Nutrients 7 (6) (2015) 4542–4554.

[16] V.G. Zanwar, S.V. Pawar, P.A. Gambhire, et al., Symptomatic improvement with gluten restriction in irritable bowel syndrome: a prospective, randomized, double blinded placebo controlled trial, Intest Res 14 (4) (2016) 343–350.

[17] M. Barone, E. Gemello, M.T. Viggiani, et al., Evaluation of Non-Celiac Gluten Sensitivity in Patients with Previous Diagnosis of Irritable Bowel Syndrome: A Randomized Double-Blind Placebo-Controlled Crossover Trial, Nutrients. 12 (3) (2020).

[18] I. Aziz, N. Trott, R. Briggs, J.R. North, M. Hadjivassiliou, DS. Sanders, Efficacy of a Gluten-Free Diet in Subjects With Irritable Bowel Syndrome-Diarrhea Unaware of Their HLA-DQ2/8 Genotype, Clin. Gastroenterol. Hepatol. 14 (5) (2016) 696–703.e691.

[19] A. Carroccio, P. Mansueto, G. Iacono, et al., Non-celiac wheat sensitivity diagnosed by double-blind placebo-controlled challenge: exploring a new clinical entity, Am. J. Gastroenterol. 107 (12) (2012) 1898–1906 quiz 1907.

[20] R.L. Wu, M.I. Vazquez-Roque, P. Carlson, et al., Gluten-induced symptoms in diarrhea-predominant irritable bowel syndrome are associated with increased myosin light chain kinase activity and claudin-15 expression, Lab. Invest. 97 (1) (2017) 14–23.

[21] D. Wild, G.G. Robins, V.J. Burley, PD. Howdle, Evidence of high sugar intake, and low fibre and mineral intake, in the gluten-free diet, Aliment. Pharmacol. Ther. 32 (4) (2010) 573–581.

[22] T. Thompson, M. Dennis, L.A. Higgins, A.R. Lee, MK. Sharrett, Gluten-free diet survey: are Americans with coeliac disease consuming recommended amounts of fibre, iron, calcium and grain foods? J. Hum. Nutr. Diet. 18 (3) (2005) 163–169.

[23] S.J. Shepherd, PR. Gibson, Nutritional inadequacies of the gluten-free diet in both recently-diagnosed and long-term patients with coeliac disease, J. Hum. Nutr. Diet. 26 (4) (2013) 349–358.

[24] S.L. Raehsler, R.S. Choung, E.V. Marietta, JA. Murray, Accumulation of Heavy Metals in People on a Gluten-Free Diet, Clin. Gastroenterol. Hepatol. 16 (2) (2018) 244–251.

[25] E. Bennett, S.A.E. Peters, M. Woodward, Sex differences in macronutrient intake and adherence to dietary recommendations: findings from the UK Biobank, BMJ Open. 8 (4) (2018) e020017.

[26] M.J. Bonder, E.F. Tigchelaar, X. Cai, et al., The influence of a short-term gluten-free diet on the human gut microbiome, Genome. Med. 8 (1) (2016) 45.

[27] G. De Palma, I. Nadal, M. Medina, et al., Intestinal dysbiosis and reduced immunoglobulin-coated bacteria associated with coeliac disease in children, BMC Microbiol. 10 (2010) 63.

[28] S. Schippa, V. Iebba, M. Barbato, et al., A distinctive 'microbial signature' in celiac pediatric patients, BMC Microbiol. 10 (2010) 175.

[29] G. De Palma, I. Nadal, M.C. Collado, Y. Sanz, Effects of a gluten-free diet on gut microbiota and immune function in healthy adult human subjects, Br. J. Nutr. 102 (8) (2009) 1154–1160.

[30] C.V. Ferreira-Halder, A.V.S. Faria, SS. Andrade, Action and function of Faecalibacterium prausnitzii in health and disease, Best. Pract. Res. Clin. Gastroenterol. 31 (6) (2017) 643–648.

[31] S. Reddel, L. Putignani, F. Del Chierico, The Impact of Low-FODMAPs, Gluten-Free, and Ketogenic Diets on Gut Microbiota Modulation in Pathological Conditions, Nutrients. 11 (2) (2019).

[32] J.F. Garcia-Mazcorro, G. Noratto, JM. Remes-Troche, The Effect of Gluten-Free Diet on Health and the Gut Microbiota Cannot Be Extrapolated from One Population to Others, Nutrients. 10 (10) (2018).

[33] M.I. Pinto-Sánchez, EF. Verdú, Non-coeliac gluten sensitivity: are we closer to separating the wheat from the chaff? Gut. 65 (12) (2016) 1921–1922.

[34] Y. Junker, S. Zeissig, S.J. Kim, et al., Wheat amylase trypsin inhibitors drive intestinal inflammation via activation of toll-like receptor 4, J. Exp. Med. 209 (13) (2012) 2395–2408.

[35] K. de Punder, L. Pruimboom, The dietary intake of wheat and other cereal grains and their role in inflammation, Nutrients. 5 (3) (2013) 771–787.

[36] G.I. Skodje, V.K. Sarna, I.H. Minelle, et al., Fructan, Rather Than Gluten, Induces Symptoms in Patients With Self-reported Non-celiac Gluten Sensitivity, Gastroenterology. (2017).

[37] C.K. Yao, P.R. Gibson, S.J. Shepherd, Design of clinical trials evaluating dietary interventions in patients with functional gastrointestinal disorders, Am. J. Gastroenterol. 108 (5) (2013) 748–758.

[38] C. Barmeyer, M. Schumann, T. Meyer, et al., Long-term response to gluten-free diet as evidence for non-celiac wheat sensitivity in one third of patients with diarrhea-dominant and mixed-type irritable bowel syndrome, Int. J. Colorectal. Dis. 32 (1) (2017) 29–39.

[39] S.M. Barratt, J.S. Leeds, D.S. Sanders, Quality of life in Coeliac Disease is determined by perceived degree of difficulty adhering to a gluten-free diet, not the level of dietary adherence ultimately achieved, J. Gastrointestin. Liver. Dis. 20 (3) (2011) 241–245.

[40] D. Paduano, A. Cingolani, E. Tanda, P. Usai, Effect of Three Diets (Low-FODMAP, Gluten-free and Balanced) on Irritable Bowel Syndrome Symptoms and Health-Related Quality of Life, Nutrients. 11 (7) (2019).

[41] A. Lenhart, C. Ferch, M. Shaw, W.D. Chey, Use of Dietary Management in Irritable Bowel Syndrome: Results of a Survey of Over 1500 United States Gastroenterologists, J. Neurogastroenterol. Motil. 24 (3) (2018) 437–451.

[42] H. Mitchell, J. Porter, P.R. Gibson, J. Barrett, M. Garg, Review article: implementation of a diet low in FODMAPs for patients with irritable bowel syndrome-directions for future research, Aliment. Pharmacol. Ther. 49 (2) (2019) 124–139.

[43] K. Murray, V. Wilkinson-Smith, C. Hoad, et al., Differential effects of FODMAPs (fermentable oligo-, di-, mono-saccharides and polyols) on small and large intestinal contents in healthy subjects shown by MRI, Am. J. Gastroenterol. 109 (1) (2014) 110–119.

[44] R. Spiller, How do FODMAPs work? J. Gastroenterol. Hepatol. 32 (Suppl 1) (2017) 36–39.

[45] G. Major, S. Pritchard, K. Murray, et al., Colon Hypersensitivity to Distension, Rather Than Excessive Gas Production, Produces Carbohydrate-Related Symptoms in Individuals With Irritable Bowel Syndrome, Gastroenterology. 152 (1) (2017) 124–133.e122.

[46] H.M. Staudacher, M.C. Lomer, J.L. Anderson, et al., Fermentable carbohydrate restriction reduces luminal bifidobacteria and gastrointestinal symptoms in patients with irritable bowel syndrome, J. Nutr. 142 (8) (2012) 1510–1518.

[47] H.M. Staudacher, M.C.E. Lomer, F.M. Farquharson, et al., Diet Low in FODMAPs Reduces Symptoms in Patients with Irritable Bowel Syndrome and Probiotic Restores Bifidobacterium Species: a Randomized Controlled Trial, Gastroenterology (2017).

[48] L. Böhn, S. Störsrud, T. Liljebo, et al., Diet low in FODMAPs reduces symptoms of irritable bowel syndrome as well as traditional dietary advice: a randomized controlled trial, Gastroenterology. 149 (6) (2015) 1399–1407.e1392.

[49] S.L. Eswaran, W.D. Chey, T. Han-Markey, S. Ball, K. Jackson, A Randomized Controlled Trial Comparing the Low FODMAP Diet vs. Modified NICE Guidelines in US Adults with IBS-D, Am. J. Gastroenterol. 111 (12) (2016) 1824–1832.

[50] K. McIntosh, D.E. Reed, T. Schneider, et al., FODMAPs alter symptoms and the metabolome of patients with IBS: a randomised controlled trial, Gut. 66 (7) (2017) 1241–1251.

[51] T.N. Hustoft, T. Hausken, S.O. Ystad, et al., Effects of varying dietary content of fermentable short-chain carbohydrates on symptoms, fecal microenvironment, and cytokine profiles in patients with irritable bowel syndrome, Neurogastroenterol. Motil. 29 (4) (2017).

[52] M.J. Zahedi, V. Behrouz, M. Azimi, Low fermentable oligo-di-mono-saccharides and polyols diet versus general dietary advice in patients with diarrhea-predominant irritable bowel syndrome: A randomized controlled trial, J. Gastroenterol. Hepatol. 33 (6) (2018) 1192–1199.

[53] K. Whelan, L.D. Martin, H.M. Staudacher, M.C.E. Lomer, The low FODMAP diet in the management of irritable bowel syndrome: an evidence-based review of FODMAP restriction, reintroduction and personalisation in clinical practice, J. Hum. Nutr. Diet. (2018).

[54] M. O'Keeffe, C. Jansen, L. Martin, et al., Long-term impact of the low-FODMAP diet on gastrointestinal symptoms, dietary intake, patient acceptability, and healthcare utilization in irritable bowel syndrome, Neurogastroenterol. Motil. 30 (1) (2018).

[55] R.M. Harvie, A.W. Chisholm, J.E. Bisanz, et al., Long-term irritable bowel syndrome symptom control with reintroduction of selected FODMAPs, World J. Gastroenterol. 23 (25) (2017) 4632–4643.

[56] S.L. Peters, C.K. Yao, H. Philpott, G.W. Yelland, J.G. Muir, P.R. Gibson, Randomised clinical trial: the efficacy of gut-directed hypnotherapy is similar to that of the low FODMAP diet for the treatment of irritable bowel syndrome, Aliment Pharmacol. Ther. 44 (5) (2016) 447–459.

[57] L. Maagaard, D.V. Ankersen, Z. Végh, et al., Follow-up of patients with functional bowel symptoms treated with a low FODMAP diet, World J. Gastroenterol. 22 (15) (2016) 4009–4019.

[58] R.H. de Roest, B.R. Dobbs, B.A. Chapman, et al., The low FODMAP diet improves gastrointestinal symptoms in patients with irritable bowel syndrome: a prospective study, Int. J. Clin. Pract. 67 (9) (2013) 895–903.

[59] HM. Staudacher, Nutritional, microbiological and psychosocial implications of the low FODMAP diet, J. Gastroenterol. Hepatol. 32 (Suppl 1) (2017) 16–19.

[60] M. Bellini, S. Tonarelli, A.G. Nagy, et al., Low FODMAP Diet: Evidence, Doubts, and Hopes, Nutrients. 12 (1) (2020).

[61] E.P. Halmos, C.T. Christophersen, A.R. Bird, S.J. Shepherd, P.R. Gibson, J.G. Muir, Diets that differ in their FODMAP content alter the colonic luminal microenvironment, Gut. 64 (1) (2015) 93–100.

[62] S.M.P. Bennet, L. Böhn, S. Störsrud, et al., Multivariate modelling of faecal bacterial profiles of patients with IBS predicts responsiveness to a diet low in FODMAPs, Gut. 67 (5) (2018) 872–881.

[63] M. Rossi, R. Aggio, H.M. Staudacher, et al., Volatile Organic Compounds in Feces Associate With Response to Dietary Intervention in Patients With Irritable Bowel Syndrome, Clin. Gastroenterol. Hepatol. 16 (3) (2018) 385–391 e381.

[64] E.P. Halmos, V.A. Power, S.J. Shepherd, P.R. Gibson, J.G. Muir, A diet low in FODMAPs reduces symptoms of irritable bowel syndrome, Gastroenterology 146 (1) (2014) 67–75 e65.

[65] S.J. Shepherd, P.R. Gibson, Fructose malabsorption and symptoms of irritable bowel syndrome: guidelines for effective dietary management, J. Am. Diet. Assoc. 106 (10) (2006) 1631–1639.

[66] M. O'Keeffe, MC. Lomer, Who should deliver the low FODMAP diet and what educational methods are optimal: a review, J. Gastroenterol. Hepatol. 32 (Suppl 1) (2017) 23–26.

[67] N. Trott, I. Aziz, A. Rej, D. Surendran Sanders, How Patients with IBS Use Low FODMAP Dietary Information Provided by General Practitioners and Gastroenterologists: A Qualitative Study, Nutrients 11 (6) (2019).

[68] R.B. Gearry, P.M. Irving, J.S. Barrett, D.M. Nathan, S.J. Shepherd, PR. Gibson, Reduction of dietary poorly absorbed short-chain carbohydrates (FODMAPs) improves abdominal symptoms in patients with inflammatory bowel disease-a pilot study, J. Crohns. Colitis. 3 (1) (2009) 8–14.

[69] A. Rej, I. Aziz, H. Tornblom, D.S. Sanders, M. Simren, The role of diet in irritable bowel syndrome: implications for dietary advice, J. Intern. Med. 286 (5) (2019) 490–502.

[70] D. Schumann, J. Langhorst, G. Dobos, H. Cramer, Randomised clinical trial: yoga vs a low-FODMAP diet in patients with irritable bowel syndrome, Aliment Pharmacol. Ther. 47 (2) (2018) 203–211.

[71] Y.A. McKenzie, R.K. Bowyer, H. Leach, et al., British Dietetic Association systematic review and evidence-based practice guidelines for the dietary management of irritable bowel syndrome in adults (2016 update), J. Hum. Nutr. Diet. 29 (5) (2016) 549–575.

[72] National Institute for Health and Clinical Excellence. Irritable. In.

[73] J. Dionne, A.C. Ford, Y. Yuan, et al., A Systematic Review and Meta-Analysis Evaluating the Efficacy of a Gluten-Free Diet and a Low FODMAPs Diet in Treating Symptoms of Irritable Bowel Syndrome, Am. J. Gastroenterol. 113 (9) (2018) 1290–1300.

[74] H.M. Staudacher, P.M. Irving, M.C.E. Lomer, K. Whelans, The challenges of control groups, placebos and blinding in clinical trials of dietary interventions, Proc. Nutr. Soc. 76 (4) (2017) 628.

[75] X.J. Wang, M. Camilleri, S. Vanner, C. Tuck, Review article: biological mechanisms for symptom causation by individual FODMAP subgroups – the case for a more personalised approach to dietary restriction, Aliment Pharmacol. Ther. 50 (5) (2019) 517–529.

[76] S. Dunn, A. Datta, S. Kallis, E. Law, C.E. Myers, K. Whelan, Validation of a food frequency questionnaire to measure intakes of inulin and oligofructose, Eur. J. Clin. Nutr. 65 (3) (2011) 402–408.

[77] A. Mari, D. Hosadurg, L. Martin, N. Zarate-Lopez, V. Passananti, A. Emmanuel, Adherence with a low-FODMAP diet in irritable bowel syndrome: are eating disorders the missing link? Eur. J. Gastroenterol. Hepatol. 31 (2) (2019) 178–182.

Index

Page numbers followed by "*f*" and "*t*" indicate, figures and tables respectively.

Printed in the United States
by Baker & Taylor Publisher Services